Springer Protocols Handbooks™

For further volumes:
http://www.springer.com/series/8623

Molecular Biological Technologies for Ocean Sensing

Edited by

Sonia M. Tiquia-Arashiro

The University of Michigan-Dearborn, Dearborn, MI, USA

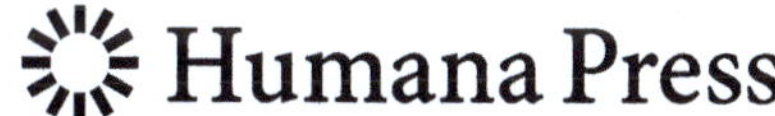

Editor
Sonia M. Tiquia-Arashiro
The University of Michigan-Dearborn
Dearborn, MI, USA

ISSN 1949-2448 ISSN 1949-2456 (electronic)
ISBN 978-1-61779-914-3 ISBN 978-1-61779-915-0 (eBook)
DOI 10.1007/978-1-61779-915-0
Springer New York Heidelberg Dordrecht London

Library of Congress Control Number: 2012940191

Printed on acid-free paper

Humana Press is a brand of Springer
Springer is part of Springer Science+Business Media (www.springer.com)

Preface

The pursuit of knowledge of the ocean has progressed in recent years due to the availability of new technologies and tools. Satellites are now equipped with sensors that can measure the optical densities of surface waters; profiling floats can collect physical and chemical data from around the world; and fluorescence detectors can provide information about the chlorophyll concentrations at depths beyond reach of satellite-based sensors. These technologies have stimulated a renewed interest in ocean exploration, and the vast quantities of physicochemical data collected have aided in the development of predictive models for some of the major ocean processes. By contrast, the integration of cell biology and genomics into oceanographic research is much less developed, even though biological (and especially microbiological) processes are fundamental for maintaining a functional global ecosystem. The study of aquatic organisms at the sub-molecular or molecular level spans from the smallest of viruses to the largest marine mammals. Issues addressed using molecular techniques are deeply rooted in questions posed by biological oceanographers: What species are present? How are they related to one another? How do they respond to alterations in the chemical and physical environment? What role do they play in biogeochemical cycling, food webs, and maintenance of global climate? Can we use specific organisms as indicators of environmental change? One of the major challenges in understanding ocean processes is the difficulty in four-dimensional sampling of a complex and dynamic habitat because of a lack of suitable microbial and biogeochemical sensors. The solution to this challenge is to develop new tools and methodologies for genomic-enabled oceanography.

The study of aquatic organisms at the molecular level is now widespread in ocean sciences. These techniques target molecular signatures (i.e., carbohydrates, lipids, proteins, and nucleic acid sequences), which provide a broad foundation for identifying specific organisms, revealing evolutionary relationships, inferring metabolic potential and function, and indicating stress due to environmental change. Subjects span from environments from all corners of the globe, including the shallowest of waters to the deep sea and cold seeps to hot vents. Target organisms range from viruses and phage to the largest marine mammals. Most recently, research on aquatic organisms has come to include genomics and associated technologies. The initial phase of the global ocean survey, an ambitious expedition to chart the ocean genome, generated an impressive number of open reading frames equivalent to half of the entire GenBank inventory of known genes. This study of marine bacteria highlighted the vast and previously unknown genetic information contained in extant marine microorganisms, from new protein families to novel metabolic processes. However, many of the open reading frames are unlike any known genes, which could encode metabolic processes that are yet to be discovered or be important in the regulation of cellular activity in the dynamic and variable marine environment. Metatranscriptomics and/or proteomics allowed us to elucidate metabolic activities under different conditions in the ocean in various organisms. The collection of large amounts of sequence information from these molecular studies paved the way for construction of DNA microarrays—high-density matrixes of DNA probes immobilized on a solid support. DNA array technology offers the possibility of detecting the presence and expression of many genes simultaneously in a single sample. They are able to identify functional groups, species, and ecotypes reliably and rapidly. Protein arrays on the other hand, complement those based on nucleic acids by providing a basis for detecting specific gene products. Applications of DNA and protein arrays are projected to enable multivariant assessments of species diversity and function in relation to changes in the chemical and physical environment on scales not possible previously. Increasing reliance on cell-free analytical techniques like DNA and protein has also probably spur application of probes in the whole-cell format, to relate genetic content and expression to identifiable, intact cells and associations.

Although application of molecular techniques has become well established in ocean science, the vast majority of the work typically occurs in the laboratory after the return of a discrete set of samples, limiting opportunities to collect molecular data in near real time. While many molecular-based analyses are described as "high-throughput," those methods are generally restricted to laboratory use and require substantial effort to process and collect. Molecular techniques can form the basis for remote instrumentation sensing technologies for marine microbial diversity and ecological function. For example, molecular biological approaches may be deployed in the ocean sensor platforms to track microorganisms or processes of interest in real time. Most of the sensors used today utilize optical techniques to derive presence, abundance, and activity of organisms. New molecular analytical-based sensors are also emerging, such as ecogenomic sensors. Ecogenomic sensors provide measures or microbial presence and function at the molecular level. Included in the latter are the Autonomous Microbial Genosensor (AMG) and Environmental Sample Processor (ESP). The AMG is a microbiological sensing buoy developed by scientists from College of Marine Science at the University of South Florida. The AMG is the first microbiological detection buoy designed that uses nucleic acid-based amplification. Currently, the AMG is designed to transmit data through a WiFi connection and is battery powered for complete autonomous operation and can be connected to a cable network system for data transmission and power. The ESP was developed by Chris Scholin and his colleagues from Monteray Bay Aquarium Research Institute in Moss Landing, California. It is an electromechanical/fluidic device that collects water samples from the ocean subsurface and allows for the application of DNA probe arrays to detect target rRNAs present in crude homogenate using sandwich hybridization assay. Previous applications of the ESP have focused on detecting harmful algae and invertebrate larvae in situ.

The availability and the reliability of commercially available instruments, methods, and supplies have made it possible to dig into the molecular underpinnings of just about everything that is "environmental." These achievements, coupled with advances in ocean observatory technology and plans to extend those networks, have fueled the idea of applying molecular sensors in remote settings. The evolution and application of ecogenomic sensors seems as natural as instruments that have brought real-time chemical, physical, and bio-optical assessments of the ocean to our desks at a mere click of button. A key to autonomous observations of microbes in the ocean is the continuing development of sensing technologies in the laboratory, transitioning sensors from the bench to the field, and integrating sensor suites into observing platforms appropriate to spatial and temporal dimensions of specific process and phenomena. More exciting advances on molecular ocean sensing are still to come as more remotely operated autonomous undersea vehicles are incorporated into oceanographic research. There is no doubt that much work remains to define the assays that will be deployed in situ and the concomitant sample collection and processing requirements. Sustained investment in the development of in situ instrumentation is essential to bring to fruition the testing of ideas and models discussed in this book.

Dearborn, MI, USA *Sonia M. Tiquia-Arashiro*

Acknowledgments

The work presented in this book collectively draws on the work of many great scientists including our distinguished authors and highly regarded academics and industrial experts from different disciplines within the science and engineering. It also owes much to the influence of many peers and colleagues worldwide, who are too numerous to mention. It goes without saying that a book project of this size could not have been successfully completed without the excellent cooperation and enormous effort on the part of the authors, many of whom have revised their manuscripts more than once.

It is also fitting to thank the dedicated publications staff of Springer Publishers for their skills in transforming our manuscripts into high-quality book, which I hope meets with your expectations as a reader. I wish to thank the Assistant Editor on Springer Science + Business Media, Melissa Higgs, for giving me an opportunity to publish this book. Finally, I hope the readers will share my joy and find this book very useful. The credit for producing this book is only partly ours; it is the blame that rests totally with us. Have a good read.

Editor ***Sonia M. Tiquia-Arashiro***

Contents

Contributors

AMPARO ALFONSO • *Departamento de Farmacología, Fisiología, Facultad de Veterinaria, USC, Lugo, Spain*

JOSEPHINE ALLER • *School of Marine and Atmospheric Sciences, Stony Brook University, Stony Brook, NY, USA*

ANA M. BOTANA • *Departamento de Farmacología, Química Analítica, Facultad de Ciencias, USC, Lugo, Spain*

LUIS M. BOTANA • *Departamento de Farmacología, Fisiología, Facultad de Veterinaria, USC, Lugo, Spain*

AMBER R. BRATCHER • *University of Maine, School of Marine Science, Orono, ME, USA*

FILIPE A. CARDOSO • *Instituto de Engenharia de Sistemas e Computadores-Microsistemas e Nanotecnologias (INESC-MN) and Institute for Nanosciences and Nanotechnologies (IN), Lisbon, Portugal*

JAMES CLEMENTS • *Clements Scientific, Brentwood, NH, USA*

LAURIE B. CONNELL • *University of Maine, School of Marine Science, Orono, ME, USA*

NEESHMA DAVE • *Department of Chemistry, Waterloo Institute for Nanotechnology, University of Waterloo, Waterloo, ON, Canada*

HARBANS S. DHADWAL • *Department of Electrical and Computer Engineering, Stony Brook University, Stony Brook, NY, USA*

LUÍS P. FONSECA • *Instituto de Engenharia de Sistemas e Computadores-Microsistemas e Nanotecnologias (INESC-MN) and Institute for Nanosciences and Nanotechnologies (IN), Lisbon, Portugal*

PAULO P. FREITAS • *Instituto de Engenharia de Sistemas e Computadores-Microsistemas e Nanotecnologias (INESC-MN) and Institute for Nanosciences and Nanotechnologies (IN), Lisbon, Portugal*

TERUO FUJII • *Center for International Research on MicronanoMechatronics (CIRMM), Institute of Industrial Science, University of Tokyo, Tokyo, Japan*

TATSUHIRO FUKUBA • *Department of Mechanical and Biofunctional Systems, Institute of Industrial Science, University of Tokyo, Tokyo, Japan; Japan Agency for Marine-Earth Science and Technology (JAMSTEC) Marine technology and Engineering Center, Marine Technology Development Department Natsushima-cho 2-15, Yokosuka-city Kanagawa, Japan*

PO-JUNG JIMMY HUANG • *Department of Chemistry, Waterloo Institute for Nanotechnology, University of Waterloo, Waterloo, ON, Canada*

KENNETH S. JOHNSON • *Monterey Bay Aquarium Research Institute, Moss Landing, CA, USA*

PAUL KEMP • *Center for Microbial Oceanography: Research and Education, School of Ocean and Earth Science and Technology, University of Hawaii, Honolulu, HI, USA*

IMRAN KHIMJI • *Department of Chemistry, Waterloo Institute for Nanotechnology, University of Waterloo, Waterloo, ON, Canada*

JUNG HEON LEE • *Department of Materials Science and Engineering, University of Illinois, Urbana, IL, USA; School of Advanced Materials Science and Engineering, Sungkyunkwan University (SKKU), Suwon, South Korea*

JUEWEN LIU • *Department of Chemistry, Waterloo Institute for Nanotechnology, University of Waterloo, Waterloo, ON, Canada*
M. CARMEN LOUZAO • *Departamento de Farmacología, Fisiología, Facultad de Veterinaria, USC, Lugo, Spain*
YI LU • *Department of Materials Science and Engineering, University of Illinois, Urbana, IL, USA; Department of Chemistry, University of Illinois, Urbana, IL, USA*
SOFIA S.A. MARTINS • *Institute for Biotechnology and Bioengineering, Centre for Biological and Chemical Engineering, Instituto Superior Técnico, Lisbon, Portugal; Instituto de Engenharia de Sistemas e Computadores-Microsistemas e Nanotecnologias (INESC-MN) and Institute for Nanosciences and Nanotechnologies (IN), Lisbon, Portugal*
VERÓNICA C. MARTINS • *Instituto de Engenharia de Sistemas e Computadores-Microsistemas e Nanotecnologias (INESC-MN) and Institute for Nanosciences and Nanotechnologies (IN), Lisbon, Portugal; International Iberian Nanotechnology Laboratory (INL), Braga, Portugal*
KATELYN M. MCKINDLES • *Department of Natural Sciences, University of Michigan, Dearborn, MI, USA*
LINDA K. MEDLIN • *Department of NanoEngineering, University of California, San Diego (UCSD), La Jolla, CA, USA; UPMC, Univ Paris 06, UMR 7621, LOMIC, Banyuls/mer, France; CNRS, UMR 7621, LOMIC, Banyuls/mer, France*
JOSEPH A. NEEDOBA • *Division of Environmental and Biomolecular Systems, and the Center for Coastal Margin Observation and Prediction, Oregon Health and Science University, Beaverton, OR, USA*
JAHIR OROZCO • *Department of NanoEngineering, University of California, San Diego (UCSD), La Jolla, CA, USA; UPMC, Univ Paris 06, UMR 7621, LOMIC, Banyuls/mer, France; CNRS, UMR 7621, LOMIC, Banyuls/mer, France*
TAWNYA D. PETERSON • *Division of Environmental and Biomolecular Systems, and the Center for Coastal Margin Observation and Prediction, Oregon Health and Science University, Beaverton, OR, USA*
JOANN RADWAY • *School of Marine and Atmospheric Sciences, Stony Brook University, Stony Brook, NY, USA*
JULIE C. ROBIDART • *Ocean Sciences Department, University of California, Santa Cruz, CA, USA*
IRINA N. SHILOVA • *Ocean Sciences Department, University of California, Santa Cruz, CA, USA*
SONIA M. TIQUIA-ARASHIRO • *Department of Natural Sciences, University of Michigan, Dearborn, MI, USA*
CARMEN VALE • *Departamento de Farmacología, Química Analítica, Facultad de Ciencias, USC, Lugo, Spain*
MERCEDES R. VIEYTES • *Departamento de Farmacología, Fisiología, Facultad de Veterinaria, USC, Lugo, Spain*
NATALIA VILARIÑO • *Departamento de Farmacología, Fisiología, Facultad de Veterinaria, USC, Lugo, Spain*
ZIDONG WANG • *Department of Materials Science and Engineering, University of Illinois, Urbana, IL, USA*
JONATHAN P. ZEHR • *Ocean Sciences Department, University of California, Santa Cruz, CA, USA*

Chapter 1

"Omics"-Enabled Microbial Sensors on Ocean Platforms

Julie C. Robidart, Irina N. Shilova, and Jonathan P. Zehr

Abstract

In order to assess the diversity and function of microbial communities most effectively, molecular assays need to be designed that target the phylogenetic markers and functional genes that are key to major ecological processes and microorganisms. A streamlined design process is presented here that designs probes for microarray and quantitative PCR (qPCR) assays, for application on the lab bench or on remote instrumentation. These assays can be used for DNA (genome) and RNA (gene transcription) studies. The pipeline described here establishes a database of environmental sequences, which is then used for the design of molecular probes for microarrays and to inform the design of qPCR assays, which is detailed here along with the assay optimization process. Finally, the process for the design of high-density microarrays is described. The qPCR protocol is currently used for assay optimization on the Environmental Sample Processor, a deployable robotic "genosensor" (http://www.mbari.org/esp). The protocols described here should advance applications in microbial oceanography using robotic instrumentation as well as traditional sampling methods.

Key words: Quantitative PCR, Microarray, Biosensors, Environmental microbiology, Ocean platforms, Robotic instrumentation, Microbial oceanography, Genosensor, Environmental sample processor

1. Introduction

With the advent of environmental metagenomics, marine microbiologists have benefited greatly from the availability of sequence data that is not dependent on cultured isolates. Since databases of environmental gene sequences have rapidly grown due to the application of high-throughput sequencing techniques, it is possible to design molecular assays to target specific processes or phylogenetic groups in the environment with greater ease than previously possible. Gene sequence databases can now be created that represent a large diversity of organisms from the environment, using publicly available data. These databases provide the starting point for the design of molecular assays to target microorganisms in the environment. Presented in this chapter is a streamlined qPCR and

Sonia M. Tiquia-Arashiro (ed.), *Molecular Biological Technologies for Ocean Sensing*, Springer Protocols Handbooks,
DOI 10.1007/978-1-61779-915-0_1, © Springer Science+Business Media New York 2012

microarray design processes to increase the probability of successful implementation with environmental samples. The assays designed by this approach can be used to examine microbial population dynamics as well as functional activities.

1.1. Quantitative PCR

Quantitative PCR has been used extensively in the field to quantify specific phylotypes (1–10), and in the lab to monitor functional gene expression in cultures or enrichments (quantitative reverse-transcription PCR or qRT-PCR) (11–16). Once designed, assays can be used for gene abundance quantification and gene expression analyses, depending on the nature of the template (DNA or cDNA, respectively). Quantification in qPCR is achieved by detection of fluorescence signal from nonspecific fluorescent dyes (i.e., SYBR Green) or sequence-specific DNA oligonucleotide probes (i.e., TaqMan) that hybridize to PCR amplicons. TaqMan chemistry has higher specificity due to hybridization of the fluorogenic probe to complementary DNA. This qPCR chemistry is described in this chapter.

1.2. Oligonucleotide Microarrays

Oligonucleotide microarrays contain hundreds to thousands of oligonucleotide probes (25–70-nt long) that are specific to genes/species and immobilized or synthesized on a slide carrier. Samples (DNA, RNA) are labeled and hybridized to the array, and hybridization signal (fluorescent or electrochemical) is quantitatively measured. Microarrays offer fast and reliable high-throughput identification of microorganisms and their activity, and have been increasingly applied to study microbial ecology (17–21) oftentimes in marine settings (7, 22–32).

Environmental microarrays can be classified into several categories including phylogenetic arrays targeting 16S-rRNA genes (PhyloChip: 18) or genomes (31, 32) and functional arrays that target a set of genes coding for ecologically relevant processes (21, 22). Functional arrays can be used for both identification of important species in a community (27) and assessment of their transcriptional activity (33). Oligonucleotide microarray designs can be spotted (e.g., (25)) or synthesized directly on the slide (or "in situ synthesis"), as are most high-density arrays (18). Spotted arrays feature a single oligonucleotide probe per target gene; these arrays are usually of relatively low throughput (hundreds of probes) and used mainly for identification of community composition (phylogenetic and functional arrays). The probe for each target is carefully designed and may target a single species or a group of closely related microorganisms. Synthesized in situ or high-density microarrays (Affymetrix, Agilent, NimbleGen) contain 4–20 probes per each target gene that are synthesized on-chip, and specificity is provided by probe redundancy. These arrays have tens to hundreds of thousands of probes and are often used for assessing transcriptional activity, but also can be used for community phylogenetic fingerprinting.

1.3. Molecular Markers

While the probability of successful application of molecular assays has increased with the availability of environmental sequence data, the design of molecular probes is still a laborious process, and much forethought should be invested into the choice of gene targets prior to design. Target genes can provide information on community phylogenetic composition, specific microbial processes (e.g., nitrogen fixation, nitrification), and nutrient limitation, or provide information on physiological status and growth rate of key microorganisms. Depending on the research questions and gene sequence conservation, the probes can be designed to differentiate among taxonomic units at various levels. Target genes can act as phylogenetic markers (to quantify specific clades within a population) when used on DNA, as proxies for a specific activity (implied from gene expression) when used on RNA, or as both. Less conserved functional genes, for example, can be used as markers for processes as well as specific clades of organisms, while it may be impossible to design a clade-specific assay for more conserved genes. Information about target genes comes from prior laboratory studies with cultured representatives, field studies, as well as (meta)genomic and metatranscriptomic data (4, 27, 34–40). If such information is not available for the organisms of interest, it can be deduced from closely related organisms. Thus, the fundamental question is: *What does this target gene represent?* Particularly, the following questions regarding the target gene should be answered.

- *How does the abundance of this gene, with contextual data (e.g., physical/chemical parameters), provide information on the ecology of the organism?*
- *How does the expression of this gene provide information on the current status (activity, physiology) of the organism(s)?*
- *What taxonomic resolution is required to provide meaningful information: i.e., are there selected sentinel organisms or specific groups of organisms?*

Table 1 provides examples of molecular assays and the information they provide. In the procedures described below, it is assumed that target genes (molecular markers) and organisms of interest have already been identified.

1.4. Gene Database

Establishing a database for probe design for environmental microarrays and degenerate PCR is critical in increasing the sensitivity of the assays. For phylogenetic studies using the 16S rRNA gene, comprehensive databases exist that include genes from cultured and uncultured microorganisms: The ARB project (ARB at http://www.arb-home.de), Ribosomal Database Project (RDP: http://www.rdp.cme.msu.edu), SILVA (http://www.arb-silva.de), Green Genes (http://greengenes.lbl.gov/), and National Center for Biotechnology Information (NCBI: http://www.ncbi.nlm.nih.gov).

Table 1
Examples of quantitative PCR assays for cyanobacteria

Organism/group	Gene	Assay interpretation
Synechococcus	Phosphonate transport, *phnD* groups	Clade-specific indicators of phosphorus stress
Synechococcus	Iron stress, *idiA*	Indicators of Fe stress
Synechococcus	Global N regulator, *ntcA* groups	Clade-specific indicators of nitrogen limitation
Synechococcus	RuBisCO, *rbcL*	Near-universal *Synechococcus* proxy for carbon fixation
Unicellular cyano, grp A	Nitrogenase	Group-specific proxy for nitrogen fixation
Crocosphaera	Nitrogenase	Group-specific proxy for nitrogen fixation
Heterocystous cyano 1	Nitrogenase	Group-specific proxy for nitrogen fixation
Trichodesmium	Nitrogenase	Group-specific proxy for nitrogen fixation

Each assay provides information regarding the ecology and/or physiology of the target organism

To represent the diversity of functional genes, homologous sequences are acquired from databases (e.g., NCBI) usually using gene annotation and key words as was used in developing the GeoChip array (20, 21). However, because of the underrepresentation of full-length gene sequences from uncultured microorganisms, ambiguity in gene annotation, and common misannotation, resulting datasets may not resemble the natural nucleotide diversity of the target genes. This is especially true for less conserved genes such as *narB,* encoding the enzyme assimilatory nitrate reductase, which was believed to be absent in *Prochlorococcus* spp. due to the bias of the isolation technique (*Prochlorococcus* strains have been isolated on media containing ammonium as N source, thus decreasing a chance of isolating the strains that are able to utilize nitrate). The presence of *narB* in *Prochlorococcus* spp. was discovered only through analyzing metagenomic data for natural populations of *Prochlorococcus* spp. (41). Another approach for acquiring natural gene diversity is through amplification of the target gene from environmental samples with degenerate PCR (22). The extent of sequence diversity in this case can be biased by the degenerate primer design, depending on the coverage of the input database used to inform the design. A more comprehensive database may be built by exploring the increasingly available metagenomic and metatranscriptomic data. Considering that the majority of key players in ocean biogeochemistry are uncultured, and their genomes remain unsequenced, the use of high-throughput environmental sequencing data in constructing a database for probe design is essential.

1.5. Molecular Assay Design

Assay design in this chapter is discussed in the order of operations in the design process: the establishment of gene database, followed by qPCR assay and microarray assay designs. These protocols assume no prior information on diversity of the target genes. Those interested in using personal gene sequence databases from the environment of interest to design probes and/or qPCR assays can begin with step 6 (*Choosing representative sequences*) of the *Establishment of gene database* protocol, or step 1, *TaqMan qPCR primer and probe design*, of the *Quantitative PCR assay design* protocol.

1.5.1. Procedural Overview: Establishing a Database for a Target Gene

The objective of the following procedure is to collect all available and relevant DNA homologous sequence diversity for the target gene. Before proceeding, the genes and organism(s), and the environment of interest, need to be defined. The amino acid sequence of a target gene is obtained from any of the available sequence databases and used as a query in a Basic Local Alignment Search Tool (BLAST, (42)) against metagenomic and metatranscriptomic nucleotide datasets in the Community Cyberinfrastructure for Advance Microbial Ecology Research & Analysis (CAMERA). The hits (or "subject sequences") are then filtered based on length and geographic location (optional), and a reciprocal BLAST is run against all proteins and genomes in NCBI or CAMERA to confirm the identity of the subject sequences. This set of environmental sequences is combined with the nucleotide sequences from sequenced genomes, and representative sequences are selected based on at least specified nucleotide dissimilarity. The representative sequences are used directly for microarray probe design or for creating an alignment, from which degenerate primers for PCR are designed (Fig. 1). This protocol has been applied to prokaryotic genes, but can be adapted to eukaryotic genes if a eukaryotic database is available, or in the near future with more sequenced eukaryotic genomes, although specifics of eukaryotic genes should be considered (e.g., introns, gene length).

1.5.2. Procedural Overview: Quantitative PCR

A variety of assays have been designed to assess the abundances and distributions of microorganisms in situ, using quantitative PCR. This process begins with the design of degenerate primers based on a database of global environmental sequences, obtained from the process outlined above. Degenerate primers should be used to amplify genes from the environment where the instrument will be deployed, from a variety of seasons, if possible. Clone libraries are generated from the amplicons and cloned gene fragments sequenced in order to assess the diversity of the gene in that environment. These sequences are used to design specific quantitative PCR primers and probes, based on clades or subclades, depending on the research question and the extent of gene conservation (i.e., subclades cannot be targeted with conserved genes). qPCR assays are optimized with ESP chemistry at the bench to create a standard curve with adequate efficiency and sensitivity, using linearized

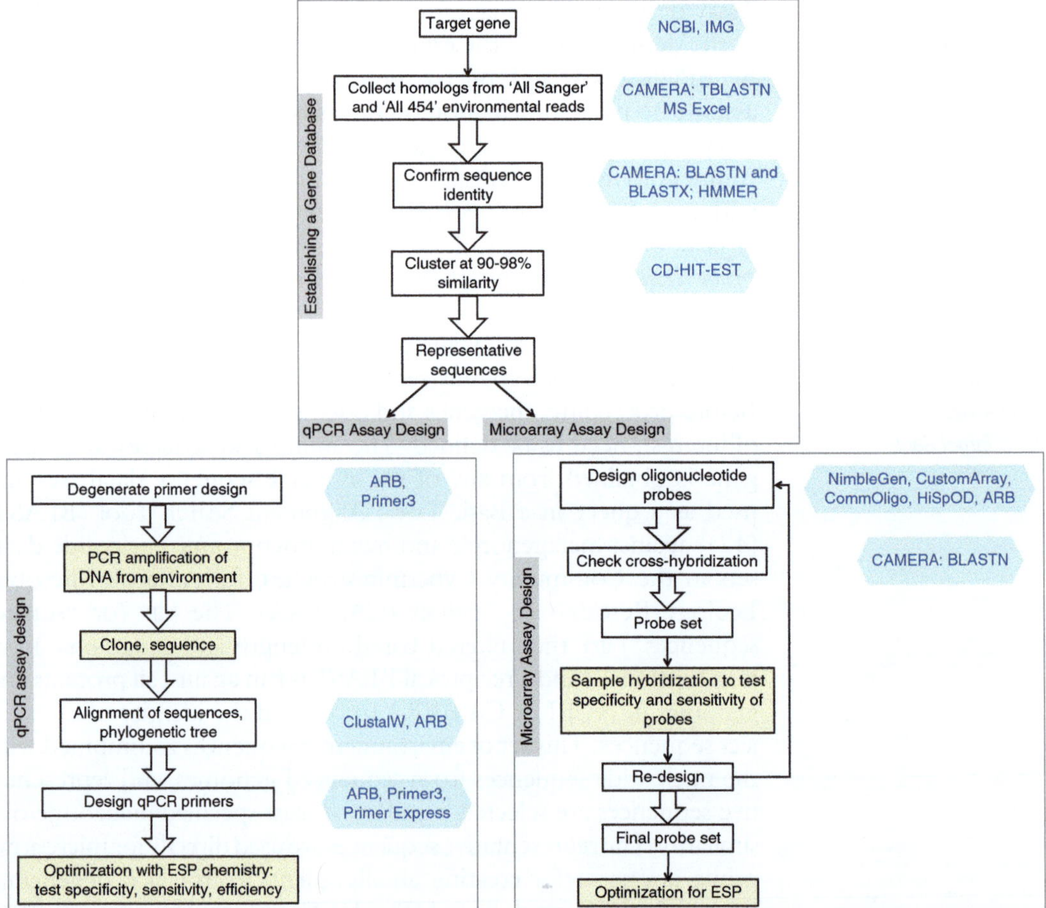

Fig. 1. Order of operations as described in this chapter. Workflow describes the three main procedures: (**a**) database generation, (**b**) qPCR design, and (**c**) microarray design. Types of software used at specific steps in molecular assay design are indicated to the *right*.

plasmids containing the gene of interest. Environmental samples from various times of the year are then queried for gene abundances, and assays are eliminated if the temporal data is uninformative (e.g., the abundance does not change over the course of the year). Assays are then checked for cross-reactivity. A 4- to 6-point standard curve should have the sensitivity to detect targets at environmentally relevant abundances, show repeatable amplification dynamics, and have an efficiency of 90–110 %; otherwise, the qPCR assay must be modified in order to achieve accurate abundance information.

1.5.3. Procedural Overview: Microarrays

The experimental scheme of developing a microarray approach has been described in detail previously (e.g., (21, 43–45)). Briefly, it consists of establishing a database of target sequences, probe design, spotting or synthesizing probes on the selected array platform, checking sensitivity and specificity of the probes by hybridizing control samples, and redesigning the array if needed (Fig. 1). A more general universal procedure for different types of arrays is described here, but specific procedural details vary between platforms. Since microarray technology on the ESP is currently in development, and the microarray platform that will be implemented is not yet known, we refer the reader to previously published protocols for specific details (43, 45). Outlined here is a general procedure for the design of a high-density oligonucleotide microarray (NimbleGen, Agilent, or CustomArray type) for identification of microbial community composition and activity (gene expression) in marine environments.

2. Reagents and Equipment

2.1. Degenerate PCR

Gradient thermocycler

PCR tubes

PCR rack

Gel electrophoresis system

Gel visualization system

1.5-ml Eppendorf tubes

PCR Master Mix

$MgCl_2$

dNTP mix

Taq DNA polymerase

Sterile water suitable for PCR

Degenerate primers

Three to five DNA extracts representing different time points from the environment of interest

Agarose

Gel loading dye

DNA ladder (spanning the size of the expected amplicon)

Gel stain

Qiagen QiaQuick Gel Extraction Kit

2.2. Clone Libraries

300-μl Multichannel pipettor.

2-ml Grow blocks for cultures.

Sterile breathable adhesive for grow blocks.

200-μl Microtiter plates.

LB media: 10 g tryptone, 5 g yeast extract, 10 g NaCl per liter milliQ water. Autoclave for 45 min on a liquid cycle.

Selection antibiotic.

TOPO TA Cloning Kit (Invitrogen, Inc.) or pGEM Vector System (Promega).

Sterile 80 % glycerol: 80 ml glycerol, 20 ml milliQ water. Autoclave for 45 min on a liquid cycle.

Plasmid Miniprep 96 Kit (Millipore).

2.3. Making qPCR Standards

1-ml Pipettor

Culture tubes

Sterile toothpicks

Microcentrifuge

Shaking incubator

Gel electrophoresis system

Gel visualization system

NanoDrop Spectrophotometer (ThermoScientific) or Qubit Fluorometer (Invitrogen Corp.)

LB media

QIAprep Spin Miniprep Kit (Qiagen)

Restriction enzyme appropriate for chosen vector sequence

Agarose

Gel loading dye

DNA ladder (spanning the size of the expected amplicon)

Gel stain

Qiagen QiaQuick Gel Extraction Kit

Sterile water

2.4. qPCR Reaction Optimization and Optimizing Sensitivity and Efficiency

Real-time PCR machine

Accuprime Supermix II

$MgCl_2$ (25 or 50 mM)

Sterile water suitable for PCR

10 mM Primers

100 mM Probe

10^5 Copies per 6 ml standard

Optical PCR tubes

PCR rack

1.5-ml Eppendorf tubes

3. Protocol

3.1. Establishing a Database for a Target Gene

A combination of Web-based tools and software packages (Table 2) is used to collect homologous sequences from available environmental databases. In the future, the workflow may be automated

Table 2
Software used in this chapter

Software or Web sites	Address
CAMERA	http://camera.calit2.net/
NCBI	http://www.ncbi.nlm.nih.gov
UCSC Genome Browser	http://genome.ucsc.edu/
CD-HIT	http://weizhong-lab.ucsd.edu/cdhit_suite
IMG	http://img.jgi.doe.gov/
ClustalW	http://www.clustal.org
Jalview	http://www.jalview.org
ARB	http://www.arb-home.de

and more widely available. Although it is possible to work with several genes simultaneously, it is advised to work with one gene of interest at a time since different genes may require different selection criteria.

1. Nucleotide sequences for the marker gene are obtained from all available genomes of targeted organisms and clones at NCBI, Integrated Microbial Genomes (IMG), or the UCSC genome browser and saved in FASTA format. These sequences will be used in step 6. See Note 1.
2. An amino acid sequence of the marker gene (any representative) is acquired in FASTA format from any of the available sequence databases (NCBI, IMG, UCSC genome browser).
3. The amino acid sequence is used as a query in two TBLASTN runs in CAMERA against the reference datasets "All Metagenomic 454" and "All Metagenomic Sanger." TBLASTN searches nucleotide databases using a protein sequence as a query. This allows the acquisition of all homologous protein sequences and returns an output of hits (subject sequences) as nucleotides. The criteria should be set for a loose search: for example, *E* value cutoff $1e^{-3}$ and 10,000 hits per query. The results are returned as tables of all subject sequences with corresponding metadata (BLAST statistics, location of the source sample, etc.). The result tables are exported to a local computer, and the subject sequences are filtered and processed.
   (a) Select unique sequences by subject sequence ID.
   (b) Select sequences that are longer than 60 amino acids (or 180 bp).
   (c) Filter by geographic location (optional).

(d) All sequences have to be in the sense (+) orientation. The strand is indicated in the result table (column "Frame"). Sequences that have "+/−" match have to be converted into the reverse complement strand.

(e) Save subject sequences in FASTA format. If no program is implemented, the FASTA file can be created from Excel by copying the column with subject sequences into any text editor program and by manually formatting the entries.

The filtering can be done in Microsoft Excel by conditional sorting of columns or by implementing Perl (http://www.perl.org) or R (The R project, http://www.r-project.org) scripts. The implementation of scripts is recommended if there are several target genes.

4. Reciprocal BLASTN and BLASTX is performed in CAMERA with the selected subject sequences (from step 3) against "All prokaryotic genomes" and "All Prokaryotic Proteins." The purpose of this step is to confirm sequence identity (BLASTX) and select sequences that originate from the organisms of interest (BLASTN). If the goal is to collect all possible homologues for the target gene, BLASTN results are used to identify taxonomy or closest related organisms for the environmental reads. The criteria for BLAST jobs are 1–3 hits per query and an E value cutoff of at least $1e^{-6}$. Outputs are saved as tables with matching sequences and metadata, and the results are filtered in MS Excel to choose sequences that hit to the target organism(s) and target annotation. For annotation, the percent cutoff for sequence similarity is based on the gene conservation and the goal, for example at least 60 % for an alignment length of at least 30 amino acids. See Note 2.

5. Environmental sequences are trimmed according to the open reading frame (ORF) start and stop positions. The reason for this is that metagenomic reads often contain genomic fragments that are not part of the gene sequence and will cause nonspecific hybridizations if left for the probe design, and will complicate sequence alignment. The start and stop positions of the ORF are deduced from the BLASTX results based on the hit position and average gene length. Custom scripts for this procedure can be written in Perl, Python, or R language. See Note 3 for an alternative way.

6. Full-length genomic and trimmed environmental gene sequences are combined into one FASTA file and clustered in CD-EST-HIT (46, 47) at 90 % nucleotide similarity for microarray and 99 % for degenerate PCR assay design. The program returns a list of longest (representative) sequences from each cluster in FASTA format. The metadata obtained from previous steps should also be selected for the representative sequences. These sequences are used for the degenerate PCR primer and microarray probe design.

3.2. Quantitative PCR Assay Design

3.2.1. Degenerate PCR

1. Using sequences from step 6 above, create a nucleotide alignment in ClustalW and import into ARB or Jalview.
2. Analyze alignments and note conserved regions[1]. Design primers with the nucleotide sequences for these regions. These rules should be followed when designing degenerate primers (see Notes 4–6):
   (a) Amplicons should be as long as possible, and must be greater than 600 base pairs (bp)[2].
   (b) Primer length should be 18–22 bp.
   (c) Degeneracy should be kept at minimum (less than four degenerate bases per primer, and do not include them on the 3′ end).
   (d) Melting temperature (T_m) should be 52–60 °C, and should be within 4 °C of the T_m of the second primer. The higher the more specific the reaction.
   (e) GC content should be 40–60 %.
   (f) Only 2–3 G or C bases should be included at the 3′ end of each primer, but three in a row should be avoided to prevent mispriming.
   (g) Avoid primer secondary structures (hairpins, primer dimers, etc.).
   (h) Avoid dinucleotide repeats (AGAGAG) and mononucleotide runs (CCCCC).
3. Amplify 3–5 environmental samples from various time points if possible, and optimize the degenerate PCR reaction. A PCR protocol to begin optimization of degenerate PCR amplification contains:
   (a) 3 μl 10× PCR buffer
   (b) 4 mM $MgCl_2$ (can decrease to zero added $MgCl_2$)
   (c) 200 μM dNTP mix
   (d) 0.2 μM each primer (can go up to 1.5 μM)
   (e) 1.5 Units Taq DNA polymerase[3]
   (f) 2 μl Template DNA (or 1:10 dilution of template DNA)
   (g) Water to a final volume of 30 μl

   For each pair of degenerate primers, the reaction needs to be optimized for annealing temperature, primer concentration (0.1–1 μM), and $MgCl_2$ concentration (2–6 mM). See Notes 7–9.

[1] If the sequences are too diverse to visualize easily, translate in frame and find more conserved regions with amino acid alignments. Target these regions of the nucleotide sequence for degenerate primer design.
[2] The less conserved the sequence, the larger the amplicon necessary.
[3] Must use *Taq* polymerase in order to generate the A-overhangs for subsequent cloning reactions.

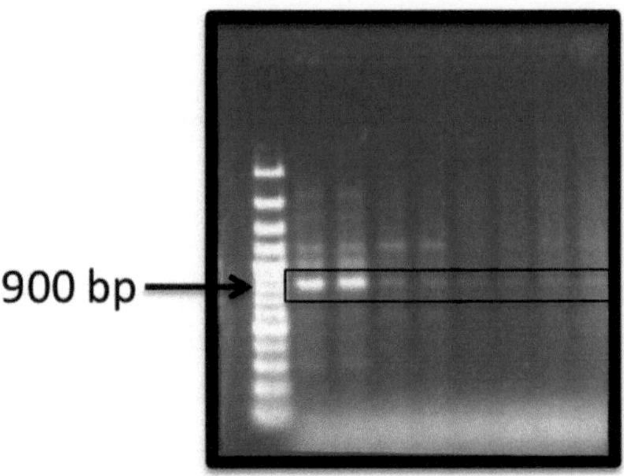

Fig. 2. Example of degenerate PCR products, visualized on an agarose gel. The desired amplification product is ca. 900 bp with this particular example. The single band at this size is gel-purified.

4. Run PCR products on a low-melting-point agarose gel. Stain and visualize the gel (Fig. 2), and gel-purify amplified bands of the correct length using a QiaQuick Gel Extraction Kit (Qiagen).

3.2.2. Clone Libraries

1. Construct libraries using the TOPO TA Cloning Kit (Invitrogen, Inc.) or pGEM Vector System (Promega Corp.) using amplicons from all designed primer sets in order to assess the complete diversity of the gene from the same samples.
2. Grow up clones overnight on 96-well grow blocks in 1.5 ml LB medium with the appropriate antibiotic.
3. Distribute 150 μl aliquots of each overnight culture to a separate microtiter plate, adding 35 μl of 80 % glycerol. Seal the plate, vortex, and store at −80 °C as an archive.
4. Extract the remaining (1 ml volume) clone libraries with the Montagé Plasmid Miniprep 96 Kit (Millipore).
5. Sequence inserts by your lab's preferred method. Maintain plate well orientation information in the submitted sequencing file, in order to return to specific clones.

3.2.3. TaqMan qPCR Primers and Probe Design

1. Verify that sequences are from targeted genes using BLAST within NCBI (42).
2. Align sequenced targets in ClustalW, and import into ARB (48) or a similar program, where a phylogenetic tree can be constructed. See Note 10. Depending on the objective, a comprehensive primer/probe set can be designed for qPCR based on all amplified sequences (which may not be possible if the gene is too diverse), or a series of assays for similar sequences in clusters, or clades.

3. qPCR primers and probes should conform to the same rules stated above, with the exception that the required amplicon length is 50–200 bp and primers are not degenerate; see Note 11. Additionally, probes must:
   (a) Be between the two PCR primers
   (b) Be situated near a primer and on the same strand
   (c) Have no guanine on the 5′ end
   (d) Contain more cytosines than guanines
   (e) Have a T_m that is 10 °C higher than that of the primers
   (f) Have a T_m near 68–70 °C
   (g) Be labeled with a fluorophore of your choice (FAM, TAMARA, or NED if using with the ESP) at the 5′ end, and Black Hole Quencher at the 3′ end

3.2.4. Making qPCR Standards

1. Add 5 ml LB media to ten culture tubes.
2. Retrieve the archive plate from the −80 °C freezer and thaw *partially* on ice.
3. Using the well information from the sequence data to orient within the plate, choose diverse clones from the archived library. For each clone, insert a sterile toothpick into the partially thawed culture from the well, and transfer toothpick to one 5-ml LB culture tube.
4. Add the lid to the tube and grow up shaking overnight at 37 °C.
5. Plasmid prep 2 × 2 ml of each culture using the QIAprep Spin Miniprep Kit (Qiagen).
6. Before digestion, aliquot 5 μl plasmid into a separate tube and keep at 4 °C for Step 8.
7. Linearize plasmids by overnight restriction digest, using a restriction enzyme appropriate to the vector sequence and terminating the reaction the following day according to the manufacturer's protocol.
8. Run the entire volume of digested plasmids on a low-melting-point agarose gel, alongside undigested plasmids and previously linearized plasmids (if available), in order to distinguish between the bands on the gel. Only linearized plasmids are gel-extracted with the Qiagen QiaQuick Gel Extraction Kit.
9. Plasmids are quantified using a NanoDrop Spectrophotometer (ThermoScientific) or Qubit Fluorometer (Invitrogen Corp.), and diluted to concentrations equivalent to $1 \times 10^{X}/6$ μl.
10. 1:10 serial dilutions of the most concentrated plasmid are performed until $1 \times 10^{0}/6$ μl using sterile water.

Table 3
Recipes for primer and probe mixes for qPCR optimization

Number	Probe (μl)	Primer (μl)	Water (μl)	Final concentration in reaction (μM)
1	0.6	45.0	2.4	1.50
2	0.6	30.0	17.4	1.00
3	0.6	15.0	32.4	0.500
4	0.6	10.0	37.4	0.333
5	0.6	3.0	44.4	0.100

3.2.5. qPCR Reaction Optimization

It is advisable to optimize the primer annealing temperature using standard PCR with an environmental sample "spiked" with a positive control. A starting point for optimization would be 7 °C below the lowest T_m of the primers. Amplification products are visualized on an agarose gel, and the annealing temperature of the most specific and efficient reaction (as evaluated by a tight, bright band) is used for optimization. Alternatively, if a temperature gradient quantitative PCR thermocycler is available, annealing temperature can be optimized during the protocol. This protocol optimizes $MgCl_2$ and primer concentrations for qPCR reactions. See Notes 12–14.

1. Dilute 10 μM primers and 100 μM probe as described here (1F to 5F, then 1R to 5R) (Table 3).
2. Make the enzyme master mix:

   480 μl Accuprime

   115.2 μl $MgCl_2$ (25 mM) or water

   19.2 μl water + *38.4 μl template* (10^5 copies per 6 μl, spiked into an environmental extract)

   652.8 μl total volume "master mix"
3. In 5 × 1.5-ml Eppendorf tubes marked 1R to 5R, dispense 28.8 μl of each R primer dilution into each tube.
4. Add 122.4 μl master mix to each of the tubes.
5. Set a 5 × 5 square of optical tubes in the PCR rack.
6. Aliquot 25.2 μl of Reverse primer dilutions (Table 4).
7. Add 4.8 μl Forward primer dilutions (Table 5).
8. Perform thermocycling reaction. A typical qPCR thermal profile is as follows: initial denaturation of 94 °C for 75 s, followed by 40–45 cycles of 94 °C for 15 s, then the annealing temperature (i.e., 7 °C below the melting temperature of the primers) for 60 s.

Table 4
Reverse primer/probe and master mix aliquot scheme

	1	2	3	4	5
A	1R	2R	3R	4R	5R
B	1R	2R	3R	4R	5R
C	1R	2R	3R	4R	5R
D	1R	2R	3R	4R	5R
E	1R	2R	3R	4R	5R

Table 5
Forward primer/probe aliquot scheme

	1	2	3	4	5
A	1F	1F	1F	1F	1F
B	2F	2F	2F	2F	2F
C	3F	3F	3F	3F	3F
D	4F	4F	4F	4F	4F
E	5F	5F	5F	5F	5F

9. Evaluate results. Reactions are evaluated based on C_T and the change in fluorescence intensity from background to maximum amplification (Δ fluorescence). Reactions with the minimum C_T are considered equivalent, unless there are differences in Δ fluorescence. We have found that the PCR modules on the ESP's Microfluidic Block (MFB) are more robust with assays with (even slightly) higher delta fluorescence on benchtop qPCR instruments. Those reactions that have the lowest C_T and largest Δ fluorescence should be chosen for further testing.

3.2.6. Optimizing Sensitivity and Efficiency

1. Reactions considered optimal based on these criteria are further evaluated by standard curve comparisons. Amplify a dilution series of linearized plasmid standards (minimum 10^0 copies per reaction, maximum 10^8 copies per reaction), in duplicate.
2. Evaluate the sensitivity of the reaction. It is imperative that the sensitivity (e.g., ten copies per reaction) is sufficient to detect the abundances of the gene quantified, in a typical sample from the environment of interest.

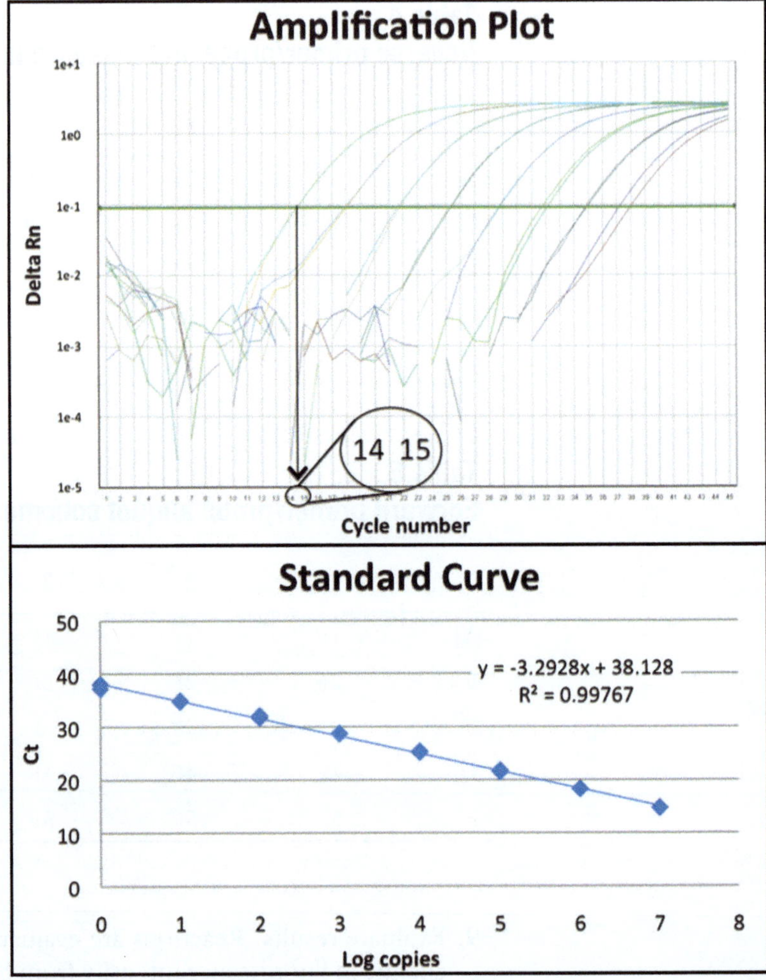

Fig. 3. Example of an optimized qPCR reaction. The C_T for each dilution of the standard is calculated at the same threshold by determining the cycle number at which each amplification plot crosses the threshold (see *arrow*). Data was generated on the Applied Biosystems 7500 Real Time PCR System. A standard curve is created by plotting the log of the copy number vs. the C_T for that amplification plot. In the example given here, the assay has the sensitivity to quantify less than ten copies per reaction. Using the slope of −3.29, the calculated efficiency of the standard curve is 101.3 %.

3. Plot the log of the copy number (e.g., for 10^4, plot 4) vs. the C_T of the reactions with that number of copies (Fig. 3). Do this for all of the amplified dilution series.
4. Fit a trendline to the dilution series data, and display equation and r^2. Use the slope of the trendline equation to calculate the standard curve efficiency:

$$\text{Efficiency} = ((10^{(-1/\text{slope})}) - 1) \times 100.$$

For accurate quantification, the efficiency must be between 90 and 110 %. See Note 15.

3.2.7. Specificity

1. For a first estimate of the specificity of your designed primers and probes, use BLAST to compare the sequences to NCBI's nonredundant sequence database (http://www.ncbi.nlm.nih.gov/tools/primer-blast/). The primer and probe sequences should align to the target gene best, and all other hits should be of low quality or nonexistent.
2. For a more accurate specificity evaluation of the qPCR assay perform dissociation curve analysis with SYBR Green chemistry (which is useful for testing primer specificity though TaqMan chemistry is used for the designed assay). Amplify samples from the environment, a negative control, and a positive control using the designed qPCR primers, minus the probe, with SYBR Green I (Biorad), according to the manufacturer's instructions and using the optimized reaction conditions. Add a dissociation curve to the end of the PCR reaction by slowly increasing the temperature from the primer annealing temperature to 96 °C, and collecting data during the ramp. As PCR products and primer dimers melt, SYBR molecules dissociate, and the instrument records a drop in fluorescence. SYBR dissociation profiles (the temperature plotted against the derivative of the fluorescence) (see Sect. 4) should have a single distinct peak in fluorescence at the melting temperature of the intended amplification product, at the same temperature as the positive control reaction, if the reaction is specific and there are no primer dimers. There should be no peak at that temperature in the no-template control (NTC).
3. If no dissociation was performed, a less sensitive but alternate method is gel visualization of amplified product. After amplification, this reaction can be loaded on an agarose gel, and visualized. A single band should be visible on the gel. See Notes 16 and 17.
4. If assays were design to target clades of microbes within a phylogenetic group, each assay should be tested for cross-reactivity across a dilution series of the non-target standards. If cross reactivity is high enough to obscure interpretation of results, it should be discarded.

3.3. Design of a High-Density Oligonucleotide Microarray (NimbleGen, Agilent, or CustomArray Type)

Oligonucleotide microarrays synthesized in situ can feature tens of thousands of probes per chip, can be made multiplex (multiple chips on one slide that can be hybridized to multiple samples), and can be easily modified. The focus of this chapter is on the design of long probes (50–60 nt), as needed for microarrays listed in the Table 6, and Affymetrix technology (that synthesizes 25-nt-long probes) is not addressed here.

1. The set of representative sequences from Sect. 3.1 is used for designing microarray probes. First, determine which sequences

Table 6
Oligonucleotide microarrays, for which probe design is considered in this chapter

Microarray	Address	Features[a]
Agilent	http://www.agilent.com	1 M, 2× 400K, 4× 180K, 8× 60K, 244K, 2× 105K, 4× 44K, 8× 15K
CustomArray	http://www.customarrays.com	4× 2K, 12K, 90K
NimbleGen	http://www.nimblegen.com	12× 135K, 385K, 4× 72K

[a]Characterized by how many chips are on one side and by number of probes: 2× 400K means duplex of 400,000 probes

require gene-specific (probes target the single sequence) or group-specific probes, Note 18.

2. Include sequences for control genes: positive and negative control (preferably, according to External RNA Control Consortium, (49)), Note 19.
3. General rules apply for both gene-specific and group-specific probes (as a starting point):
   (a) Probe melting temperature: 85–90 °C
   (b) GC range: 40–65 %
   (c) Length range: 55–60 nt
   (d) Temperature threshold for secondary structure: 65 °C
   (e) Temperature threshold for cross-hybridization: 65 °C
   (f) No runs of the same nucleotides longer than 5 nt
   (g) Number of probes per gene: 10
   (h) Maximum distance from 3′-end: 1,500 bp, Note 20
4. For group-specific probe design, align sequences in each group using ClustalW, and use ARB (Note 21), CommOligo 2.0 (46), or HiSpOD (50) for probe design. Group-specific probes should have ≥96 % of sequence identity to the target, ≥35-base of continuous stretch, and ≤−60 kcal/mol of free energy within the group (51).
5. For gene-specific probe design, use one of the following software in addition to software listed for group-specific probes: OligoArray 2.1 (52), OligoWiz (53), YODA (54), eArray (https://earray.chem.agilent.com/earray/). Gene-specific probes should have >=90 % of sequence identity to the target, >=20-base continuous stretch of identical nucleotides, and <=−35 kcal/mol free energy (55). Also, custom design of gene-specific probes for high-density arrays is offered at NimbleGen, Agilent, or CustomArrays.
6. Standard control probes (random oligonucleotides) are generated by a microarray supplier, Note 22.

7. All probes including control are tested for cross-hybridization in silico by running BLASTN against:
   (a) The whole database of target sequences
   (b) "Non-identical nucleotide database" in CAMERA
   (c) "All Sanger reads" and "All 454 reads" in CAMERA
   (d) Ribosomal genes in RDP

   The criteria for a nonspecific hit are >90 % sequence identity, >20-bases continuous stretches, or <−35 kcal/mol free energy (20). Probes that have nonspecific hits should be eliminated.
8. Hybridization behavior of probes is predicted with CalcOligo (56), http://www.calcoligo.info/, which returns weighed mismatches of the probes against the entire sequence database used for probe design. Based on the result matrix, reference sequences representing the desired diversity can be selected to use in the array validation (45).
9. The oligonucleotide probes are synthesized directly on the slide by the manufacturer (NimbleGen, CustomArrays, Agilent).
10. Specificity, sensitivity, and validation control:
    - Hybridize test samples of extracted nucleic acids from at least some cultures of targeted microorganisms, if available, and from natural environmental samples varying in concentrations of nucleic acid, Notes 23 and 24.
    - Validate microarray detection of selected genes with (RT)-qPCR.
11. Redesign is based on the results of specificity and sensitivity tests. Probes that yielded cross-hybridization, random probes that yielded high hybridization signal, and negative and positive probes that yielded erroneous results are eliminated.

4. Typical Protocol Results

4.1. Generation of a Gene Database

The generation of a gene database is described using the example of the *phnD* gene that codes for a phosphonate-binding protein in an ABC-type high-affinity transporter. The target organisms are the picocyanobacteria *Prochlorococcus* and *Synechococcus*. The product of the gene is responsible for scavenging phosphonate compounds, which are used as a phosphorus source in phosphate-limited open ocean environments.

These specific results were obtained at the time of writing, but as databases evolve the numbers of reads in these results will change. Nucleotide sequences for *phnD* from all available genomes of *Prochlorococcus* and *Synechococcus* and an amino acid sequence from *Synechococcus* WH8102 were acquired from ProPortal (http://proportal.mit.edu/), with a total of 26 nucleotide sequences.

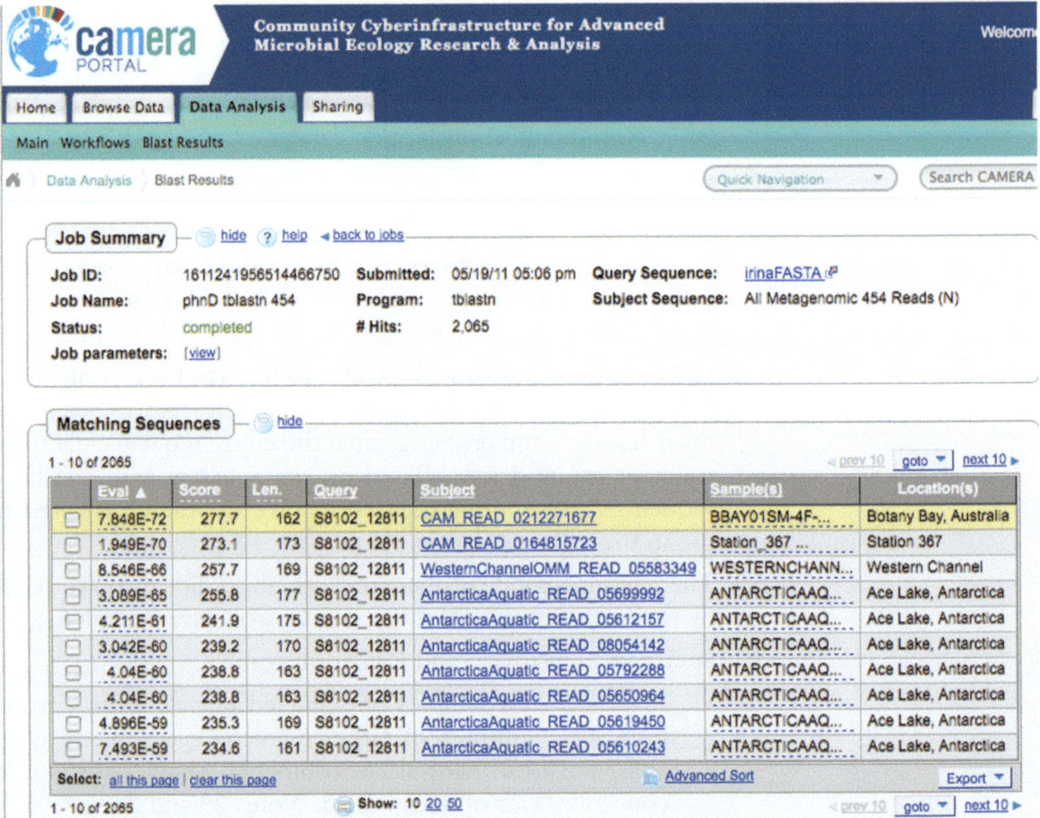

Fig. 4. TBLASTN results for PhnD against "All 454 Metagenomic Reads" in CAMERA.

TBLASTN in CAMERA with PhnD from *Synechococcus* WH8102 yielded 1806 reads from "All Metagenomic Sanger reads" and 2065 reads from "All Metagenomic 454 reads" (Fig. 4). The CSV files of results with metadata were exported into Excel to filter by "Habitat," leaving only unique hits (subject sequences) that originated from the marine environment, with a sequence length of at least 60 amino acids and a percent similarity to the query of at least 50 % (Fig. 5). This resulted in a total of 990 sequences. Next, reciprocal BLASTN ("All Prokaryotic Genomes") and BLASTX ("All Prokaryotic Proteins") in CAMERA were performed, with the FASTA file of subject sequences combined from 454 and Sanger TBLASTN filtered results. The results of BLASTX were filtered by the following criteria:

(a) Annotation of the hit (The "Subject Seq" column should have "phosphonate," "transporter," and/or "binding")

(b) Length of the hit (a protein length of about 300 amino acids)

(c) Percent similarity to the hit of at least 40 % over a length of 20 amino acids

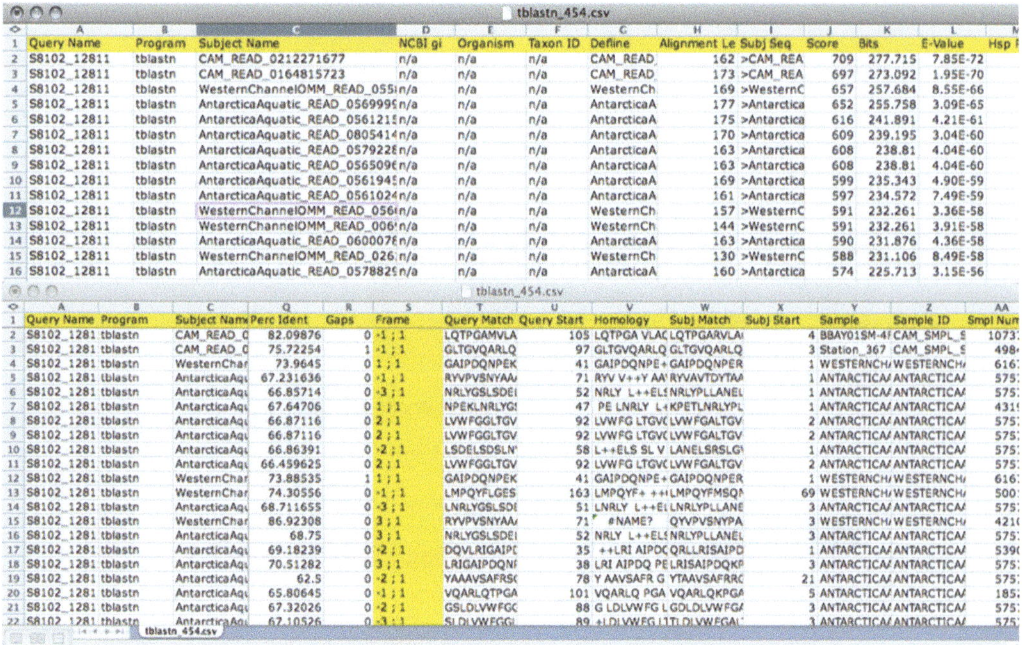

Fig. 5. TBLASTN results obtained from CAMERA and exported into Excel. The results ("Subject Name") are then filtered by Alignment Length, Percent Similarity, and/or E-Value, Habitat. Note that alignment length is shown in amino acid residues. On the *bottom panel*, the "Frame" column shows that if there is a minus sign, the subject sequences need to be converted into the reverse-complement strand.

The results of BLASTN were filtered to leave sequences that hit *Prochlorococcus* or *Synechococcus* with at least 75 % similarity on the nucleotide level over 60-nt alignment length, which corresponded to the highest *E*-value of 6.33*E*–06. The IDs for selected sequences (Column "Query Name") were saved as a separate column, and nucleotide sequences for the selected IDs were obtained from the TBLASTN result table. The sequences were trimmed at 5′, 3′-ends according to hit position from BLASTX (Fig. 6), and sequences that are at least 180 nt were selected. At this step, total 706 sequences were confirmed to be *phnD* genes and to originate from marine picocyanobacteria. The FASTA file of selected sequences was combined with sequences of *phnD* gene obtained from the genomes, and CD HIT EST was used to cluster the sequences at 90 % nt similarity with the rest of parameters left with default settings (the resulting statistics are shown in Fig. 7). This resulted in 184 representative sequences.

4.2. Results for Quantitative PCR Assay Design

While databases of environmental gene sequences enhance the ability to amplify diverse target sequences from the environment, degenerate PCR remains the most difficult procedure in molecular assay design. It is best to use a minimal number of cycles in the PCR to amplify a band of the correct size when visualized on the

blastX_sel.csv

	A	T	U	V	W	X	Y	Z	AA
1	Query Name	Query Mat	Query Star	Homology	Subj Match	Subj Start	Sample	Sample ID	Smpl Num
13	JCVI_READ_1104596112765	GKGFTKELTS	839	G+GFTKEL SV	GEGFTKELKS	242	n/a	n/a	n/a
14	JCVI_READ_1099679481156	LMPEYFLNDA	818	LMPEY+LNDA	LMPEYYLNDA	158	n/a	n/a	n/a
15	HF_READ_04951163	KTKNLRFTFG	791	K KNLRFTFGS	KLKNLRFTFG	136	n/a	n/a	n/a
16	JCVI_READ_1105499837202	KRVRFSGSHD	788	KRV FSGSHD	KRVGFSGSHD	177	n/a	n/a	n/a
17	HF_READ_04941671	KILDMFNAKK	766	KILDMFNAKK	KILDMFNAKK	269	n/a	n/a	n/a
18	HF_READ_05016014	KKAGFSGSHD	733	KKAGFSGSHD	KKAGFSGSHD	173	n/a	n/a	n/a
19	JCVI_READ_1103242873253	EKSSFSGSHD	732	#NAME?	KKAGYSGSHD	173	n/a	n/a	n/a
20	JCVI_READ_1105333460831	AQRDIDTGIH	708	AQR+ID V	AQREIDAEFT	115	n/a	n/a	n/a
21	HF_READ_04975994	QRTSNLELFW	692	#NAME?	KRTSNLELFW	212	n/a	n/a	n/a
22	HF_READ_04993995	WENTLKNNP	641	WEN LKNNPK	WENNLKNNP	203	n/a	n/a	n/a
23	JCVI_READ_1103242720382	NLKNFLLSIYL	637	NLKNFL+S YL	NLKNFLISSYL	3	n/a	n/a	n/a
24	JCVI_READ_1108829626926	LKVGAIPDQN	637	LKVGAIPDQN	LKVGAIPDQN	29	n/a	n/a	n/a
25	JCVI_READ_1100018129339	LKVGAIPDQN	606	LKVGAIPDQN	LKVGAIPDQN	29	n/a	n/a	n/a
26	JCVI_READ_1103180395350	VLKVGAIPDQ	599	VLKVGAIPDQ	VLKVGAIPDQ	28	n/a	n/a	n/a
27	HF_READ_05035586	PVFSNPKVLK	584	PVFSNPKVLK	PVFSNPKVLK	21	n/a	n/a	n/a
28	JCVI_READ_1105333597449	LKVGAIPDQN	576	LKVGAIPDQN	LKVGAIPDQN	29	n/a	n/a	n/a
29	JCVI_READ_570904	NPKVLKVGAI	567	NPKVLKVGAI	NPKVLKVGAI	25	n/a	n/a	n/a

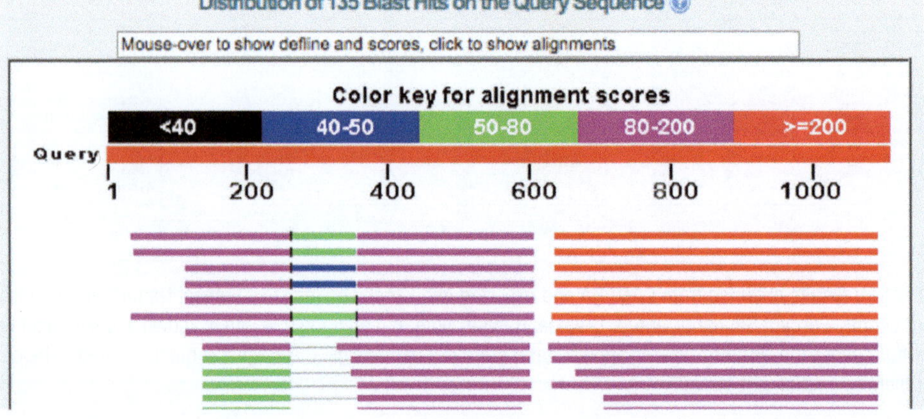

Fig. 6. *Top*: BLASTX results in Excel: JCVI_READ_1103242720382 is an example of an environmental sequence that has to be trimmed at 5′-end. *Bottom*: The BLASTX results for this sequence were visualized in NCBI BLASTX. The sequence contains parts of two genes, and the second one is *phnD*. Thus, 5′-end of this sequence was trimmed at 637 nt position.

agarose gel (Fig. 2). Since nontarget genes will likely amplify in degenerate PCR, the band of the correct size should be extracted, avoiding surrounding pieces of the gel.

Presuming that the gene is not 100 % conserved, a diversity of sequences should be obtained from the environment. This may require a few sequencing runs. The optimal scenario is a 96-well plate containing a few target genes from nontarget organisms. In theory, this would mean that the diversity of targets amplified from the environment was large enough to span from the target organism to nontarget organisms so that any potential targets in the sample would have been amplified.

The amplification plot and standard curve should look similar to Fig. 3, though the sensitivity in this example is rare. It is often the case that the lower concentration standards skew the efficiency of the PCR reaction. If this is the case, oftentimes increasing the primer and $MgCl_2$ concentrations can help. Also, one should not expect 100% consistent amplification of the 10^0 (and sometimes 10^1) standard concentration(s), and if the efficiency of the curve is not 90–110 %, eliminating that standard from the curve will likely

Raw output
Download all files
Browse clusters by size
Browse clusters by length
Distribution of clusters

```
Sequence type   DNA
No. sequences   184
Longest sequence        1200
Shortest sequence       190
Average length  690
Total letters   126981
Total N letters 7
Total non N     126974
Sequences with N        4
```

Raw output
Download all files
Browse clusters by size
Browse clusters by length
Distribution of clusters

```
>Cluster 0
0       1200nt, >HF_READ_04858478... *
1       844nt, >JCVI_READ_1103359321337... at +/94%
2       854nt, >HF_READ_04990679... at +/93%
3       935nt, >JCVI_READ_1091141761184... at +/91%
4       970nt, >JCVI_READ_1092959703068... at +/92%
5       553nt, >CAM_READ_0234777981... at +/93%
6       327nt, >HF_READ_02109625... at +/96%
7       468nt, >CAM_READ_0231598421... at +/96%
8       303nt, >HF_READ_02684535... at +/97%
>Cluster 1
0       452nt, >CAM_READ_0236104633... at +/96%
1       281nt, >HF_READ_02750685... at +/98%
2       1033nt, >JCVI_READ_1103466314991... at +/91%
3       910nt, >HF_READ_05499600... at +/90%
4       960nt, >JCVI_READ_1091140934553... at +/93%
5       1178nt, >HF_READ_04853245... *
6       911nt, >HF_READ_05602508... at +/90%
7       844nt, >JCVI_READ_87756... at +/91%
```

Fig. 7. CD-HIT-EST output. *Top*: Statistics of the output *phnD* representative sequences. The input was 732 sequences. *Bottom*: Clusters at 90 % similarity are shown, where a *an asterisk* indicates the representative sequence for the cluster.

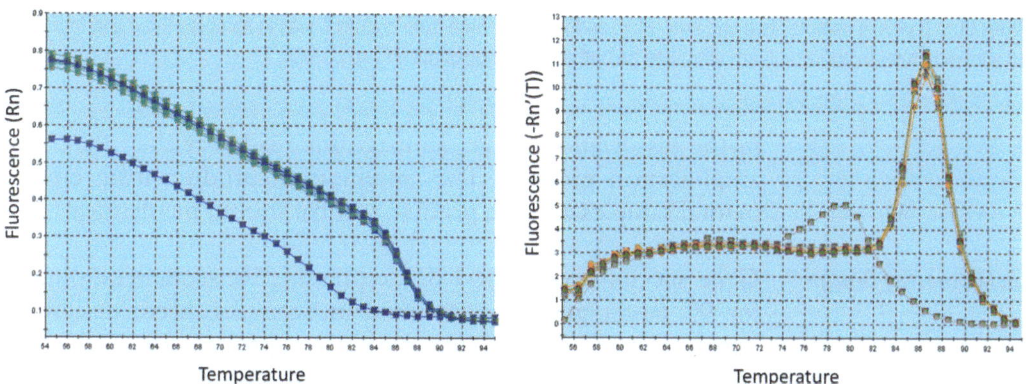

Fig. 8. Example of qPCR dissociation curves. The *first plot* is the fluorescence plotted by cycle number. The *second plot* is the usual output from qPCR software. The *first peak* in that plot is the NTC reaction, which should not have a major peak corresponding to the cycle number of the other reactions' peaks. The remaining samples are examples of what a positive control and several environmental samples should look like. There is only one peak in these profiles, indicating a specific amplification reaction. Curve was generated on a Stratagene M×3000P.

produce a better curve. Good target sensitivity must be achieved, however, so amplifying 5×10^1 standards when 10^1 is not quantifiable, for example, will provide additional sensitivity information.

Figure 8 is an example of a dissociation curve, used to determine the reaction's specificity. While the NTC shows a melt peak

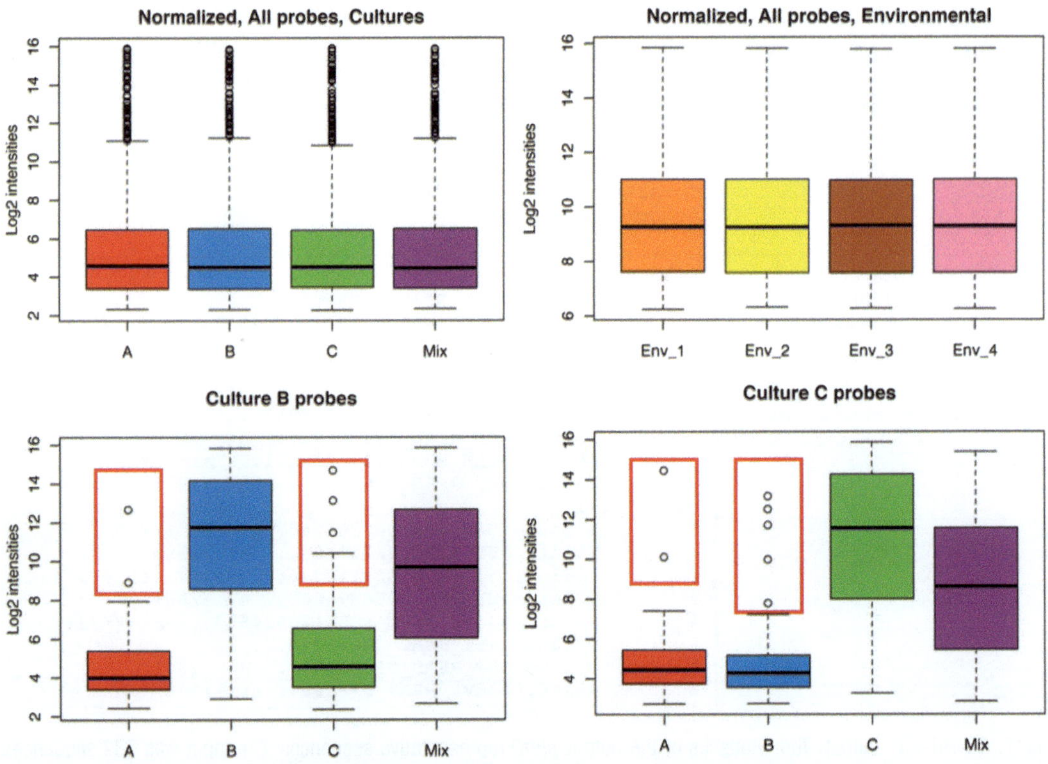

Fig. 9. Example of quality and specificity control for environmental microarray. *Top*: Distribution of normalized hybridization signal intensities for all probes (Samples from cultures A, B, C, and Mix of A, B, C; and four environmental samples). *Bottom*: Distribution of hybridization signal intensities for probes specific to culture B only (*left*) or culture C only (*right*). Transcriptome from culture B was in samples two and four (Mix), and transcriptome from culture C was in samples three and four. Probes that showed cross-hybridization are identified as outliers in the boxplots (in *red boxes*).

in the dissociation curve, its magnitude is lower than that of the amplification product. The amplification product from environmental samples and the positive control should be identical.

4.3. Microarray Design

Gene-specific probes for 1,371 representative sequences, obtained for 97 target genes as described in Sect. 3.1, were designed at NimbleGen (Roche). Generating 10 probes for each target sequences plus control probes resulted in 14,950 probes total. The probes were checked for cross-hybridization by BLASTN against the original dataset, the ribosomal gene database in RDP, and the "Non-identical nucleotide sequences" database in CAMERA. Probes (160 total) that had hit to a nontarget sequence with >95 % nt similarity over a 47-nt length were omitted. Five replicates of each probe (a set of unique probes is called a BLOCK) were randomly printed on the 4× 72K chip (one slide contains four chips of 72,000 probes,). To test for specificity, samples from targeted cultures and a mix of these samples were hybridized to the microarray at NimbleGen (Fig. 9).

5. Notes

5.1. Notes for Establishing a Gene Database

1. If the gene is highly conserved and the number of sequences available from genomes and clones is sufficient to represent the extent of diversity, the search in metagenomic and metatranscriptomic is likely redundant. Proceed with step 6.
2. Alternatively, the functional identity of hits can be confirmed with the sequence-region-consensus program HMMER (57). For this, a profile hidden Markov model is built using hmmbuild (57) on the alignment of selected full-length gene sequences (seed sequences) with ClustalW (58). This approach allows confident identification of an environmental sequence based on the conserved domains versus annotation-based confirmation from BLASTX results; thus, this approach is preferable if seed sequences can be identified.
3. If the gene model is built (as in Note 2), the environmental sequences are aligned to the seed alignment in ARB, and 5′, 3′-ends of the subject reads are trimmed in agreement with the alignment.

5.2. Notes for Quantitative PCR Assay Design

4. If comprehensive pair of primers cannot be designed according to these rules, it is necessary to design a number of primer sets for which these rules do apply. The easiest method is to evaluate a phylogenetic tree of the gene to choose sequences to use for primer design based on their phylogenetic clustering (i.e., eliminate more divergent sequences because the closer the clustering, the easier the primer design).
5. Design several alternate versions of degenerate primers if possible to maximize the chances of success with degenerate PCR. Additionally, amplifying from a diversity of samples from the target environment increases the chances of obtaining a comprehensive representation of sequences, so much time should be dedicated to this part of the protocol.
6. If a stretch of similar sequences has a potential to act as a primer region, but the resulting primers are too short to use in PCR (the T_m is too low), it is possible to add a stretch of ca. 5 bp that are not homologous to the target sequence, to the 5′ end of the primer, to optimize the T_m without affecting the amplification.
7. A touchdown PCR is often used to increase the specificity of degenerate PCR.
8. It is best practice to include a positive control reaction from a cultured organism that contains the gene of interest, and a negative control reaction from an organism that does not.

9. A temperature gradient thermocycler allows efficient optimization of annealing temperature. Use the same PCR conditions in a single run, with annealing temperature as the only variable, changing by 2 °C increments.
10. Care should be taken to recover all the diversity of the target gene in the environment. Strategies to increase diversity in the clone library include using a lower number of cycles in the degenerate PCR and using a higher diversity of environmental samples (representing different times of the year, for example).
11. Aside from manual primer design (as described above), a primer design program such as Primer Express (Applied Biosystems) or Primer3 (59), as well as ARB (also in (45)), can be used for primer and probe design, taking care to verify that the designed primers and probes anneal to all target sequences.
12. Accuprime Supermix II (Invitrogen Corp.) contains $MgCl_2$. The protocol below can be performed once with 3 mM additional $MgCl_2$ by using $MgCl_2$ rather than 115.2 μl water in the master mix recipe. If reactions are not suitable under these conditions, $MgCl_2$ can be increased to a maximum of 4 mM or it can be eliminated from the reaction.
13. Try with 0.4 μM probe (these instructions) and separately with 0.2 μM probe if necessary.
14. Avoid freeze/thaw cycles with the probe and the standards.
15. If the efficiency does not fall between 90 and 110 %, return to the optimization procedure and increase the added $MgCl_2$ concentration in the reactions to 4 mM (it will be necessary to use 50 mM stock $MgCl_2$). Optimize primer concentrations with 4 mM $MgCl_2$ per reaction. If adequate standard curve efficiencies are not achieved under these new conditions, you should return to your less optimized reaction conditions, but avoiding the least successful conditions. In the end, it may be necessary to redesign the primers and/or probe.
16. If multiple amplicons are created with the primers, the efficiency of the reaction may be affected with environmental samples. Sequence a clone library based on "low cycle number amplifications" from several environmental samples in order to assess the diversity of the nonspecific amplicons. The probes should not cross-hybridize to the unspecific amplicons, and the targeted gene should be the dominant amplification product. It may be necessary to redesign your assay to avoid cross-reactivity.
17. A reference for optimization and a guideline for publication requirements can be found in the Minimum Information for Publication of Quantitative Real-Time PCR Experiments (MIQE) Guidelines (60). This publication includes information on best practices in order to obtain reliable quantitative

results, and a table of the minimum information required for publication of qPCR data, to be used by both authors and reviewers.

5.3. Notes for Microarray Design

18. The choice to design group-specific probes is optional and depends on the sequences and the goal.
19. It is generally recommended that control probes constitute 10 % of all probes on the array (61).
20. This is due to the 3′-end bias of double-stranded cDNA synthesis.
21. Instructions for using ARB for probe design can be found in documentation for ARB (http://www.arb-home.de) and in (45).
22. Control random oligonucleotides can yield relatively high signal from environmental samples. In this case, the threshold of nonspecific hybridization is set as a mean of 5 % of the lowest signals (33).
23. Samples are extracted by a preferred method of extraction. For extraction of both DNA and RNA from one sample, we recommend the guanidinium thiocyanate/acid phenol–chloroform method (62, 63) with addition of bead-beating step for efficient cell lysis. The extraction is followed by purification by ethanol precipitation or using commercial columns (such as Qiagen). Since phenol inhibits reverse transcription, make sure to extract with chloroform at least three times after using phenol. RNA samples need to be DNase treated. For concentrating DNA or RNA, use ethanol precipitation. Total RNA required for processing and hybridization is 10–30 μg, which is often impossible to obtain from some samples (open ocean environmental samples, for example). Therefore, DNA or RNA from environmental samples should be amplified, and in this case, the test samples should also be amplified (not necessary if environmental samples are not amplified before hybridization). Synthesis of double-stranded cDNA from RNA is done using the SuperScript double-stranded cDNA synthesis kit (Invitrogen, Carlsbad, CA). Quality of DNA/RNA should be assessed using a NanoDrop spectrophotometer and gel electrophoresis or Bioanalyzer (Agilent) before and after amplification, and for RNA, after cDNA synthesis. The A260/A280 and A260/A230 ratios measured with NanoDrop should be of at least 1.8. RNA should appear intact when analyzed on the gel (ratio of 23S to 16S ~2). Quantities of DNA and cDNA for hybridization: 1–5 μg of DNA/cDNA. Labeling and hybridization of the samples to the chip and getting raw signals are performed at a microarray facility. For data normalization and analysis, Bioconductor (http://www.bioconductor.org) is

recommended. The data and annotation from the microarray should follow standards developed by the Minimum Information About a Microarray Experiment (64, 65).

24. If cultured representatives are not available, directly amplified genes or transcripts can be used for testing cross-reactivity and sensitivity.

6. Future Directions

Environmental molecular assays are most robust when a significant amount of effort is invested in their design. This initial investment results in minimal time spent troubleshooting in subsequent research applications. The more robust assays result in straightforward transitions to robotic implementation.

The Environmental Sample Processor (ESP; described in Scholin, Doucette, Chap. 9 and (66)) is a sample acquisition and delivery system amenable to various downstream analyses, and is therefore the optimal platform for molecular assay implementation. qPCR on the ESP has been used in Monterey Bay to quantify clades of cyanobacteria (*Synechococcus*), ammonia-oxidizing Archaea (*Thaumarchaeota*), and alpha proteobacteria (*Roseobacter* and SAR11 clades) at near-shore station M0 (67, 68). ESP deployments are coupled with chemical and physical sensors to provide environmental data to evaluate ecological factors that drive population dynamics. While the ESP Generation 2 platform provides superior temporal resolution of microbial processes, it is currently limited to five target assays per sample; so a consensus of the best indicators of ecosystem dynamics should be agreed upon. While whole cell sample archives can be used to interrogate RNA (69), the instrument does not allow quantification of messenger RNA in real time. ESP Generation 3 technology will likely have the capability for a larger suite of qPCR detection assays and microarray capabilities, allowing detection of functional genes and their expression.

The high-density oligonucleotide microarray designed based on sequence diversity obtained from metagenomic and metatranscriptomic studies and targeting about 1,350 sequences was used to assess gene expression in natural cyanobacterial populations in South Pacific Ocean (70). This design approach was extended to build a comprehensive marine microbial functional microarray as a result of a scientific community effort (MicroTOOLS workshop II, sponsored by the Gordon and Betty Moore Foundation). The major challenge in designing a comprehensive marine microbial chip is limited knowledge on natural sequence diversity for marine phytoplankton in general as most data comes from surface oceans and particularly underrepresentation of data for eukaryotes in comparison to prokaryotes. One of the advantages using oligonucleotide microarray technologies such as NimbleGen and CustomArray is that the design can be easily modified. The availability of data will increase in the near future; thus, an automated process of gene database establishment is needed to consider constant database updates. Implementation of explorative probes (71, 72) may help to account for the lack of sequence diversity information, but this approach could only be used for a limited number of genes. With the development and optimization of a high-density microarray technology for marine microbial communities in the lab, the implementation of microarray on ESP generation 3 becomes feasible. In addition to common microarray challenges such as probe specificity, defining the control probes, hybridization issues, and data acquisition and analysis, there are ESP-specific challenges. Due to stability issues, the choice of detector for signal hybridization detection is one of the main keystones in implementing microarray on ESP, and electrochemical signal may be preferred. Another challenge is amplification of the nucleic acid samples at sea; this is a multistep process and requires a set of enzymes that lose activity when not stored at −20 °C. Despite the difficulties, the efforts of implementation of a well-designed microarray will be justified by the high-resolution (both in time and space) data on changes in microbial population composition and activity—view on marine microbial ecology that is not possible with present frequency of sampling.

Acknowledgements

The authors would like to thank Christina Preston, Kendra Turk, Tracy Cote, and Jim Tripp for help with protocol development. This work was funded by the MEGAMER (Microbial Environmental Genomics Applications, Modeling, Experimentation and Remote Sensing) grant by the Gordon and Betty Moore Foundation and the Center for Microbial Oceanography: Research and Education.

References

1. Okano Y, Hristova KR, Leutenegger CM, Jackson LE, Denison RF, Gebreyesus B et al (2004) Application of real-time PCR to study effects of ammonium on population size of ammonia-oxidizing bacteria in soil. Appl Environ Microbiol 70:1008–1016
2. Randa MA, Polz MF, Lim E (2004) Effects of temperature and salinity on Vibrio vulnificus population dynamics as assessed by quantitative PCR. Appl Environ Microbiol 70:5469–5476
3. Thompson JR, Randa MA, Marcelino LA, Tomita-Mitchell A, Lim E, Polz MF (2004) Diversity and dynamics of a North Atlantic Coastal Vibrio Community. Appl Environ Microbiol 70:4103–4110
4. Church MJ, Short CM, Jenkins BD, Karl DM, Zehr JP (2005) Temporal patterns of nitrogenase gene (nifH) expression in the oligotrophic North Pacific Ocean. Appl Environ Microbiol 71:5362–5370

5. Noble RT, Griffith JF, Blackwood AD, Fuhrman JA, Gregory JB, Hernandez X et al (2006) Multitiered approach using quantitative PCR to track sources of fecal pollution affecting Santa Monica Bay, California. Appl Environ Microbiol 72:1604–1612
6. Blackstone GM, Nordstrom JL, Bowen MD, Meyer RF, Imbro P, DePaola A (2007) Use of a real time PCR assay for detection of the ctxA gene of vibrio cholerae in an environmental survey of Mobile Bay. J Microbiol Methods 68:254–259
7. Smith CJ, Nedwell DB, Dong LF, Osborn AM (2007) Diversity and abundance of nitrate reductase genes (narG and napA), nitrite reductase genes (nirS and nrfA), and their transcripts in estuarine sediments. Appl Environ Microbiol 73:3612–3622
8. Tai V, Palenik B (2009) Temporal variation of Synechococcus clades at a coastal Pacific Ocean monitoring site. ISME J 3:903–915
9. Mosier AC, Francis CA (2011) Determining the distribution of marine and coastal ammonia-oxidizing archaea and bacteria using a quantitative approach. Methods Enzymol 486:205–221
10. Paerl RW, Johnson KS, Welsh RM, Worden AZ, Chavez FP, Zehr JP (2011) Differential distributions of Synechococcus subgroups across the California current system. Front Microbiol 2:1–22
11. Holtzendorff J, Marie D, Post AF, Partensky F, Rivlin A, Hess WR (2002) Synchronized expression of *ftsZ* in natural Prochlorococcus populations of the Red Sea. Environ Microbiol 4:644–653
12. Wagner VE, Bushnell D, Passador L, Brooks AI, Iglewski BH (2003) Microarray analysis of Pseudomonas aeruginosa quorum-sensing regulons: effects of growth phase and environment. J Bacteriol 185:2080–2095
13. Treusch AH, Leininger S, Kletzin A, Schuster SC, Klenk HP, Schleper C (2005) Novel genes for nitrite reductase and Amo-related proteins indicate a role of uncultivated mesophilic crenarchaeota in nitrogen cycling. Environ Microbiol 7:1985–1995
14. Mock T, Samanta MP, Iverson V, Berthiaume C, Robison M, Holtermann K et al (2008) Whole-genome expression profiling of the marine diatom Thalassiosira pseudonana identifies genes involved in silicon bioprocesses. Proc Natl Acad Sci 105:1579–1584
15. Ilikchyan IN, Mckay RML, Zehr JP, Dyhrman ST, Bullerjahn GS (2009) Detection and expression of the phosphonate transporter gene phnD in marine and freshwater picocyanobacteria. Environ Microbiol 11:1314–1324
16. Tetu SG, Brahamsha B, Johnson DA, Tai V, Phillippy K, Palenik B et al (2009) Microarray analysis of phosphate regulation in the marine cyanobacterium Synechococcus sp. WH8102. ISME J 3:835–849
17. Rhee SK, Liu X, Wu L, Chong SC, Wan X, Zhou J (2004) Detection of genes involved in biodegradation and biotransformation in microbial communities by using 50-mer oligonucleotide microarrays. Appl Environ Microbiol 70: 4303–4317
18. Brodie EL, Desantis TZ, Joyner DC et al (2006) Application of a high-density oligonucleotide microarray approach to study bacterial population dynamics during uranium reduction and reoxidation. Appl Environ Microbiol 72:6288–6298
19. DeSantis TZ, Brodie EL, Moberg JP, Zubieta IX, Piceno YM, Andersen GL (2007) High-density universal 16 S rRNA microarray analysis reveals broader diversity than typical clone library when sampling the environment. Microb Ecol 53:371–383
20. He Z, Deng Y, Van Nostrand JD et al (2010) GeoChip 3.0 as a high-throughput tool for analyzing microbial community composition, structure and functional activity. ISME J 4: 1167–1179
21. He Z, Gentry TJ, Schadt CW et al (2007) GeoChip: a comprehensive microarray for investigating biogeochemical, ecological and environmental processes. ISME J 1:67–77
22. Taroncher-Oldenburg G, Griner EM, Francis CA, Ward BB (2003) Oligonucleotide microarray for the study of functional gene diversity in the nitrogen cycle in the environment. Appl Environ Microbiol 69:1159–1171
23. Tiquia SM, Wu L, Chong SC, Passovets S, Xu D, Xu Y, Zhou J (2004) Evaluation of 50-mer oligonucleotide arrays for detecting microbial populations in environmental samples. Biotechniques 36:664–670
24. Moisander PH, Morrison AE, Ward BB, Jenkins BD, Zehr JP (2007) Spatial-temporal variability in diazotroph assemblages in Chesapeake Bay using an oligonucleotide *nifH* microarray. Environ Microbiol 9:1823–1835
25. Moisander PH, Shiue L, Steward GF, Jenkins BD, Bebout BM, Zehr JP (2006) Application of a nifH oligonucleotide microarray for profiling diversity of N_2-fixing microorganisms in marine microbial mats. Environ Microbiol 8:1721–1735
26. Ward BB, Eveillard D, Kirshtein JD, Nelson JD, Voytek MA, Jackson GA (2007) Ammonia-oxidizing bacterial community composition in estuarine and oceanic environments assessed

using a functional gene microarray. Environ Microbiol 9:2522–2538

27. Bulow SE, Francis CA, Jackson GA, Ward BB (2008) Sediment denitrifier community composition and nirS gene expression investigated with functional gene microarrays. Environ Microbiol 10:3057–3069
28. Wu L, Kellogg L, Devol AH, Tiedje JM, Zhou J (2008) Microarray-based characterization of microbial community functional structure and heterogeneity in marine sediments from the Gulf of Mexico. Appl Environ Microbiol 74: 4516–4529
29. Wu L, Liu X, Schadt CW, Zhou J (2006) Microarray-based analysis of subnanogram quantities of microbial community DNAs by using whole-community genome amplification. Appl Environ Microbiol 72:4931–4941
30. Wu L, Thompson DK, Li G, Hurt RA, Tiedje JM, Zhou J (2001) Development and evaluation of functional gene arrays for detection of selected genes in the environment. Appl Environ Microbiol 67:5780–5790
31. Rich VI, Konstantinidis K, DeLong EF (2008) Design and testing of 'genome-proxy' microarrays to profile marine microbial communities. Environ Microbiol 10:506–521
32. Rich VI, Pham VD, Eppley J, Shi Y, DeLong EF (2011) Time-series analyses of Monterey Bay coastal microbial picoplankton using a 'genome proxy' microarray. Environ Microbiol 13:116–134
33. Smith MW, Herfort L, Tyrol K et al (2010) Seasonal changes in bacterial and archaeal gene expression patterns across salinity gradients in the Columbia River coastal margin. PLoS One 5:e13312
34. Webb EA, Moffett JW, Waterbury JB (2001) Iron stress in open-ocean cyanobacteria (Synechococcus, Trichodesmium, and Crocosphaera spp.): identification of the IdiA protein. Appl Environ Microbiol 67:5444–5452
35. Lindell D, Post AF (2001) Ecological aspects of ntcA gene expression and its use as an indicator of the nitrogen status of marine *Synechococcus* spp. Appl Environ Microbiol 67:3340–3349
36. Sebastian M, Ammerman JW (2009) The alkaline phosphatase PhoX is more widely distributed in marine bacteria than the classical PhoA. ISME J 3:563–572
37. Jayakumar DA, Francis CA, Naqvi SWA, Ward BB (2004) Diversity of nitrite reductase genes (*nirS*) in the denitrifying water column of the coastal Arabian Sea. Aquat Microb Ecol 34: 69–78
38. Francis CA, Roberts KJ, Beman JM, Santoro AE, Oakley BB (2005) Ubiquity and diversity of ammonia-oxidizing archaea in water columns and sediments of the ocean. Proc Natl Acad Sci 102:14683–14688
39. Varaljay VA, Howard EC, Sun SL, Moran MA (2010) Deep Sequencing of a dimethylsulfoniopropionate-degrading gene (dmdA) by using PCR primer pairs designed on the basis of marine metagenomic data. Appl Environ Microbiol 76:609–617
40. LaRoche J, McKay RM, Boyd P (1999) Immunological and molecular probes to detect phytoplankton responses to environmental stress in nature. Hydrobiologia 401:177–198
41. Martiny AC, Kathuria S, Berube PM (2009) Widespread metabolic potential for nitrite and nitrate assimilation among *Prochlorococcus* ecotypes. Proc Natl Acad Sci 106:10787–10792
42. Altschul SF, Madden TL, Schaffer AA, Zhang JH, Zhang Z, Miller W et al (1997) Gapped BLAST and PSI-BLAST: a new generation of protein database search programs. Nucl Acids Res 25:3389–3402
43. Bodrossy L, Stralis-Pavese N, Murrell JC, Radajewski S, Weilharter A, Sessitsch A (2003) Development and validation of a diagnostic microbial microarray for methanotrophs. Environ Microbiol 5:566–582
44. Loy A, Bodrossy L (2006) Highly parallel microbial diagnostics using oligonucleotide microarrays. Clin Chim Acta Int J Clin Chem 363:106–119
45. Stralis-Pavese NAG, Sessitsch A, Bodrossy L (2011) Analysis of methanotroph community composition using a *pmoA*-based microbial diagnostic microarray. Nat Protoc 6:609–624
46. Li W, Godzik A (2006) Cd-hit: a fast program for clustering and comparing large sets of protein or nucleotide sequences. Bioinformatics 22:1658–1659
47. Huang Y, Niu B, Gao Y, Fu L, Li W (2010) CD-HIT Suite: a web server for clustering and comparing biological sequences. Bioinformatics 26:680–682
48. Ludwig W, Strunk O, Westram R, Richter L, Meier H, Yadhukumar et al (2004) ARB: a software environment for sequence data. Nucl Acids Res 32:1363–1371
49. Baker SC, Bauer SR, Beyer RP et al (2005) The external RNA controls consortium: a progress report. Nat Methods 2:731–734
50. Dugat-Bony E, Missaoui M, Peyretaillade E et al (2011) HiSpOD: probe design for functional DNA microarrays. Bioinformatics 27:641–648
51. He Z, Wu L, Li X, Fields MW, Zhou J (2005) Empirical establishment of oligonucleotide probe design criteria. Appl Environ Microbiol 71:3753–3760

52. Rouillard JM, Zuker M, Gulari E (2003) OligoArray 2.0: design of oligonucleotide probes for DNA microarrays using a thermodynamic approach. Nucl Acids Res 31:3057–3062
53. Wernersson R, Nielsen HB (2005) OligoWiz 2.0–integrating sequence feature annotation into the design of microarray probes. Nucl Acids Res 33:W611–W615
54. Nordberg EK (2005) YODA: selecting signature oligonucleotides. Bioinformatics 21: 1365–1370
55. Liebich J, Schadt CW, Chong SC, He Z, Rhee SK, Zhou J (2006) Improvement of oligonucleotide probe design criteria for functional gene microarrays in environmental applications. Appl Environ Microbiol 72:1688–1691
56. Kibbe WA (2007) OligoCalc: an online oligonucleotide properties calculator. Nucl Acids Res 35:W43–W46
57. Eddy SR (1998) Profile hidden Markov models. Bioinformatics 14:755–763
58. Thompson JD, Higgins DG, Gibson TJ (1994) CLUSTAL W: improving the sensitivity of progressive multiple sequence alignment through sequence weighting, position-specific gap penalties and weight matrix choice. Nucl Acids Res 22:4673–4680
59. Rozen S, Skaletsky H (1999) Primer3 on the WWW for general users and for biologist programmers. Bioinformatics Methods 132: 364–386
60. Bustin SA, Benes V, Garson JA, Hellemans J, Huggett J, Kubista M et al (2009) The MIQE guidelines: minimum information for publication of quantitative real-time PCR experiments. Clin Chem 55:611–622
61. Russel S, Roslin LAM (2009) Russel, R. Microarray technology in practice. Elsevier, Amsterdam
62. Chomczynski P, Sacchi N (2006) The single-step method of RNA isolation by acid guanidinium thiocyanate-phenol-chloroform extraction: twenty-something years on. Nat Protoc 1: 581–585
63. Farrell RE (1998) RNA methodologies. A laboratory guide for isolation and characterization. Academic, Amsterdam
64. Brazma A (2009) Minimum Information About a Microarray Experiment (MIAME)—successes, failures, challenges. Sci World J 9: 420–423
65. Brazma A, Hingamp P, Quackenbush J et al (2001) Minimum information about a microarray experiment (MIAME)—toward standards for microarray data. Nat Genet 29: 365–371
66. Preston C, Harris A, Ryan JP, Roman B, Marin R, Jensen S et al (2011) Application of quantitative PCR on a coastal mooring. PLoS One 6(8):e22522
67. Robidart JC, Preston CM, Paerl RW, Turk KA, Mosier AC, Francis CA et al (2012) Seasonal dynamics of Synechococcus and Thaumarchaeal populations resolved in real time with in situ instrumentation. ISME J 6(3):513–523
68. Varaljay VA, Robidart JC, Preston CM, Scholin CA, Moran MA (in prep) *In situ* time series of bacterial dimethylsulfopropionate degradation potential in Monterey Bay
69. Ottesen EA, Marin R III, Preston CM, Young CR, Ryan JP, Scholin CA et al (2011) Metatranscriptomic analysis of autonomously collected and preserved marine bacterioplankton. ISME J 5(12):1881–1895
70. Shilova IN, Tripp HJ, Robidart J et al. Marine Microbial microarray (MicroTOOLs project): development and application to the assessment of gene expression in the surface open ocean microbial communities. In prep.
71. Militon C, Rimour S, Missaoui M et al (2007) PhylArray: phylogenetic probe design algorithm for microarray. Bioinformatics 23: 2550–2557
72. Terrat S, Peyretaillade E, Goncalves O et al (2010) Detecting variants with metabolic design, a new software tool to design probes for explorative functional DNA microarray development. BMC Bioinformatics 11:478

Chapter 2

Development of a Capillary Waveguide Biosensor Analytical Module for Use with the MBARI Environmental Sample Processor

Harbans S. Dhadwal, JoAnn Radway, Josephine Aller, Paul Kemp, and James Clements

Abstract

The Monterey Bay Aquarium Research Institute's Environmental Sample Processor (ESP) is well established as an innovative sampling and instrument platform for sensors designed for in situ monitoring of microorganisms in the ocean. The platform can be deployed for periods up to 3 months and real-time data can be remotely downloaded at any time. A daughter platform, the microfluidic block (MFB), is used as an interface between the ESP and analytical modules to permit rapid deployment of new sensor technologies. We have developed a capillary waveguide biosensor (CWB) which employs nucleic acid hybridization for detection and quantification of specific microorganisms, and have integrated it with the MFB for use with the ESP. An important aspect of the CWB is the use of a combined capillary waveguide/hybridization surface that can be regenerated for repeated use. This chapter describes design issues related to the integration of the CWB and MFB and the development of coating protocols to maximize the operational life of the capillary.

Key words: Capillary waveguide, Fluorescence, Hybridization, Nucleic acid, RNA, MBARI environmental sample processor, Capillary waveguide biosensor, Microfluidic block, Capture probe, Ocean sensing

1. Introduction

Many biological processes occur on temporal and spatial scales that require near-continuous monitoring to fully understand their dynamics. Although a few biological variables (e.g., chlorophyll concentration) can be monitored via satellite or moored instruments, for many others no methods are available to collect data at appropriate temporal and spatial scales. The dynamics of microbial processes are particularly difficult to study except by intensive, ship-based studies. However, shipboard studies are expensive and

Sonia M. Tiquia-Arashiro (ed.), *Molecular Biological Technologies for Ocean Sensing*, Springer Protocols Handbooks,
DOI 10.1007/978-1-61779-915-0_2, © Springer Science+Business Media New York 2012

therefore occur too infrequently to capture more than snapshots of the true variability of microbial dynamic processes. In recent years, increasing attention has been focused on the need to develop automated, in situ methods to address the need for long-term, cost-effective measurements of microbial properties and processes, e.g., microbial abundance, biomass, productivity, and metabolism. Some of these properties are amenable to direct measurements (e.g., cell counts, concentrations of the biochemical products of metabolic pathways), while others can be examined through proxy variables such as the presence, abundance, and expression of specific genes. Automated, in situ sampling has enormous potential benefit for water quality management, human health, and the seafood industry, both economically and with respect to food safety. Applications include detection of human, fish, and shellfish pathogens in seawater, including water at aquaculture facilities, and detection and monitoring of harmful and nuisance species in a diverse range of environments. In conjunction with physical and chemical monitoring, perhaps based on the same monitoring platforms, real-time information on microbial populations will expand our understanding of (and ability to predict) microbially mediated processes in the ocean and their responses to stresses such as climate change or hypoxia. To provide this type of information, researchers at the Monterey Bay Aquarium Research Institute (MBARI) have developed the Environmental Sample Processor (ESP), which is now being commercially sold by McLane Laboratories (http://www.mclanelabs.com). The ESP's capabilities include large volume water sample collection, biomass concentration, and extraction of nucleic acids. Samples may be stored for later retrieval, or delivered to downstream analytical modules developed by other researchers as well as by MBARI.

Several analytical modules are already in development, including a filter-based sandwich hybridization assay to detect toxic microalgae, and an ELISA assay for the algal toxin domoic acid (1, 2). MBARI scientists have also developed a PCR module to amplify specific gene targets from a DNA sample delivered by the ESP core (1). For in-depth description of the ESP the reader is directed to the MBARI URL at http://www.mbari.com/.

Detection methods based on nucleic acids are often employed because they can be used to infer the presence and activity of specific organisms and metabolic pathways. Most such methods are based on nucleic acid hybridization, and those techniques involving direct detection of target nucleic acids (rather than amplification) generally rely upon the use of surface-bound probes to which the targets hybridize. The hybridized products may be detected by a variety of techniques, including fluorescent labeling, chemiluminescence, and plasmon resonance. Fluorescent labeling is perhaps the best established and is the mainstay of microarray-based systems. We have developed a capillary waveguide biosensor (CWB) which draws upon microarray technology in its use of surface-bound oligonucleotide capture probes and fluorescence-based detection. Because direct chemical labeling of the target with fluorophore is not amenable to use in the field, we employ a secondary hybridization with a fluorescently labeled detection probe. To eliminate some of the instrument-dependent fluctuations inherent in a fluorescence measurement, we (3) proposed an instantaneous normalization of the fluorescent signal by the unfiltered excitation signal, overcoming difficulties attendant on extracting concentration data from photon counts.

Recognizing that in the early stages of instrument development a complete ESP is not necessary, MBARI engineers designed an intermediate platform, the Micro Fluidic Block (MFB), to encourage development of novel analytical modules. The MFB provides functions for full control of rotary valves and syringes needed for moving reagents through the reaction chamber, as well as the computer platform needed for developing the software codes for control of the new instrument. This chapter focuses on the procedures necessary for integrating the CWB with the MFB for use on the ESP platform. We also describe a crucial part of the CWB development process, the attachment of capture probes to the interior surface of a capillary. In addition, we have optimized probes and hybridization conditions, and demonstrated the ability of the capillary to be regenerated for sequential measurements.

1.1. The Micro Fluidic Block

The MFB (shown as delivered in Fig. 1) is modified to be operated independently but is otherwise identical to the ESP's fluid handling system subsequent to sample collection. It is powered by a 12 V 15 A sealed lead acid battery (Powersonic PS-12120), which is continuously charged using a Xenotronix HPX-30 charger. The MFB is designed around the ARM host board, which provides a serial port and an Ethernet connection to the outside environment. The host board supports the Linux operating system, and the programming language is Ruby v1.6.8. Control software for the MFB and user-developed codes for the analytical module reside on the host board. A 2 GB compact flash card provides additional data storage; however, data can be downloaded from a remote computer at any time via a Web-based interface. The host board also supports two USB 1.1 peripherals. A gateway from the host board to the Dwarf stack is established through the use of a core board, which connects to the former via the RS232 cable and to the latter through the I2C bus connector. The Dwarf stack comprises a core board, rotary valve board and a sensor board. The host board uses the I2C network to issue instructions to the core board, which directly controls the solenoid valves (eight on the current MFB), as well as two heaters. The core board also passes RS232 from the host board to the analytical module being developed. A connection between the host board and the analytical module can also be

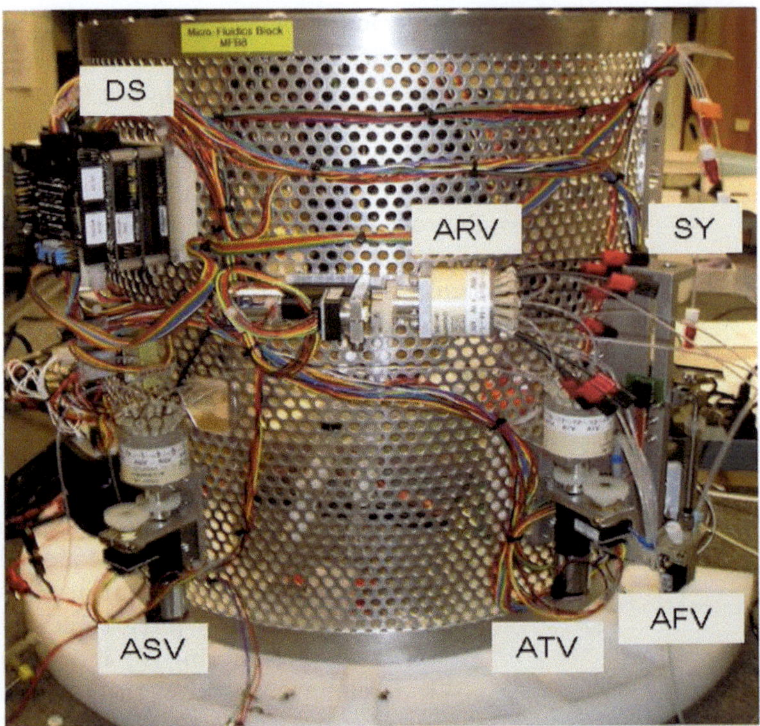

Fig. 1. Unmodified microfluidic block (MFB). *DS* Dwarf stack, *ARV* analytic reagent valve, *SY* syringe, *SPE* solid phase extraction, *ASV* analytic sample valve, *ATV* analytic top valve, *AFV* analytic fluid valve.

established through the USB port. The rotary valve can control up to four DC brushed motors and their encoders. The sensor board can also process data from two temperature sensors and one pressure sensor.

Figure 2 shows the complete fluidics pathway available on the MFB. Reagents can be pumped into the reaction chamber through various ports of the analytic reagent valve (ARV). These valves are controlled by Ruby functions supplied with the MFB and resident on the ARM host board. Use of these functions from user-developed software code will be described later. The input port of the capillary reaction chamber is connected to port 14 of the analytical top valve (ATV) and the output to the center port of the analytical sample valve (ASV). The analytic fluidic valve (AFV) is used to switch between the carrier fluid and reagents. Use of the carrier fluid, rather than air, for shuttling reagents and samples through the fluid system is recommended, as it allows purging of air bubbles from the fluid lines. In order to prevent mixing of neighboring fluids, small air pockets can be introduced in the fluid pathway.

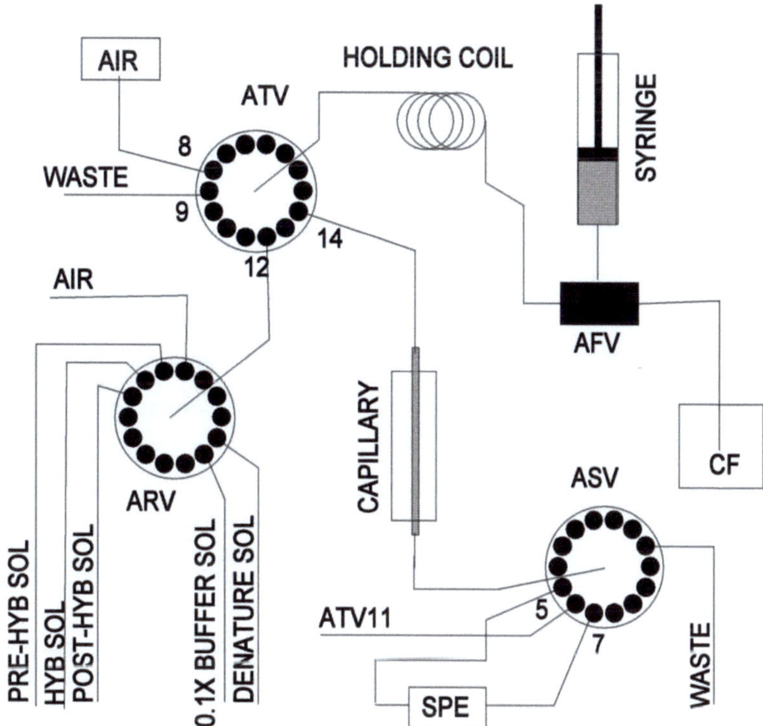

Fig. 2. Fluid pathways of the MFB. *ATV* analytic top valve, *ARV* analytic reagent valve, *AFV* analytic fluid valve, *ASV* analytic sample valve, *SPE* solid phase extraction, *CF* carrier fluid.

In the actual ESP system, a nucleic acid extract is supplied from the lysate coil, which is connected to the output of the solid phase extraction heater. The analytical module will draw the DNA slug from the lysate coil and move it into the capillary for detection.

Figure 3 illustrates the various interconnects between the MFB and CWB. During the development phase, the local host computer connects to the ARM host on the MFB via a hardwire Ethernet cable. This connectivity is used to upload user Ruby script files to the MFB for execution and for downloading the data files back to the local machine. However, the code can also be developed on the local machine by creating the **esp** environment on the hard drive. The RS232 port on the AM can be connected to the MFB through either the emulated serial port on the Dwarf stack or the USB1.1 connector on the host board using a RS232 to USB dongle.

1.1.1. MFB Installation

The first step upon delivery of the MFB is to establish communication between the local host machine and the MFB core board, as directed in the accompanying user manual. This is achieved by using a wired Ethernet connection from the local computer to the ARM processor host board. The host board will integrate itself into any network that supports both DHCP and DDNS. The pre-configuration file assigns a hostname to the MFB, for example,

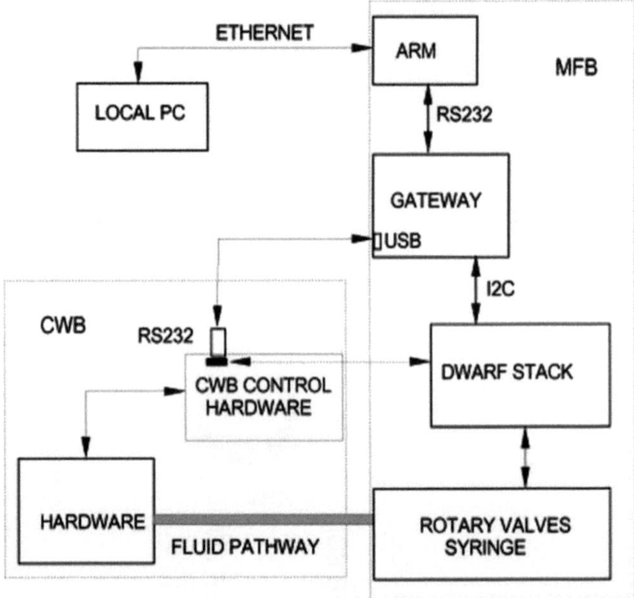

Fig. 3. Schematic of the CWB-AM, showing the interconnects between the MFB and CWB.

MFB8. At start-up a unique IPv4 address is assigned to this hostname. From the local machine running the Linux OS, the command *ssh mfb@MFB8* will prompt for a password (supplied by the vendor). For networks supporting DDNS, the MFB can only be identified by its numeric IP address of the form 192.168.1.204. This address is typically assigned by the network router and may change when the MFB is powered down. It should also be noted that some routers assign the IP on a lease arrangement. In other words, at the expiration of the lease period the IP will have to be reassigned. Occasional problems with connectivity may be traced back to this issue.

The MFB is accessed by running the **esp** application that controls the MFB from the host's (*MFB8*) Linux command prompt. Typically, a new terminal window is opened and the following sequence of commands sets up the **esp** environment from which the MFB's various valves and syringes can be tested.

ssh~dir@192.168.1.104	[~ is a space character, dir: folder on MFB]
cd~esp2/bin	[change directory]
.~ESPenv~mfb~MFB8	[opens the **esp** environment]
su	[login as super user]
AFV.to~: holdingcoil	[MFB command to move AFV to holding coil]

Operation of various valves, as described in the MFB user manual, can be verified in the **esp** environment.

1.2. The Capillary Waveguide Biosensor

Figure 4 shows a desktop version of the CWB, prior to integration with the MFB. The heart of the CWB is a silica capillary, which serves as the reaction chamber for target detection, and as an optical element for delivering the excitation energy to the surface-bound hybridized products and for collecting the fluorescence emission. For details on the CWB the reader is referred to our earlier papers (3, 4) and a recent chapter (5). The essential features include an opto-fluid connector, which serves as an input fluid port and also as a laser delivery system. Light from a laser source is launched directly into the annulus wall of the capillary, which has the optical properties of a multimode optical fiber. A small fraction of the guided optical modes penetrate into the core region of the liquid-filled capillary, creating an evanescent wave field, which excites the fluorochrome label attached to immobilized hybridization products. The penetration depth is typically smaller than the excitation wavelength. Some of the photons emitted from the excitation volume are trapped within the capillary wall, through the mechanism of tunneling, and can be collected at the capillary ends, while some are emitted into the lumen of the capillary and still others are radiated out of the curved outer surface. As fluorescence detection is noncoherent, theoretically the signal could be enhanced through a collection of all possible emissions; however, in practice the optical system for achieving this task is not easily realizable (see Note 1).

The CWB uses a 1 mm silica optical fiber to collect photons emitted from the curved surface. The distal end of the pickup fiber

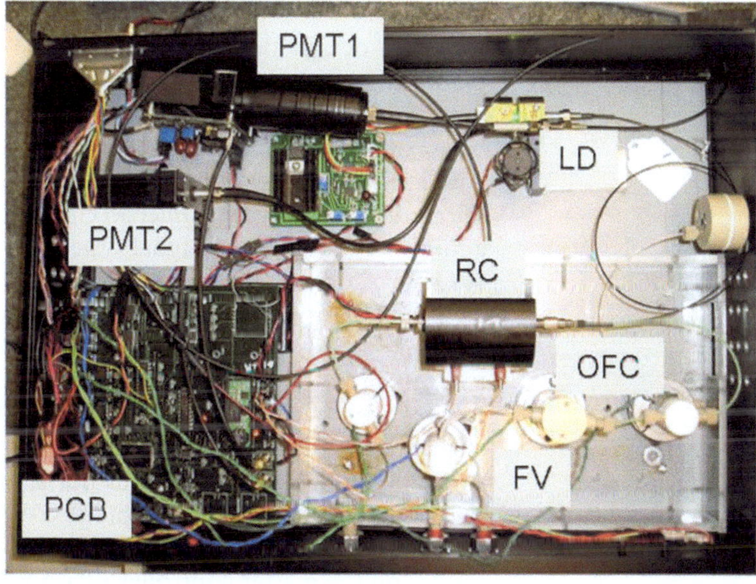

Fig. 4. Earlier (desktop) version of the capillary waveguide biosensor (CWB). *FV* fluid valves, *PMT1(2)* photomultipliers, *RC* reaction chamber, *OFC* opto/fluid connector, *PCB* printed circuit board, *LD* laser diode.

is mated with a 1×2 fiber optic multiplexer, having a power splitting ratio of 200:1. Photon fields emanating from the two fiber arms are filtered to separate the emission and excitation components before being detected by two single photon counting photomultipler modules.

2. Development of the Capillary Waveguide Biosensor Analytic Module

As discussed in Sect. 1.1, the MFB platform allows the user to adapt an existing (or new) sensor in order to create an analytic module (AM) that can be merged with the ESP at the front (upstream) end. Design of the AM must accommodate constraints imposed by the ESP platform, of these, autonomous functioning is critical to its success. It is expected that the AM will be activated by a single function call from within the ESP application. Size and energy constraints will fundamentally limit the scope of the AM's functionality. For example, heaters, being the most energy consuming components, will require special attention, and consideration of the data transfer rate over the RS232 serial interface will limit overly ambitious imaging-based sensors.

As depicted in Fig. 5, the capillary waveguide biosensor analytic module (CWB-AM) has three distinct functional subsystems: (1) a fluid management system, (2) a data acquisition system, and (3) a temperature controller. The MFB platform takes care of the first. Design of the remaining two subsystems is central to an efficient merging of the two technology platforms.

2.1. Data Acquisition System

2.1.1. Microcontroller

The first design decision concerns the selection of the control platform for the modified CWB, and involves a trade-off between functionality, cost, and complexity of development. Solutions based on the use of a single-board computer or a microprocessor-based embedded system are unnecessary for the CWB-AM. We selected the Silicon Labs microcontroller (#634-C8051F120DK), a fully integrated mixed-signal system on a chip. It has eight 8-pin I/O ports, one 12-bit 9-channel analog-to-digital converter, three 12-bit digital-to-analog converters, 128 kB of programmable flash memory for in-system programming, 8,448 bytes of RAM, and two 24-bit counters with a minimum sample time of 15.625 ms. An external 1 MB RAM (IDT #IDT71V124) was added to the system to give the capability to store count data for a 100 s duration. An in-depth design discussion is not appropriate here. Briefly, the microcontroller forms the hub of the CWB-AM platform. Firmware for controlling and checking the status of the subcomponents resides on the master controller's flash RAM, and can be upgraded in the system at any time. For interested readers the microcontroller code can be obtained from the corresponding author.

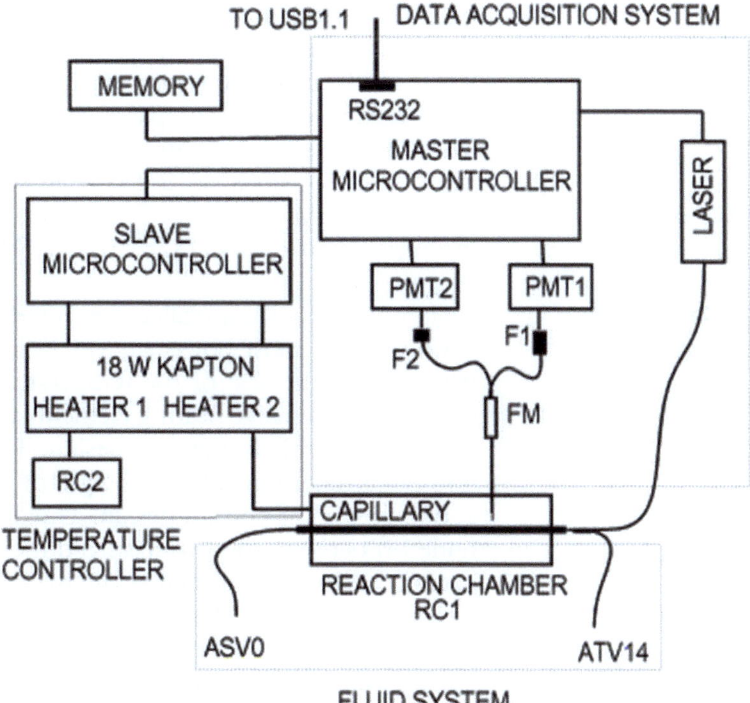

Fig. 5. Schematic of the CWB. *PMT1(2)* single photon counting modules, *F1* emission filter set, *F2* neutral density filter, *RC2*: second reaction chamber, *FM* 1 × 2 fiber optic multiplexer.

2.1.2. Detection System

The detection system comprises both electronic and optical components. A well-designed optical system will have an optimal filter set that matches the excitation and emission spectra of the fluorescent dye complex being used with the detection probe. However, angular sensitivity of these components is usually overlooked. A few degrees off normal incidence can significantly broaden the filter bandwidth, resulting in an increased background signal. For fiber-based detection schemes such as this one, care should be taken in designing the excitation and emission optics, particularly when the numerical aperture exceeds 0.2. The 1 mm collection fiber, with a NA of 0.5, requires collimation optics, which increases its physical size (see Note 2).

An earlier model of the CWB used a compact DPSS laser, with a footprint of less than 200 mm^2; however, we later sacrificed size for better power stability and the capability for digital control. It was replaced by a 30 mW, temperature-stabilized, solid-state 532 nm laser (World Star Tech TECGL-05-TTL), with power stability better than 2 %. For single photon counting systems, there are two possible solutions, one based on a photomultiplier (PMT) and the other on an avalanche photodiode detector (APD). The latter is more costly but has a higher quantum efficiency at the operating wavelength. However, it requires active cooling to keep the dark counts at ~150 cps, whereas the PMT has a dark count less than 50 cps at room temperature. APD-based systems have an active area of 0.3 mm^2 compared with PMTs with an area of 5 mm^2. The detector size is not an issue for small diameter fibers, but for the 1 mm pick-up fiber, PMT-based system are the only choice. The CWB-AM uses two PMT-based single photon counting modules, Hamamatsu H8259-01 (see Note 3).

2.2. Temperature Controller

The temperature controller presented a major challenge, as it consumes the most power and its design is critical for extended battery life. The size of the temperature housing, the insulation material, the resistive heating element, and the control circuitry all play a critical role in the final design. The size of the temperature housing is determined by the length of the capillary, which is fixed at 65 mm for the current design. The diameter of the housing is a trade-off between thermal mass and the higher energy required to heat it. The mass is needed as the hybridization times could be of the order of 60 min. The CWB uses custom designed 12 V 18 W Kapton flat resistive heaters. A system requiring rapid thermal cycling (e.g., PCR) can use a much smaller thermal mass and thereby use smaller heating elements.

Most constant temperature controllers use a PID feedback loop to obtain the desired temperature profile. The PID coefficients are empirically tuned to the thermal load. The feedback loop can be implemented with either analog or digital signals. Digital feedback is more robust and flexible and allows downloading of new PID coefficients for remote applications, but requires additional electronic components. We decided on a digital PID algorithm, modified to be run on a microcontroller that did not have a floating point processor. The PID algorithm was further modified by turning off the feedback loop until the measured temperature was within 5 °C of the set point. This modification reduced the thermal time constant. A slave microcontroller was added to the system for independent temperature control. Typically, one temperature-controlled reaction chamber is needed for the CWB; however, a second temperature controller was added in case thermal denaturation of dsDNA in sample extracts was required. As this may require temperatures close to 90 °C, it cannot be performed in the capture probe coated capillary. Both temperature housings are controlled by the slave microcontroller using a 2×1 multiplexer, and each chamber has a steady-state stability of ±1 °C with a maximum attainable temperature of 100 °C. The two chambers can be operated at different set temperatures.

2.2.1. Energy Usage Considerations

In a tethered deployment with cabled power, the duration of deployment of the CWB will be determined primarily by probe and hybridization chemistry, i.e., by the longevity of capture probes and reagents. However, long-term untethered operation under battery power requires a detailed exploration and understanding of energy usage in the coupled MFB/CWB system. The entire MFB and CWB system is powered by a 12 V, 15 Ah rechargeable lead acid battery. Because developers of new analytical modules have no control over the MFB's energy consumption, analytical modules must be designed to be as energy efficient as possible. The CWB requires the use of alternating hybridization and regeneration (target removal) steps; in the current system the former typically is performed at 52 °C for 15 min and the latter at 65 °C for 5 min. Elevated temperature increases the specificity of hybridization and the efficiency of capillary regeneration, but comes at the cost of increased energy consumption. The optimal hybridization temperature varies with the length and sequence of probes as well as the composition of the buffer fluid in which hybridization takes place.

A Hall current sensor with a 1 ms response time was designed into the hardware to allow instantaneous measurement of energy consumption under different usage scenarios. For example, a measurement cycle with a 30-min hybridization at 60 °C followed by regeneration at 80 °C gave a battery consumption of 0.5 Ah (data not shown). Under these extreme conditions, the 15 Ah rechargeable battery powering the MFB and CWB would support up to 30 target detection cycles. Clearly, it is desirable to decrease energy consumption as much as possible. Strategies for achieving the lower energy budget include reducing the mass of the reaction chamber, lowering the hybridization and regeneration temperatures, and finding more efficient methods of heating the fluid inside the capillary. The latter could be done by employing integrated heating elements close to the surface of the capillary. We are already using a very efficient insulator, Pyrogel 2250 (www.aerogel.com), which reduced the heating time from 120 s (1/4 in. rubber) to 15 s. However, the use of a low thermal conductivity insulator meant that rapid temperature reductions in turn became difficult. Since forced cooling was not a viable option due to its energy cost, we added cooling channels to pump relatively cool water (e.g., ambient temperature water in temperate or colder environments) through the housing.

As a final test, the entire CWB-AM system, running with maximum load, was placed inside a refrigerator to replicate the temperature environment of an immersed ESP in cold waters. Figure 6 shows the temperature profile of the two temperature-stabilized reaction chambers (these data was taken with ¼ in. rubber insulation). All other electronic components, including the laser and the PMT, continued to operate as normal. Such tests are crucial if considerable downtime and cost are to be avoided during field trials.

2.3. Integration of the CWB and MFB

Figure 7 shows a photograph of the completed MFB-CWB platform, ready for testing with the ESP. In its present configuration, the CWB comprises a number of distributed components mounted on the aluminum curved plate. These components include two

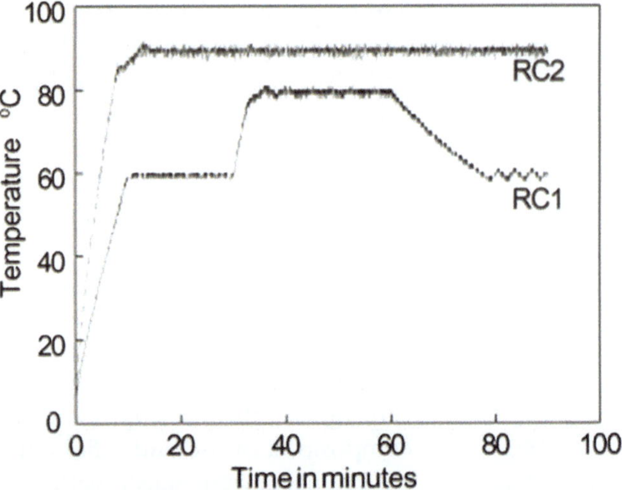

Fig. 6. Environment temperature test. The complete CWB was placed inside a refrigerator at 4 °C. Temperature profiles of the two reaction chambers RC1 and RC2 are shown.

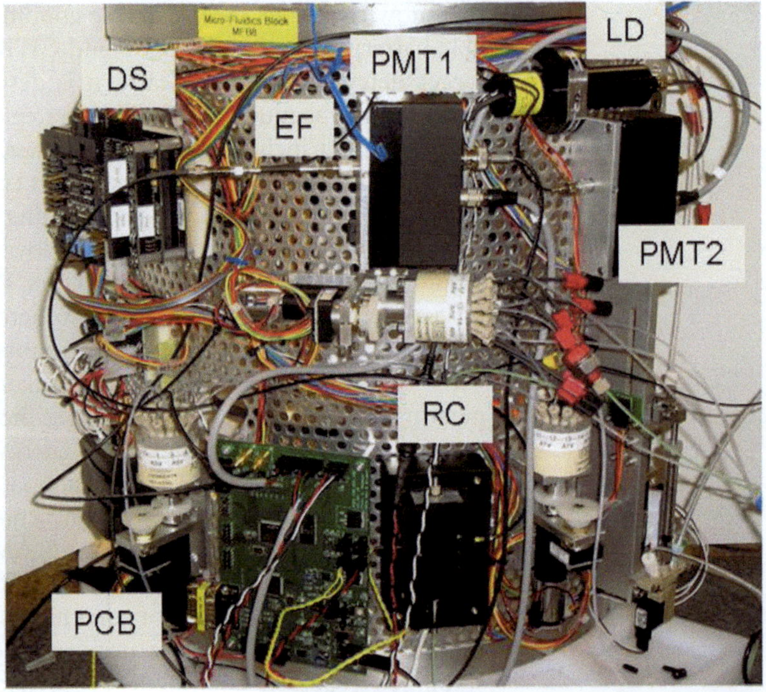

Fig. 7. MFB-CWB integrated platform. *PMT1(2)* single photon counting modules, *EF* emission filters, *RC* reaction chamber with capillary, *OFC* opto/fluid connector, *PCB* printed circuit board, *DS* Dwarf stack, *LD* laser diode assembly.

photomultipliers with optical filters, a DPSS laser source, the hardware control board, and the capillary reaction chamber.

Each target detection cycle in the CWB involves a complex sequence of tasks. As a part of the integration process, the MFB rotary valves and syringes have to be controlled from the user-written Ruby script files. This feature required that the MFB function commands be wrapped using pipes. For example, the MFB command for moving the ARV to port 12 is *"ARV.to 12"*, this command can be called from the user script by opening a pipe and sending commands as follows:

espClient=IO.popen "espclient", "a+" [once at the top of the program]

...

espClient.puts "ARV.to 12"

puts espClient.gets("\0").chomp "\0"

...

espClient.close [at the end of the script file]

At the end of the measurement cycle, a data file containing the raw fluorescent photon counts for each cycle is stored on the MFB ARM board memory. The files can be downloaded to the local machine at any time.

3. Capillary Preparation, Probe Selection, and Hybridization Protocol Development

The capillary's inner surface must allow a firm attachment of oligonucleotide probes and be capable of regeneration, i.e., the bound target must be removable without altering capture probes or their attachment to the surface. These considerations will determine the capillary's usable lifespan. In order to better understand the attachment chemistry of the capture probe to the silica surface, an extensive study was undertaken using a microarray reader.

3.1. Capillary Preparation

Hybridization/waveguide chambers for the CWB were constructed from TSP fused silica capillary tubing (1 mm ID, 1.3 mm OD; Polymicro, Inc.). The construction process involved cleaning, surface activation, silanization, final assembly, and capture probe application. Adequate cleaning and activation of the silica surface are especially critical to achieving good results.

3.1.1. Cleaning

The outer polymide coating was scraped from capillary tubing and the tubing was cut into 65 mm lengths. Capillaries were cleaned by soaking them in anhydrous ethanol (Cat# E1028-500MLGL, Spectrum Chemical, Gardena CA) for 20 min to overnight. Alcohol was blown out of the capillaries with filtered, compressed oil-free nitrogen (see Note 4). After drying (approximately 30 s per capillary), capillaries were flushed with ultraclean (18 MΩ or equivalent) water for a few seconds and drying was repeated.

3.1.2. Activation

Two methods of preparing the capillary surface for silanization have been tested. Method A was to bake the capillaries by wrapping them in cleaned aluminum foil (previously cleaned by heating to 130 °C for 30 min in an oven) and heating them to 100 °C for 2 h or longer. Higher temperatures may be beneficial, but 100 °C is the minimum to remove a "bound water" layer from the surfaces to be treated. In Method B, an ozone generator was used and a vacuum pump was employed to suction ozone through the capillaries. This removes not only the water, but also any organic contamination on the surfaces.

3.1.3. Silanization

Immediately after activation, capillaries were filled with a solution of 1–2 % 3-glycidoxypropyltrimethoxysilane (Gelest, SIG 5840.0) in anhydrous ethanol. After 1–5 min, liquid was removed by blowing with compressed nitrogen. Silanized capillaries were placed into a clean container for 24 h at 35 °C. Care was required to keep atmospheric debris away from the surfaces, although some reaction with atmospheric water vapor occurs.

3.1.4. Assembly

A half-inch stainless steel ferrule was mounted, using a 5 min epoxy, on each end of the capillary. The ferrules permit polishing of the capillary end surfaces and are necessary for ease of handling. Capillary end surfaces were manually polished to an optical flatness of $\lambda/2$ and polishing byproducts were removed by blowing with a gas duster. Capillaries were stored at room temperature under low vacuum (vacuum-sealed freezer bags) to help prevent contamination of surfaces.

3.1.5. Capture Probe Application

To remove residual particles from the polishing process, a syringe and custom-built injection fitting were used to flush capillaries 3× with 1 ml of spotting solution ($NaH_2PO_4 \cdot H_2O$, 3.4 mM; Na_2HPO_4, 146.5 mM; SDS, 5 % w/w). Capillaries were then injected with spotting solution containing 20 μM amino-modified capture probe. Capture probes modified with a fluorophore were used in tests of coating evenness and quality. Filled capillaries were incubated at room temperature overnight in a horizontal position over saturated NH_4Cl/KNO_3 to maintain humidity (6). After 16–18 h, capillaries were flushed with approximately 10 ml of 0.1× SSC, (saline sodium citrate; prepared from sterile 20× SSC; (7)). They were then dried with a gas duster and stored in vacuum-sealed freezer bags until they were used.

3.2. Probes

Two types of oligonucleotide probe are needed to bind and detect rRNA or rDNA in the CWB. These are as follows:

1. Capture probe. Capture probes bind specifically to targeted DNA or RNA; for example, to rRNA or rDNA from an organism

being monitored. For best adhesion to the capillary's silane layer, they should contain a functional group (amino group in this instance) that reacts with a functional group of the silane being used. A spacer (e.g., C_{12}) between the amino group and the oligonucleotide is desirable to facilitate free access of target molecules to the surface-bound probe.

2. Detection probe. These bind to a different site on the target nucleic acid and incorporate a fluorophore. They need not be specific, since specificity is conferred by the capture probe. Detection probes may also be used in the absence of sample RNA or DNA ("background controls") to evaluate background levels of fluorescence due to nonspecific binding between capture and detection probes.

Two other categories of oligonucleotide were needed for process development:

3. Positive control. A fluorescently labeled oligonucleotide complementary to the capture probe was used to evaluate capture probe performance (independently of the detection probe) under various conditions.
4. Labeled capture probe. A capture probe with an additional fluorophore modification was used to visualize capture probe binding to a silane layer, allowing comparison of coating methods.

Although oligonucleotide sequences can be obtained from published sources or designed in silico, empirical testing is essential. Oligonucleotides and probes used in the study are shown in Table 1. Sequences were derived from ProbeBase (8) except where noted, and probes were custom synthesized by OligosEtc Inc. or Integrated DNA Technologies. We designed two new detection probes (FVdet1 and FVdet2) based on published *Vibrio* sp. sequences and information derived from screening candidate probes using Integrated DNA Technologies' online utility OligoAnalyzer 3.1 (http://www.idtdna.com/analyzer/applications/oligoanalyzer/). The new detection probes were designed for close binding site proximity to capture probe Amn-GV, and for low tendency to form a heterodimer with the capture probe. The former allows detection even if RNA is partially fragmented (either intentionally or as a result of extraction and processing), while the latter is an important consideration in preventing high background control values. Typical concentrations in hybridization solutions were 100 nM for positive controls, and 200 nM for detection probes used in background controls or for labeling of RNA. All probes were diluted in 0.1 μm filtered, RNase-free (1 h autoclaved), deionized and distilled water.

Table 1
Oligonucleotide capture probes, detection probes, and targets used in the study

Name	Sequence	Target region	Notes	References
Amn-Vibcap	5′-Amino-C12-ACC ACC TGC ATG CGC TTT-3′	*Vibrio* sp., 16s RNA bases 572–589	Modified "VIB572a"; Vibrio capture probe; ProbeBase # pB-01188	Huggett et al. (13)
Amn-Vibcap-Cy3	5′-Amino-C12-ACC ACC TGC ATG CGC TTT-Cy3-3′	*Vibrio* sp., 16s RNA bases 572–589	Capture probe labeled with Cy3 for testing probe attachment; ProbeBase # pB-01188	Huggett et al. (13)
Amn-GV	5 -Amino-C12-AGG CCA CAA CCT CCA AGT AG-3′	*Vibrio* sp., 16s rRNA bases 841–860	Modified "GV" *Vibrio* capture probe; ProbeBase # pB-00611	Eilers et al. (14), Giuliano et al. (15)
Amn-Vibrl	5′-Amino-C12-CGC TGG CAA ACA AGG ATA AG-3′	Vibrionaceae, 16s rRNA 1113–1132	Modified "Vibrl" *Vibrio* capture probe; ProbeBase # pB02258	Kyselkova et al. (16)
A532-Vibtar	5′-Alexa Fluor 532-AAA GCG CAT GCA GGT GGT-3′	N/A	Complementary to Amn-Vibcap; used for positive controls	This study
Cy3-G Vtar	5′-Cy3-CTA CTT GGA GGT TGT GGC CT-3′	N/A	Complementary to Amn-GV; used for positive controls	This study
A532-Vibdet	5′-Alexa Fluor 532-AGG CCA CAA CCT CCA AGT AG-3′	*Vibrio* sp., 16s rRNA bases 841–860	Modified "GV"; *Vibrio* detection probe; ProbeBase # pB-00611	Eilers et al. (14), Giuliano et al. (15)
Cy3-EUB338	5′-Cy3-GCT GCC TCC CGT AGG AGT-3′	Most bacteria, 16s RNA bases 338–355	Modified "EUB338"; *Vibrio* detection probe; ProbeBase # pB-00159	Amann et al. (17)
FVdet1-A546	5′-CTA ATC CTG TTT GCT CCC CAC-Alexa Fluor 546-3′	*Vibrio* sp., 16s rRNA bases 788–808, likely others	Detection probe designed for proximity to Amn-GV	This study
FVdet2-A546	5′-CGT TTA CGG CGT GGA CTA C-Alexa Fluor 546-3′	*Vibrio* sp., 16s rRNA bases 817–836, likely others	Detection probe designed for proximity to Amn-GV	This study

3.3. Growth of Bacteria and Preparation of RNA

3.3.1. Cultivation and Harvesting

Vibrio harveyi BAA-1116 (American Type Culture Collection) was maintained on Difco Marine Broth 2216 (Becton Dickinson, 279110) solidified with 1.2 % agar (Becton Dickinson, 214010). Stock cultures were maintained at 30 °C and transferred monthly. Cell slurries were obtained by inoculating spread plates on the same medium and incubating overnight at 30 °C. Cells were scraped from the plates with an alcohol-dipped, flamed 25 × 40 mm cover slip and washed into a few ml of liquid Difco Marine Broth. The suspension was washed by centrifugation, resuspended in sterile 2 % NaCl, and its cell concentration determined by the epifluorescence method of Hobbie et al. (9) as modified by Watson et al. (10). After being pelleted by centrifugation and the supernatants discarded, aliquots (usually on the order of 10^9 cells) of packed cells were stored frozen at −80 °C.

3.3.2. RNA Extraction

An RNeasy Mini Kit (Qiagen, 74104) was used to extract RNA from cell pellets. The extraction procedure was modified to incorporate lysozyme digestion according to procedures described for bacteria by Qiagen (11). Quantity and quality of extracted RNA was evaluated by means of a Nanodrop ND-1000 (Thermo Scientific) and by agarose gel electrophoresis. RNA was aliquoted, immediately stored at −80 °C, and typically used at a concentration of 20–22 μg/mL hybridization solution (see Note 5). Total RNA extracted from bacterial cells is generally considered to contain 90–95 % ribosomal RNA.

3.3.3. RNA Fragmentation

Secondary structure of rRNA/rDNA affects its accessibility to probes (12); this problem might be exacerbated when the short probes are bound and the ca. 1,500 nt RNA is free in solution. To help eliminate secondary structure, RNA was chemically fragmented in some experiments according to the $ZnCl_2$ protocol described by Kreatech, Inc. (http://www.kreatech.com/Portals/kreatech/downloads/labeling/40_RNA%20fragmentation_v2.0doc.pdf). This method is claimed to yield fragments averaging 60–200 nt.

3.4. Optimization of Hybridization and Denaturation Protocols

3.4.1. Use of Slides as a Model System

3-Glycidoxypropyltrimethoxysilane-coated microarray slides (Corning, CLS40041) were used to optimize probes and hybridization conditions, since they allowed a large number of hybridizations to be performed simultaneously and their use enabled the concurrent optimization of 3-glycidoxypropyltrimethoxysilane coating protocols for capillaries. The spotting and hybridization procedures described by the slide manufacturer (8) were modified by the use of the same nonproprietary spotting solution used for capillaries, and by the omission of distilled water and 1× SSC washes (which proved unnecessary in preliminary tests). Slides were spotted using a manual glass slide replicator (V&P Scientific, VP478A) and a custom-built single-position indexing unit.

Hybridizations were carried out under lifter slips (Thermo Scientific, 25X401-2-4772). Elevated incubation temperatures were maintained for slides by means of a humidified multislide chamber (Genetix Ltd.) and a hybridization oven (Hybaid H9270), and for solutions by means of a water bath (GeneMate HSP1000). Hybridization times of 15 min—overnight at optimal temperature showed no consistent differences in fluorescence intensity (data not shown). For convenience, slides were hybridized for 1 h and capillaries for 15 min. All solutions were prepared using deionized, distilled water that had been 0.2 μm filtered and autoclaved for 1 h to inactivate RNases. Glassware and disposable supplies were selected to be RNase-free or were autoclaved for 1 h, while care was taken to select RNase-free reagents for all stages of hybridization. Slides were visualized and fluorescence data were acquired using a GenePix 4000B microarray scanner (Molecular Devices).

3.4.2. Hybridization in the CWB

Solutions, temperatures, and times for a typical hybridization cycle are shown in Table 2. The hybridization solution contains 15 % formamide (see Note 6) and therefore the effective hybridization temperature is 61 °C. Buffer rinses entail flushing the previous solution from the feed line and capillary using 0.1× SSC, then refilling with 0.1× SSC. Fluorescence data are gathered after refilling the capillary (i.e., after Rinse 1 and Rinse 2). Hybridization of RNA with detection and capture probes is carried out simultaneously, while positive control hybridizations are carried out with a labeled target complementary to the capture probe, and background control hybridizations receive the detection probe but no RNA. It should be noted that each cycle of target detection requires two measurements, i.e., before and after hybridization. This is necessary to prevent false-positive results.

Table 2
Hybridization and denaturation protocol used in the CWB

Step	Solution	Temperature (°C)	Incubation (min)
Denature	0.1× SSC + urea (50 % w/w)	65	5
Rinse 1	0.1× SSC	52	0
Prehybridize	5× SSC, 0.01 % BSA (w/w), 0.1 % SDS (w/w)	52	15
Hybridize	5× SSC; 15 % formamide (v/v); 0.1 % SDS (w/w); labeled target oligo, labeled detection probe, or RNA + detection probe as required	52	15
Posthybridize	2× SSC, 0.1 % SDS (w/w)	52	5
Rinse 2	0.1× SSC	52	0

4. Results of Capillary Tests and Hybridization Optimization Studies

As a first step in preparing the CWB for target detection, it was necessary to check the system under controlled conditions using a standard reference material (SRM). For this purpose an uncoated capillary was mounted into the reaction chamber and loaded with a solution of Alexa Fluor 532-labeled positive control solution (A532-Vibtar; Table 1) at an arbitrary concentration. The CWB-AM was run remotely from a Linux host machine. A Ruby script file with an abbreviated measurement cycle was uploaded into the MFB's ARM processor for execution. The CWB was instructed to load a buffer solution into the capillary and record the background fluorescence signal N_B for 30 s. Subsequently, the SRM was loaded into the capillary and the fluorescent signal N_S was recorded for 30 s. The difference between these two signals provides an indication of the operational status of the CWB. A difference below the preset limits (5) indicates that the CWB should not be operated. Figure 8 shows a summary of the results from 20 cycles

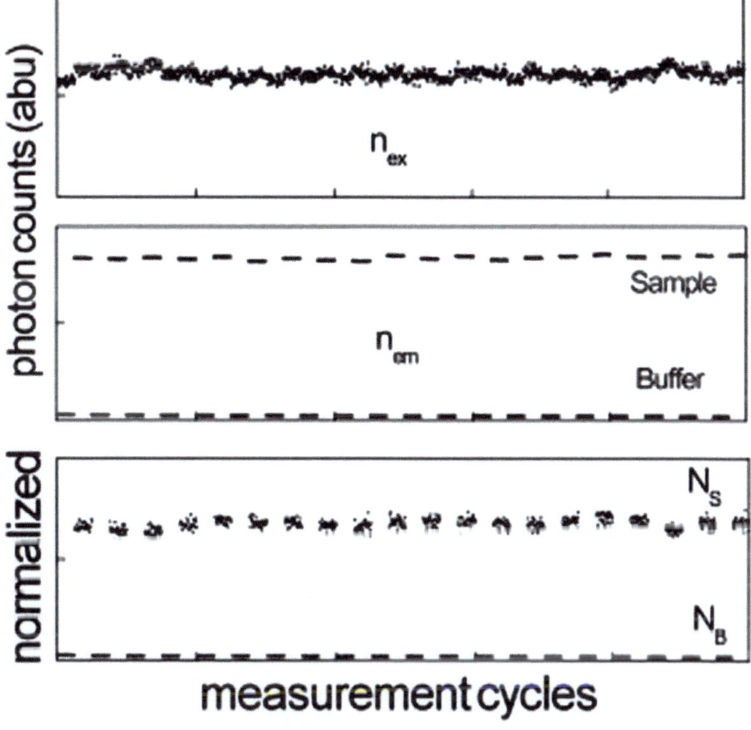

Fig. 8. System test with an uncoated capillary, loaded with a standard reference material (Alexa532 dye solution). *Top panel*: raw photon count data measured with PMT2. The signal count n_{ex} is proportional to the excitation energy; *middle panel*: raw photon count data at the output of the emission filter, measured with PMT2. This is the fluorescence emission signal n_{em}; bottom panel: normalized data ($=n_{em}/n_{ex}$). N_S and N_B correspond to sample and background, respectively.

Fig. 9. Visualization of probe attachment inside capillaries, silanized after baking (**a**) or ozone treatment (**b**). Capillaries were coated with Cy3-labeled capture probe (Amn-Vibcap-Cy3) and were visualized using a microarray scanner and a slotted polycarbonate holder.

of measurements. The top panel is the raw (scaled) photon counts corresponding to the optical excitation power n_{ex}, at the point of measurement. The middle panel is the raw photon counts corresponding to the fluorescence emission n_{em}. The bottom panel is the normalized signal, n_{em}/n_{ex} (N_B for the background solution and N_S for the dye solution). As discussed by Dhadwal et al. (4), this normalization eliminates the machine to machine variability that plagues fluorescence measurements. The results show that the system is working correctly and within specifications. From a practical view, the background fluorescence signal determines the limit of detection (see Note 7).

In order to stably immobilize a capture probe on the interior surface of the capillary, the surface of the capillary needs to be silanized. Proper cleaning and activation of the silica surface proved critical to this procedure. Figure 9 shows capillaries in which surface activation was carried out by Method A (baking) and Method B (ozone treatment) as described in Sect. 3.1.2. Visualization of the coating by application of a labeled capture probe clearly showed that the probe attached more densely and evenly to capillaries prepared by Method B, which was adopted for use with the CWB.

Hybridization temperature and buffer composition both influence the specificity of hybridization, and can be conveniently expressed as an "effective temperature," a function of the actual temperature and the effects of formamide concentration. The effective temperature must be high enough to minimize mismatches and nonspecific binding. Too high an effective temperature will reduce the number of target molecules bound to probes, including both binding of targets to capture probes and binding of detection probes to immobilized targets. Figure 10 shows a typical

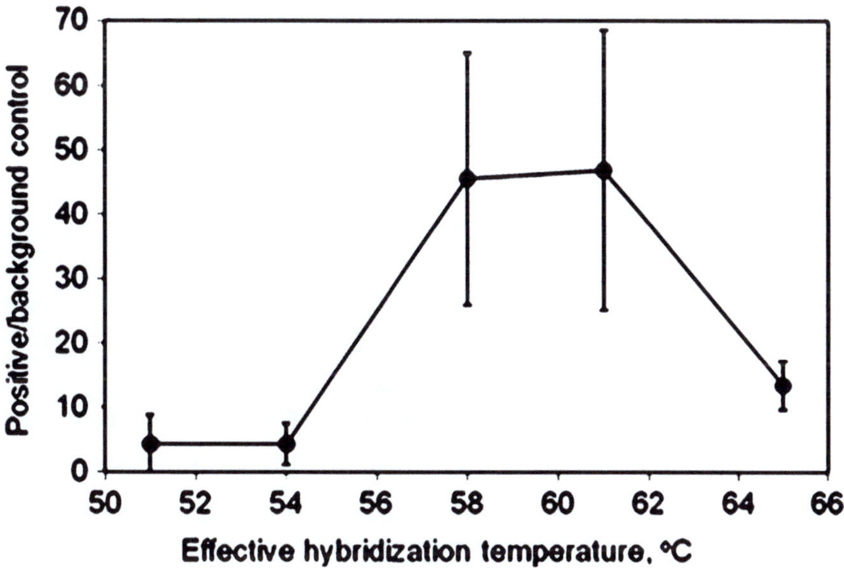

Fig. 10. Optimization of hybridization temperature. Effective hybridization temperature was varied by varying actual temperature (45, 52, or 59 °C) and formamide concentration (10 % or 15 %; 1 % formamide is equivalent to a 0.6 °C temperature increase). Overnight hybridizations were performed on microarray slides spotted with capture probe Amn-Vibcap. Values represent the ratio between fluorescence intensity of positive and background controls, where positive control slides received a complementary labeled oligonucleotide (A532-Vibtar) and background controls received detection probe A532-Vibdet in the absence of RNA as a measure of nonspecific hybridization (noise). *Error bars* represent standard deviations.

experiment in which hybridization temperature was optimized by maximizing the ratio between fluorescence intensity of positive and background control slides. The results show an optimal hybridization temperature range between 58 and 61 °C. The rather high standard deviation in the data points is due to spot variability associated with the use of a manual arrayer and normalization of the data with the low background control signal.

The use of microarrays expedited the testing of candidate probe pairs for the detection of *Vibrio*. Figure 11 shows tests of two capture probes and three detection probes. The best results (lowest background control values combined with highest detection of RNA) were achieved using capture probe sequence GV and detection probe FVdet2. Little difference was observed between fragmented and intact RNA. This finding suggests that access to both binding sites is not markedly impeded by RNA secondary structure, a conclusion supported by the effectiveness of GV as a FISH probe (10, 11) and the close proximity of the two binding sites. Our findings also indicate that, as expected from their proximity, the GV and FVdet2 binding sites are seldom separated by fragmentation of the RNA.

Methods tested for the removal of positive control oligonucleotide (to regenerate the capture probe for subsequent hybridizations)

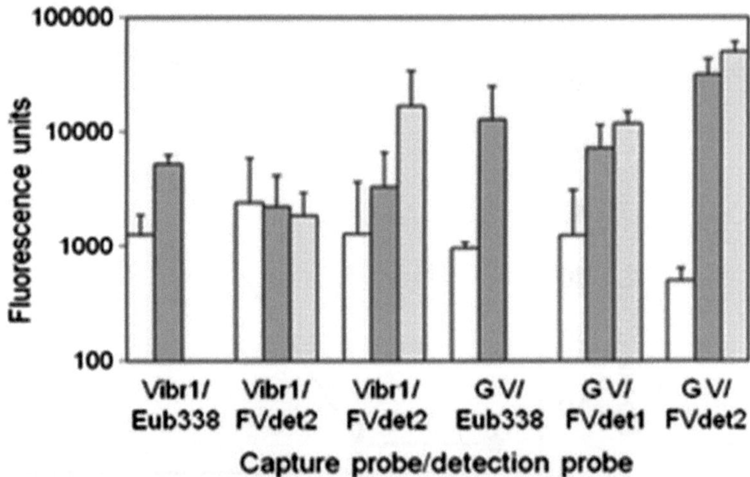

Fig. 11. Effect of capture and detection probe sequences on fluorescence signal from *Vibrio* RNA. Hybridizations were performed for 1 h on microarray slides. *Unshaded bars*: background control with detection probe but no RNA, *hatched bars*: detection probe plus unfragmented RNA, *stippled bars*: detection probe plus $ZnCl_2$-fragmented RNA. Fragmented RNA bound to our capture probes was not detected with detection probe Eub338 because of the large distance between binding sites (data not shown). 2.35–2.55 μg/slide total RNA was used. *Error bars* represent standard deviations of 60 spots. Data for the Cy3-labeled Eub338 are corrected for comparison with the two Alexa Fluor 546-labeled probes.

included formamide (up to 90 %), 0.1× SSC at temperatures up to 95 °C, and urea (Fisher Scientific BP169-500; 50 % w/w in 0.1× SSC) (data not shown). A 50 % urea treatment at 65 °C gave best results and was selected for use in the CWB (see Note 8).

The optimized probe system and hybridization/denaturation protocol (Table 2) were then tested using the CWB. Figure 12 shows sample data obtained over four cycles using a labeled target oligonucleotide. The capillary had been previously used and filled with 0.1× SSC for storage. One cycle was required to reestablish stable performance, after which the difference between normalized fluorescence values before and after hybridization remained constant. To date, over 20 cycles have been achieved using a single capillary, without apparent loss of performance (see Note 9). We are confident that we have found a coating procedure that will preserve the integrity of the capture probe for many more cycles. The final number will be established after the complete system is fully operational.

Development of a field-deployable biosensor draws upon fields ranging from optical engineering, to chemical engineering, to molecular biology. Using an interdisciplinary approach we have achieved reproducible and automated detection of an oligonucleotide target. The system is currently in final testing stage and the next step will be to demonstrate detection of extracted RNA in the CWB, as has already been done using our probes and protocols on microarrays.

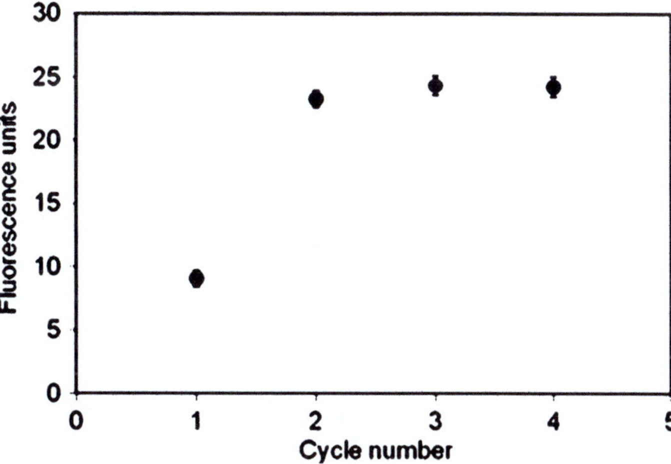

Fig. 12. Consecutive hybridizations in the CWB. Repeated hybridizations and denaturations were performed in automated mode, using a capillary coated with Amn-GV capture probe and Cy3-GVtar as the target. The capillary had been previously used for approximately 20 hybridization cycles and filled with 0.1× SSC for storage. Values represent the difference between mean normalized fluorescence intensity measured before and after hybridization at each cycle. Sixty sets of data points were acquired at each measurement time; error bars represent standard deviations for the difference between hybridized and unhybridized mean values. One cycle was needed to restore normal functioning of the capillary, after which values stabilized.

5. Notes

1. The detection limit of a fluorescence instrument is determined by the background signal, which is minimized by carefully selecting compounds that make up the solution at zero concentration and by reducing stray reflections through creative optical design and by the use of a holographic notch filter to exclude excitation energy from bleeding into the emission band.
2. Spectral response of the emission filters is based on normal incidence. A spread of a few degrees will lead to suboptimal filtering, causing an increase in the background signal. Thus, for fiber-based optical receivers, care must be taken to collimate the output of the optical fiber.
3. As discussed in Note 1, the detection limit is determined by the background signal, which will be at its lowest value when it approaches the dark count of the single photon counting system. Photomultiplers have dark counts of the order of 30 cps at room temperature, whereas avalanche photodiodes can achieve dark counts of 150 cps when actively cooled.

4. Any oil contamination in the compressed gas will coat and contaminate the interior surfaces of the capillaries.
5. RNA is exceedingly susceptible to degradation, and precautions should be taken to minimize exposure to room temperatures and to RNases. If partially used, an RNA aliquot should be discarded rather than refrozen. RNA solutions should contact only RNase-free solutions and equipment.
6. Deionized formamide is best aliquoted and stored at −20 °C, preferably under an inert gas. It should not be used if a yellowish color develops.
7. Dhadwal (5) described a target detection error rate, which computes the probability of detection based on the signal-to-noise ratio of both the zero concentration and the target solutions. He showed that it was possible to further reduce the detection limit by increasing the data accumulation time and using time-averaged signals for comparison.
8. We do not recommend this use of microarray slides for tests of probe regeneration, since drying of slides for examination in the microarray scanner makes subsequent denaturation of the probe–target complex exceedingly difficult (Eli Hatchwell, personal communication). The number of cycles that can be tested using slides is thus limited by the number of replicates that can be run (since slides must be sacrificed at each hybridization/denaturation cycle), and some air exposure is nonetheless unavoidable.
9. The CWB allows the entire process, including data acquisition and storage of the capillary between experiments, to be carried out with liquid inside the capillary, avoiding the problems described in Note 8.

6. Future Directions

We intend to explore the dynamic range of the system and its ability to quantify *Vibrio*, a genus which includes numerous human and animal pathogens, against a background of other organisms. The development of robust probe attachment capabilities, capillary regeneration methods, and field-deployable instrumentation provides a foundation upon which probe systems for other target organisms of economic and ecological importance can be constructed and optimized according to the procedures we have outlined.

Acknowledgments

We appreciate the funding support from NSF through the EAGER award OCE0929238. We also express our thanks to Dr. Eli Hatchwell for giving us access to microarray equipment at the Stony Brook Medical Center's Genomic Core Facility and for many fruitful discussions of hybridization kinetics. Discussions on silanes with Qingzhi Zhu of the School of Marine and Atmospheric Sciences were instrumental in our choice of attachment chemistry. This work certainly would not have been possible without the support of Chris Scholin and his staff at MBARI. Harban S. Dhadwal thanks his graduate students William Kaunds, Jason Yen, and Manish Muttreja for working tirelessly on the different components of hardware and software.

References

1. Scholin C, Doucette G, Jensen S, Roman B, Pargett D, Marin R, Preston C, Jones W, Feldman J, Everlove C, Harris E, Alvarado N, Massion E, Birch J, Greenfield D, Vrijenhoek R, Mikulski C, Jones K (2009) Remote detection of marine microbes, small invertebrates, harmful algae, and biotoxins using the environmental sample processor (ESP). Oceanography 22(2):158–167
2. Paul J, Scholin C, Engh G, Perry M (2007) In situ instrumentation. Oceanography 20(2): 70–78
3. Dhadwal HS, Mukherejee B, Kemp P, Aller J, Liu J, Radway J (2007) A dual detector capillary waveguide biosensor for detection and quantification of hybridized targets. Anal Chem Acta 598:147–154
4. Dhadwal HS, Kemp P, Aller J, Dantzler MM (2004) A capillary waveguide nucleic acid sensor. Anal Chem Acta 501(12):205–217
5. Dhadwal HS (2010) Capillary Waveguide Biosensor Platform. In: Zourob M, Lakhtakia A (eds) Optical guided-wave chemical and biosensors, Springer series on chemical sensors and biosensors, vol 8. Springer, Heidelberg, pp 221–257
6. Sambrook K, Fritsch EF, Maniatis T (1989) Molecular cloning: a laboratory manual (2 ed.). Cold Spring Harbor Press, Cold Spring Harbor
7. Corning Incorporated (2008) Corning epoxide coated slides instruction manual. CLS-CS-022REV2, Corning Incorporated, Corning, NY, 13 pp. http://catalog2.corning.com/lifesciences/media/pdf/corning_epoxide_instruction_manual.pdf.
8. Loy A, Maixner F, Wagner M, Horn M (2007) ProbeBase - an online resource for rRNA-targeted oligonucleotide probes: new features 2007. Nucl Acids Res 35:D800–D804
9. Hobbie JE, Daley RJ, Jasper S (1977) The use of nuclepore filters for counting bacteria by fluorescence microscope. Appl Environ Microbiol 33:1225–1228
10. Watson SW, Novitsky TJ, Quinby HL, Valois FW (1977) Determination of bacterial number and biomass in the marine environment. Appl Environ Microbiol 33:940–946
11. Qiagen (2005) RNAprotect bacteria reagent handbook. Qiagen, 51 pp. http://www.qiagen.com/literature/render.aspx?id=179.
12. Fuchs BM, Wallner G, Beisker W, Schwippl I, Ludwig W, Amann R (1998) Flow cytometric analysis of the *in situ* accessibility of *Escherichia coli* 16S rRNA for fluorescently labeled oligonucleotide probes. Appl Environ Microbiol 64:4973–4982
13. Huggett MJ, Crocetti GR, Kjelleberg S, Steinberg PD (2008) Recruitment of the sea urchin *Heliocidaris erythrogramma* and the distribution and abundance of inducing bacteria in the field. Aquat Microb Ecol 53:161–171
14. Eilers H, Pernthaler J, Glöckner F, Aman R (2000) Culturability and in situ abundance of pelagic bacteria from the North Sea. Appl Environ Microbiol 66:3044–3051
15. Giuliano L, De Domenico M, De Domenico E, Hofle MG, Yakimov MM (1999) Identification

of culturable oligotrophic bacteria within naturally occurring bacterioplankton communities of the Ligurian Sea by 16S rRNA sequencing and probing. Microb Ecol 37:77–85

16. Kyselkova M, Kopecky J, Frapolli M, Defago G, Sagova-Mareckova M, Grundmann GL, Moenne-Loccoz Y (2009) Comparison of rhizobacterial community composition in soil suppressive or conducive to tobacco black root rot disease. ISME J3:1127–1138
17. Amann RI, Binder BJ, Olson RJ, Chisholm SW, Devereux R, Stahl DA (1990) Combination of 16S rRNA-targeted oligonucleotide probes with flow cytometry for analyzing mixed microbial populations. Appl Environ Microbiol 56:1919–1925

Chapter 3

Microfabricated Flow-Through Device for In Situ Gene Analysis

Tatsuhiro Fukuba and Teruo Fujii

Abstract

A miniaturized and self-contained in situ genetic analyzer Integrated In Situ Analyzer-Gene (IISA-Gene) was developed to conduct PCR-based targeted sequence detection using microfluidic technology. The core element of the in situ analyzer is a microfluidic device that conducts cell lysis, DNA purification, PCR, and optical detection of amplified DNA fragments. In this device, a standard three-step SYBR green PCR amplification with a pair of oligonucleotide primers is conducted using environmental microbial DNA templates eluted from glass beads packed in a microchannel. The microfluidic device is integrated with pumping components and control electronics for real-field deployments. Here, the performance of the microfluidic device and results of in situ operation of the IISA-Gene in deep-sea hydrothermal areas are presented.

Key words: PCR, Microbes, Gene detection, Microfluidic device, Genetic analyzer, Three-step amplification, Microfluidic device, In situ molecular analysis, IISA-Gene system, 16S rRNA gene

1. Introduction

Microbial communities formed in various ocean environments have vast diversity on their functions and compositions. These communities play important ecological roles as final decomposers and primary producers of organic substances in complex marine food webs. This is especially true in deep-sea environments, where chemoautotrophic or chemolithoautotrophic microbes support unconventional, dense, and unique ecosystems by converting inorganic compounds into organic compounds using the oxidization energy of electron donors in surrounding environments. For example, H_2S and CH_4 are widely known and biogeochemically important electron donors supplied from hydrothermal vent fields or hydrocarbon seepage areas (1, 2). Habitats of extremophiles,

Sonia M. Tiquia-Arashiro (ed.), *Molecular Biological Technologies for Ocean Sensing*, Springer Protocols Handbooks, DOI 10.1007/978-1-61779-915-0_3, © Springer Science+Business Media New York 2012

which are microbes adapted to physically or geochemically extreme environments, may extend to areas far beneath the seafloor that are subject to extraordinarily high temperatures and pressures (3, 4). Elucidation of the phylogenetic diversity, biogeochemical roles, and total biomass of marine microbes including extremophiles is one of the most important issues involved in estimation of global cycles of materials and exploration of the frontiers of life.

Standard PCR-based highly sensitive and specific detection of microbial (bacterial and archaeal) genomic DNA partial sequences that encode ribosomal RNA or functional genes is traditional, but it is still one of the major, useful, and powerful molecular biological methods used in the field of environmental microbiology. For example, the existence of genetic fingerprinting molecules related to a series of enzymes for methane oxidation such as the particulate methane monooxygenase gene (*pmoA*) (5) indirectly, but strongly, suggests the prosperity of methanotrophs supported by methane supplied from their habitat. Species-, genus-, phylogenetic group-, and domain-specific 16S rRNA sequences or the genomic DNA sequences encoding them are also recognized as useful genetic probes (6–8). Fingerprinting genes can be used in the exploration for novel natural resources such as methane (gas) hydrates (9) or hydrothermal ore deposits (10, 11) in ocean environments because specific geochemical settings that allow the formation and accumulation of these resources will regulate the members and the properties of environment-associated microbial communities (11, 12). Furthermore detailed environmental assessments should be conducted for actual development and use of novel resources (13). Long-term and continuous monitoring of the microbial community will provide fundamental information about the impact of resource development on local ecosystems. In addition, growing industrial and medical demands on the applications of marine genetic resources (14) reinforce the importance of novel and useful survey tools to detect fingerprinting genes.

Development of in situ analysis tools that have the ability to conduct the entire analytical process for gene detection will probably lead to meaningful breakthroughs in traditional sampling-based methodologies. For instance, detailed analysis of the spatiotemporal distribution and diversity of fingerprinting genes is highly meaningful for elucidation of the environmental properties of their habitat. The use of in situ analyzers will improve reliability and spatiotemporal resolution for environmental microbiological surveys. To date, few studies have been conducted to develop in situ molecular biological tools for oceanography applications. These studies have investigated the usefulness of such tools for evaluation of the oceanic environment to detect marine invertebrate larvae (15), harmful phytoplankton (16, 17), and bacterioplankton (18). The next key achievement is to develop highly miniaturized and functionally integrated analyzers to mount them on a wide range of ocean observation platforms including small remotely operated vehicles (ROVs), autonomous underwater vehicles (AUVs), and buoy systems.

Application of "microfluidic devices" is a promising solution for meeting this requirement. Microfluidic devices (also known as micro total analysis systems (μTAS) or lab-on-a-chip) have been recognized as an essential element in realizing functionally integrated tiny devices that can be applied to the fields of clinical diagnosis, medical chemistry, and industrial chemistry. Some completed tools based on microfluidic devices for quantitative determination of dissolved metals (19) or inorganic nutrients (20, 21) have been developed for oceanography applications. However, to conduct complicated molecular biological analyses, reliable functional units for temperature control and sophisticated pumping are necessary. Accordingly, our group has developed prototypes of some microfluidic devices designed for in situ chemical (22), biochemical (23), and PCR-based molecular biological analyses (24).

Here, a completely integrated PCR-based molecular biological analyzer (Integrated In Situ Analyzer-Gene, IISA-Gene) with a built-in microfluidic device is presented. This system employs a polydimethylsiloxane (PDMS)-glass microfluidic device that conducts cell lysis, DNA purification, PCR, and optical detection of PCR products, and is integrated with miniature pumping components. The adoption of a simple flow-through PCR technique makes it possible to conduct complicated molecular biological analyses in situ.

2. IISA-Gene System

2.1. System Overview

The IISA-Gene system is a self-contained in situ molecular biological analyzer that uses a microfluidic device for PCR-based detection of targeted gene fragments in ocean environments. The IISA-Gene consists of an analysis unit and a control unit (Fig. 1). The analysis unit can conduct sample intake from environments, microbial cell lysis, DNA purification and concentration, PCR, detection, and recovery of PCR products. Analysis steps from cell lysis to PCR amplification and optical detection are conducted in a reusable microfluidic device. The temperature of the microfluidic device is controlled by integrated heaters and temperature sensors. Fluidic operations including sample intake, reagent supply, and switching are conducted by stepping motor- and DC motor-driven pumps, and solenoid valves integrated in the analysis unit. All reagents are stored in reagent tanks that have an indirect pumping function (see Note 1). All fluidic operations and molecular biological processes are conducted under ambient hydrostatic pressure in the oil-immersed analysis unit. A pair of optical fibers is connected

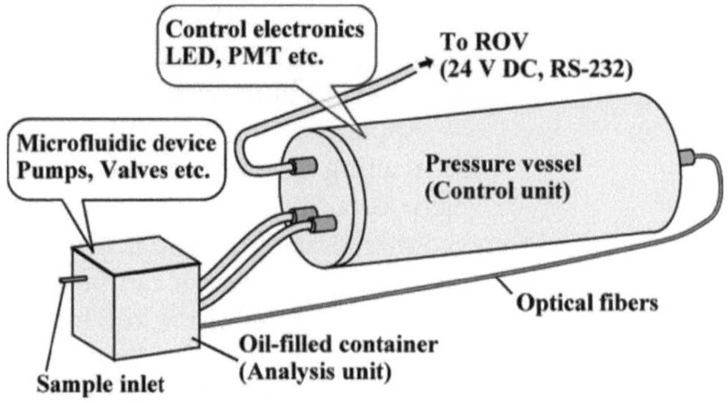

Fig. 1. Schematic illustration of the IISA-Gene.

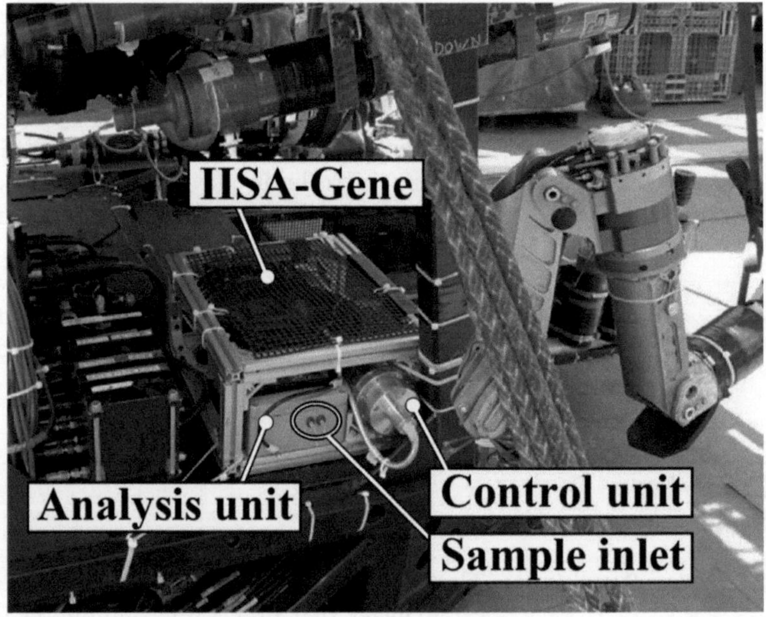

Fig. 2. The IISA-Gene mounted on the ROV "HYPER-DOLPHIN".

to the microfluidic device for fluorescence-based detection of PCR products. Optical components such as an LED as a light source and a photomultiplier tube (PMT) as a photodetector are enclosed in a cylindrical pressure vessel with control electronics. The analysis unit and control unit are fixed on a metal frame and mounted on an underwater vehicle for in situ operation (Fig. 2). An operator of the IISA-Gene system can remotely monitor the condition of the systems, and data are measured using a PC on a surface vessel.

2.2. Operational Procedure

A typical operational procedure is shown in Fig. 3. In this procedure, a seawater sample is introduced into the system using one of the pumps in the analysis unit. The sample is then processed to obtain PCR amplicons in the microfluidic device, as described

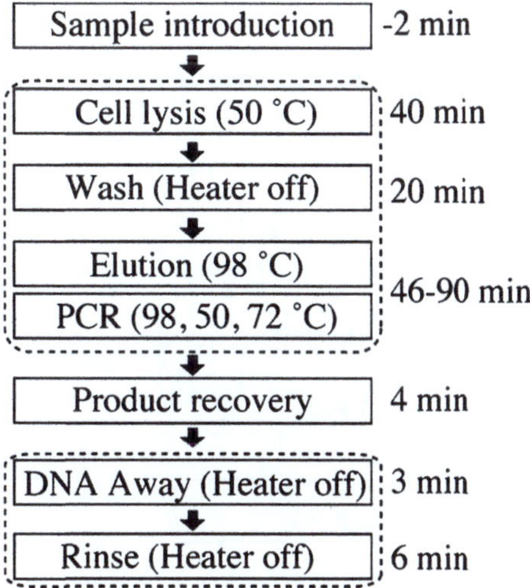

Fig. 3. Operational procedure of the IISA-Gene for typical single analysis. Adapted from Fukuba et al. (26) with permission.

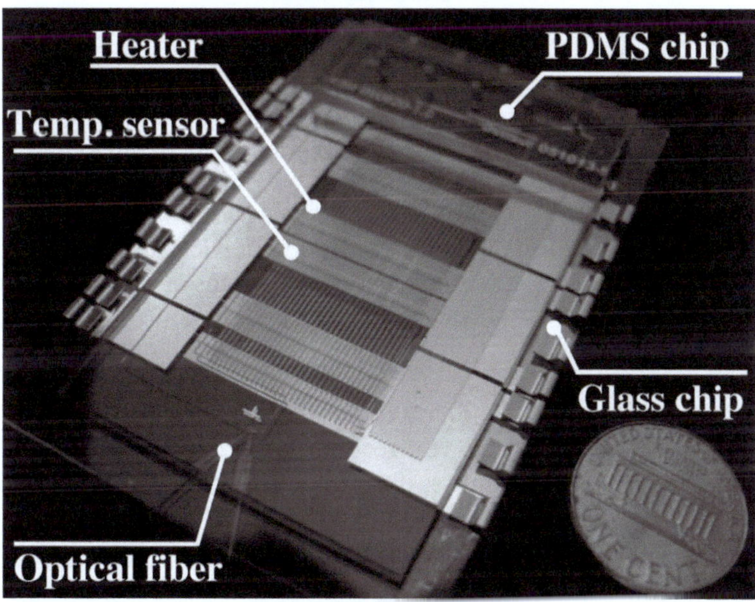

Fig. 4. Photograph of the microfluidic device for the IISA-Gene.

below. After the PCR step, the amplicon can be recovered for further onboard analysis. The microfluidic device is treated using DNA Away ™ (Molecular BioProducts Inc., USA) (25) prior to the next operation. The total time for a single whole analysis sequence is about 130–180 min depending on the elapsed time for PCR.

2.3. Microfluidic Device

The microfluidic device (Fig. 4) is the core functional component of the IISA-Gene (see Note 2). The microfluidic device is composed of a PDMS microfluidic chip with a microchannel and a glass

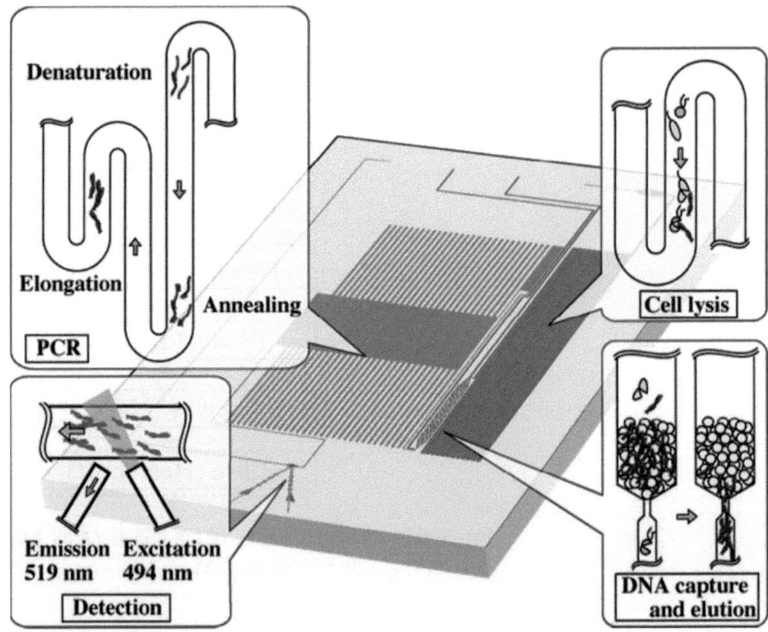

Fig. 5. Functions of the microfluidic device for PCR-based microbial gene detection.

chip with integrated heaters and temperature sensors. The microchannel is designed to perform cell lysis, DNA extraction, flow-through PCR, and optical detection of the amplicon. PDMS, a transparent silicone elastomer, is one of the most widely used materials for fabrication of microfluidic chips containing micro- to nanoscale structures by means of a simple replica-molding method (27). The glass chip is used to control the device temperature for the cell lysis and PCR steps. The thin film heaters and temperature sensors are fabricated by a conventional wet etching of metal layers deposited on the glass substrate (50×75 mm). The heaters are designed to realize well-defined uniform temperature distributions on the glass substrate for the cell lysis and flow-through PCR steps.

The functions of the microfluidic device are summarized in Fig. 5. First, a 500 μL seawater sample (see Note 3) mixed with a cell lysis buffer passes through a serpentine microchannel for microbial cell lysis. Since the cell lysis buffer contains a chaotropic agent (guanidium thiocyanate), microbial genomic DNA is captured onto the glass beads packed in the microchannel (see Note 4). The glass beads are then cleaned with a washing buffer to remove the cell lysis buffer (see Note 5). Next, the captured DNA is directly eluted into a PCR reagent mixture containing SYBR Green I dye. The flow-through PCR is subsequently conducted in a long serpentine microchannel (100-μm wide, 3,050-mm long) designed to perform 30 cycles of thermal cycling with a residence time ratio of 8:2:2:3:10 for initial denaturation, followed by 30 cycles of denaturation, annealing, extension, and final extension (see Note 6). The flow-through (or continuous-flow) PCR (28, 29) is suitable for in situ instrumentation because of its simplicity with respect to temperature and fluid control (24). Amplicons are optically detected at the end of the microchannel using a pair of optical fibers for excitation and fluorescence detection.

3. Performance Evaluations

3.1. PCR in the Microfluidic Device

To evaluate the performance of the microfluidic device for flow-through PCR, manually extracted and purified microbial genomic DNA was used as a template. The genomic DNA was prepared from frozen stock methanotroph (*Methylosinus trichosporium* OB3b) cells by a conventional protocol (see Note 7). The PCR reagent mixture contained 1× PCR buffer (TAKARA Bio Inc., Japan), 1× SYBR Green I (Invitrogen, USA), dNTPs (0.2 mM each), *Taq* DNA polymerase (0.075 U/μL, *TaKaRa Taq* Hot Start Version, TAKARA Bio Inc.), forward and reverse oligonucleotide primer (0.5 μM each), and genomic DNA (0.8 ng/μL final concentration). To prevent the adsorption of PCR mixture contents onto the PDMS surface, a 5 % (vol/vol) of Tween 20 (ICN Biomed. Inc., USA) was added to the PCR mixture. Flow-through PCR was then conducted under various flow-rate conditions (e.g., various times were used for 30 cycles of PCR). A primer pair for the particulate methane monooxygenase (pMMO) gene (pmofl-pmor) (30) was used for the functional gene detection trial. The temperature conditions for the PCR were 95, 50, and 72 °C for denaturation, annealing, and elongation, respectively. A control experiment was conducted using a desktop PCR thermal cycler (TaKaRa Thermal Cycler PERSONAL, TAKARA Bio Inc.). The PCR conditions for the control experiment were 2 min of initial denaturation followed by 30 s each for denaturation, annealing, and elongation, and a 10-min final elongation step (approx. 60 min total). The temperature setting was the same as that used for the microfluidic device. Figure 6 shows the results of agarose-gel electrophoresis analysis of the amplified pMMO gene fragments obtained using the microfluidic device. Under rapid PCR conditions (20 and 30 min for 30 cycles), the amount of specific amplicons (approx. 330 bp) was smaller than the amount obtained in the control experiment. In contrast, almost the same levels of amplicons were provided when 45 and 60 min were used for 30 cycles. A long and nonspecific band observed in the control experiment (~2,000 bp) was not observed in the amplicons from the microfluidic device. These results indicate that use of the microfluidic device for the IISA-Gene is capable of amplifying functional gene fragments from microbial genomic DNA with sufficiently high specificity and efficiency.

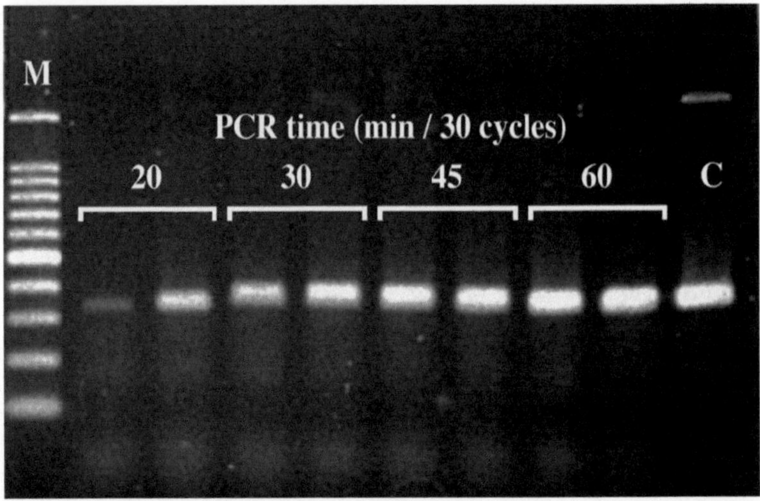

Fig. 6. Results of pMMO gene amplification using the microfluidic device. *Lane M*: 100 bp DNA ladder, *C*: control experiment.

On the IISA-Gene system, all analysis procedures should be conducted under ambient hydrostatic pressure because the microfluidic device is placed in an oil-filled analysis unit for pressure equalization. Here, the PCR performance was evaluated under high hydrostatic pressure conditions assuming practical operation in deep-sea environments. To accomplish this, the microfluidic device and a pumping apparatus were immersed in fluorinated oil (Fluorinert FC-43, 3 M, USA) and placed in a high-pressure chamber to apply 10 and 30 MPa of hydrostatic pressure. The eubacterial universal 16S rRNA gene was amplified using *Escherichia coli* genomic DNA as a template. The 968f–1401r (31) primer pair was used to amplify approximately 430 bp DNA fragments. The PCR conditions were the same as described above, except for the flow-rate setting (40 min for 30 cycles). Control experiments were conducted under atmospheric pressure (0.1 MPa) using the same microfluidic device. The PCR reagent mixture was composed of 1 × PCR buffer, dNTPs (0.5 mM each), *Taq* DNA polymerase (0.1 U/μL), forward and reverse oligonucleotide primers (1 μM each), 5 % Tween 20, and genomic DNA (0.4 ng/μL final concentration). The amplicons were analyzed by gel electrophoresis and the results were visualized by ethidium bromide staining (Fig. 7). Even at 10 and 30 MPa, no PCR inhibition as a result of elevated hydrostatic pressure was observed. The sizes of the amplicons were consistent with those of the control experiments. These results indicate that PCR can be conducted as usual, even in deep-sea environments at depth of up to 3,000 m.

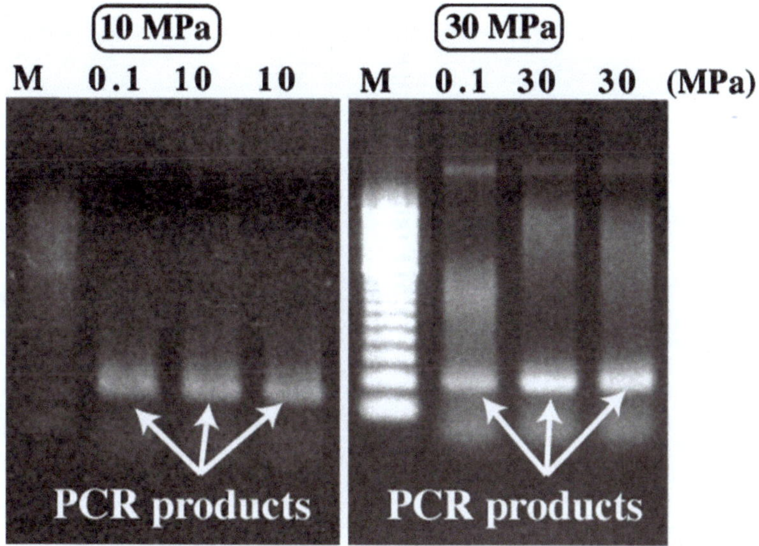

Fig. 7. Results of PCR conducted under high-pressure conditions. *Lane M*: 200 bp DNA ladder.

3.2. Performance of IISA-Gene

To evaluate the performance of the completed IISA-Gene system, *M. trichosporium* OB3b cells were processed to simulate targeted gene amplification from natural samples. Guanidium thiocyanate-based cell lysis buffer prepared as described by Boom et al. (32) is used in the processing. 1× PCR buffer for *Taq* DNA polymerase is used as the washing buffer. The PCR mixture is composed of 1× PCR buffer, dNTPs (0.2 mM each), *Taq* DNA polymerase (0.075 U/μL), forward and reverse oligonucleotide primers (0.5 μM each), 2× SYBR green I (see Note 8), and 5 % of Tween 20. To amplify the eubacterial universal gene, a primer pair specific for the 16S rRNA gene (341f–926r) (33) was used for this evaluation. The temperature conditions for the flow-through PCR are 98, 50, and 72 °C for each PCR step (46 min for 30 cycles). The results of the analysis conducted using 500 μL of 10^4 cells/mL are shown in Fig. 8. The PMT output voltage values that correspond to the fluorescence intensity from the amplicons showed a clear increase when the microbial cells were analyzed, whereas the negative control showed no increase. Similarly, analysis using a sample composed of 10^3 cells/mL provided negative amplification (data not shown). Therefore, the detection limit for the 16S rRNA gene using the IISA-Gene was estimated to be around 10^4 cells/mL. As shown here, the capability of the IISA-Gene for PCR-based molecular biological analysis using microbial cells as a sample was successfully demonstrated.

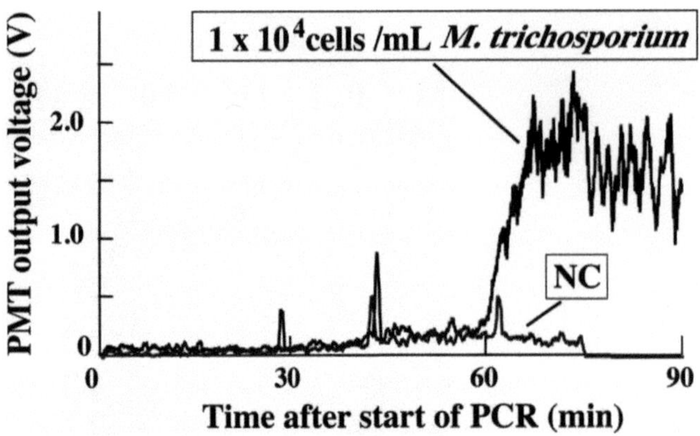

Fig. 8. Result of 16S rRNA gene amplification using microbial cells as the sample. *NC*: Negative control without microbial cells. (Adapted from Fukuba et al. (26) with permission).

4. In Situ Deployments

The completed IISA-Gene system was operated at a hydrothermal site in Hatoma Knoll (34, 35), Okinawa Trough, Japan, using the ROV "HYPER-DOLPHIN" and the research vessel "NATSUSHIMA" (Japan Agency for Marine and Earth Science and Technology: JAMSTEC, Japan) in June 2008. The IISA-Gene was mounted on a payload-bay of the ROV and was operated a few meters above an active hydrothermal vent at a depth of 1,471 m. The in situ analysis was performed twice at the same position for 16S rRNA gene and pMMO gene detection using the 341f–926r and A189–A682 (36) primer pairs, respectively. The PCR conditions were the same as those described in Sect. 3.2. The results are summarized in Fig. 9. The microbial 16S rRNA gene was successfully detected by PCR-based in situ analysis using the IISA-Gene. In contrast, PCR targeted to amplify the pMMO gene showed a negative result. These results imply that detectable numbers of microbial cells (more than 10^4 cells/mL) were present, but that methanotrophic microbes are not a major member of the microbial community in the hydrothermal fluid of Hatoma Knoll.

5. Notes

1. The tank for the reagent has two compartments separated by a deformable membrane for the indirect pumping using 1× PCR buffer as a displacement fluid. Capacity of the tank is approximately 2 mL for regular uses, and it can be modified depending on the frequency of in situ analysis. All of the tank materials should be autoclaved to avoid contaminations prior to filling of the reagents.

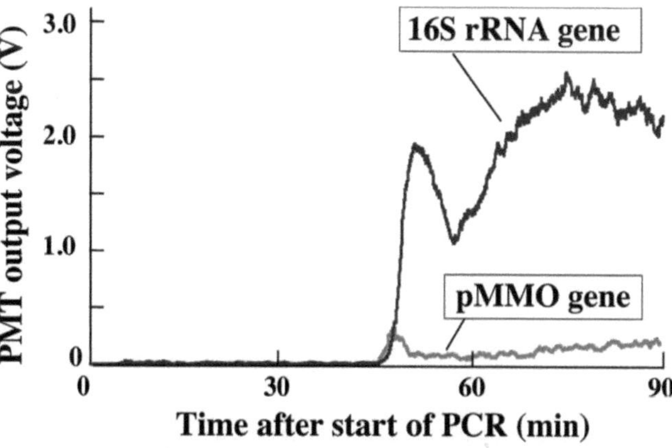

Fig. 9. Results of 16S rRNA gene and pMMO gene amplification during the dive #853 in the Hatoma Knoll hydrothermal site, Okinawa Trough, Japan. (Adapted from Fukuba et al. (26) with permission).

2. All of the materials for the microfluidic device are autoclavable. The microfluidic device should be autoclaved prior to integrating with IISA-Gene system. After the autoclave, the PDMS material becomes cloudy white color because of penetration of water vapor in the material. The PDMS returns to a transparent state by keeping it in a dry condition for a few hours.
3. The volume of the sample is defined by the total volume of a coiled tube as a sample loop. The volume of the loop is variable depends on a requirement.
4. Approximately 5 mg of glass beads (30–50 μm in diameter) are packed in the microchannel. The glass beads should be washed, and small particles of broken beads should be removed before the packaging of the glass beads to avoid a clogging of the microchannel.
5. The glass beads with template DNA should be washed at around room temperature condition. Washing the glass beads under the high temperature condition (e.g., more than 90 °C) will cause loss of the template DNA.
6. PCR conditions such as the number of thermal cycle and the residence time ratio are defined by a design of the microfluidic device. The total time for the PCR (regulated by flow rate for the continuous-flow PCR) and the temperature setting are variable.
7. A commercially available kit (DNeasy Blood & Tissue Kit, QIAGEN, Germany) was used for the preparation of the positive control DNA.
8. Addition of Tween 20 is essential to use SYBR green I in 2×concentration in PCR mixture to improve the fluorescence signal intensity. No PCR products will be obtained without Tween 20 in the case of 16S rRNA gene and pMMO gene.

6. Future Directions

The development and evaluation of the IISA-Gene in this study demonstrated that microfluidic devices have significant potential for use in functionally integrated and portable in situ analysis tools. The next essential mission to realize higher spatiotemporal resolution during in situ gene detection is improvement of the throughput of the system. In addition, for further miniaturization of the IISA-Gene system, well-miniaturized or on-chip pumping components should be used instead of the stepping motor-driven pumps and solenoid-actuated valves used as pumping components. Application of a miniature and integrated continuous pumping system that has on-chip microvalves and flow regulators (37) will contribute to improvements in the throughput and size of the IISA-Gene system. Realization of an IISA-Gene system with quantitative PCR capability is possible by applying a simple and novel quantitative PCR method using a flow-through PCR microfluidic device (38). By integrating a reverse transcription function on the microfluidic device, in situ gene expression analysis using microbial RNA as a sample will also become possible in the future.

Acknowledgements

The authors are grateful to the crew of the R/V NATSUSHIMA and the operating team of the ROV HYPER-DOLPHIN (JAMSTEC) for their helpful assistance during scientific cruise NT08-11. This work was supported by a grant-in-aid for scientific research from the Japanese Ministry of Education, Culture, Sports, Science, and Technology (MEXT). The Engineering Advancement Association of Japan (ENAA) also supported this study.

References

1. Cavanaugh CM, Gardiner SL, Jones ML, Jannasch HW, Waterbury JB (1981) Prokaryotic cells in hydrothermal vent tube worm *Riftia pachyptila* Jones: possible chemoautotrophic symbionts. Science 213(4505):340–342
2. Karl DM, Wirsen CO, Jannasch HW (1980) Deep-sea primary production at the Galapagos hydrothermal vents. Science 207(4437):1345–1347
3. Gold T (1992) The deep, hot biosphere. Proc Natl Acad Sci U S A 89(13):6045–6049
4. Stetter K (1999) Extremophiles and their adaptation to hot environments. FEBS Lett 452(1–2):22–25
5. McDonald I, Murrell J (1997) The particulate methane monooxygenase gene pmoA and its use as a functional gene probe for methanotrophs. FEMS Microbiol Lett 156(2):205–210
6. Gulledge J, Ahmad A, Steudler P, Pomerantz W, Cavanaugh C (2001) Family-and genus-level 16S rRNA-targeted oligonucleotide probes for ecological studies of methanotrophic

bacteria. Appl Environ Microbiol 67(10): 4726–4733

7. Amann R, Binder B, Olson R, Chisholm S, Devereux R, Stahl D (1990) Combination of 16S rRNA-targeted oligonucleotide probes with flow cytometry for analyzing mixed microbial populations. Appl Environ Microbiol 56(6):1919–1925
8. Karner M, DeLong E, Karl D (2001) Archaeal dominance in the mesopelagic zone of the Pacific Ocean. Nature 409(6819):507–510
9. Sloan E (2003) Fundamental principles and applications of natural gas hydrates. Nature 426(6964):353–363
10. Halbach P, Nakamura K, Wahsner M, Lange J, Sakai H, Käselitz L, Hansen RD, Yamano M, Post J, Prause B, Seifert R, Michaelis W, Teichmann F, Kinoshita M, Märten A, Ishibashi J, Czerwinski S, Blum N (1989) Probable modern analogue of Kuroko-type massive sulphide deposits in the Okinawa Trough back-arc basin. Nature 338(6215):496–499
11. Zierenberg RA, Fouquet Y, Miller DJ, Bahr JM, Baker PA, Bjerkgar T, Brunner CA, Duckworth RC, Gable R, Gieskes J, Goodfellow WD, Groschel-Becker HM, Guerin G, Ishibashi J, Iturrino G, James RH, Lackschewitz KS, Marquez LL, Nehlig P, Peter JM, Rigsby CA, Schultheiss P, Shanks WC, Simoneit BRT, Summit M, Teagle DAH, Urbat M, Zuffa GG (1998) The deep structure of a sea-floor hydrothermal deposit. Nature 392(6675): 485–488
12. Zhang C, Pancost R, Sassen R, Qian Y, Macko S (2003) Archaeal lipid biomarkers and isotopic evidence of anaerobic methane oxidation associated with gas hydrates in the Gulf of Mexico. Org Geochem 34(6):827–836
13. Awashima Y, Saito H, Hoaki T, Sawada S, Fukasawa T, Sukizaki S, Ishihara Y, Kano K, Arata N (2008) Development of monitoring system on methane hydrate production. OCEANS 2008 - MTS/IEEE Kobe Techno-Ocean, Kobe, Japan, pp 1–7
14. Niehaus F, Bertoldo C, Kähler M, Antranikian G (1999) Extremophiles as a source of novel enzymes for industrial application. Appl Microbiol Biotechnol 51(6):711–729
15. Jones W, Preston C, Martin R III, Scholin CA, Vrijenhoek R (2008) A robotic molecular method for in situ detection of marine invertebrate larvae. Mol Ecol Resour 8(3):540–550
16. Greenfield D, Marin R III, Jensen S, Massion E, Roman B, Feldman J, Scholin C (2006) Application of Environmental Sample Processor (ESP) methodology for quantifying Pseudo-nitzschia australis using ribosomal RNA-targeted probes in sandwich and fluorescent *in situ* hybridization formats. Limnol Oceanogr Methods 4:426–435
17. Fries D, Paul J, Smith M, Farmer A, Casper E, Wilson J (2007) The Autonomous microbial genosensor, an *in situ* sensor for marine microbe detection. Microsc Microanal 13(S02):514–515
18. Scholin C, Doucette G, Jensen S, Roman B, Pargett D, Marin R III, Preston C, Jones W, Feldman J, Everlove C, Harris A, Alvarado N, Massion E, Birch J, Greenfield D, Vrijenhoek R, Mikulski C, Jones K (2009) Remote detection of marine microbes, small invertebrates, harmful algae, and biotoxins using the environmental sample processor (ESP). Oceanography 22(2):158–161
19. Chapin T, Jannasch H, Johnson K (2002) *In situ* osmotic analyzer for the year-long continuous determination of Fe in hydrothermal systems. Anal Chim Acta 463(2):265–274
20. Jannasch H, Johnson K, Sakamoto C (1994) Submersible. Osmotically pumped analyzer for continuous determination of nitrate in situ. Anal Chem 66(20):3352–3361
21. Thouron D, Vuillemin R, Philippon X, Lourenço A, Provost C, Cruzado A, Garçon V (2003) An autonomous nutrient analyzer for oceanic long-term *in situ* biogeochemical monitoring. Anal Chem 75(11):2601–2609
22. Fukuba T, Provin C, Okamura K, Fujii T (2009) Development and evaluation of microfluidic device for Mn ion quantification in ocean environments. IEEJ Trans SM 129(3): 69–72
23. Aoki Y, Fukuba T, Yamamoto T, Fujii T (2009) Design optimization and evaluation of a bioluminescence detection part on a microfluidic device for in situ ATP quantification. IEEJ Trans SM 129(3):73–76
24. Fukuba T, Yamamoto T, Naganuma T, Fujii T (2004) Microfabricated flow-through device for DNA amplification -towards *in situ* gene analysis. Chem Eng J 101(1–3):151–156
25. Kemp B, Smith D (2005) Use of bleach to eliminate contaminating DNA from the surface of bones and teeth. Forensic Sci Int 154(1):53–61
26. Fukuba T, Miyaji A, Okamoto T, Yamamoto T, Kaneda S, Fujii T (2011) Integrated *in situ* genetinc analyzer for microbiology in extreme environments. RSC Adv 1:1567–1573
27. Fujii T (2002) PDMS-based microfluidic devices for biomedical applications. Microelec Eng 61–62:907–914

28. Nakano H, Matsuda K, Yohda M, Nagamune T, Endo I, Yamane T (1994) High speed polymerase chain reaction in constant flow. Biosci Biotech Biochem 58(2):349–352
29. Kopp M, Mello A, Manz A (1998) Chemical amplification: continuous-flow PCR on a chip. Science 280(5366):1046–1048
30. Cheng Y, Halsey J, Fode K, Remsen C, Collins M (1999) Detection of methanotrophs in groundwater by PCR. Appl Environ Microbiol 65(2):648
31. Nubel U, Engelen B, Felske A, Snaidr J, Wieshuber A, Amann RI, Ludwig W, Backhaus H (1996) Sequence heterogeneities of genes encoding 16S rRNAs in Paenibacillus polymyxa detected by temperature gradient gel electrophoresis. J Bacteriol 178(19):5636–5643
32. Boom R, Sol C, Salimans M, Jansen C, Wertheim-van Dillen P (1990) Rapid and simple method for purification of nucleic acids. J Clin Microbiol 28(3):495–503
33. Kato C, Li L, Tamaoka J, Horikoshi K (1997) Molecular analyses of the sediment of the 11000-m deep Mariana Trench. Extremophiles 1(3):117–123
34. Tsuchida S, Watanabe K, Ishibashi J, Miyake H, Watabe H, Yamaguchi T, Kitajima T, Nakano A, Matsumura M, Watanabe H (2000) Preliminary report of a biological, geological, and geochemical survey on hydrothermalism at the Hatoma and Minna Knolls. JAMSTEC J Deep Sea Res 17:34–42
35. Takai K, Inagaki F, Nakagawa S, Hirayama H, Nunoura T, Sako Y, Nealson KH, Horikoshi K (2003) Isolation and phylogenetic diversity of members of previously uncultivated ε-Proteobacteria in deep sea hydrothermal fields. FEMS Microbiol Lett 218(1):167–174
36. Holmes AJ, Costello A, Lidstrom ME, Murrell JC (1995) Evidence that participate methane monooxygenase and ammonia monooxygenase may be evolutionarily related. FEMS Microbiol Lett 132(3):203–208
37. Kinoshita H, Atsumi T, Fukuba T, Fujii T (2010) Active micro flow-rate reguration technique based on soft membrane deformation using miniaturized electroosmotic pumps. 14th international conference on miniaturized systems for chemistry and life sciences, Groningen, The Netherlands. pp 390–392
38. Fukuba T, Hiraga M, Takamatsu A, Provin C, Yamamoto T, Fujii T (2008) Simple method for quantitative PCR using flow-through PCR device. 12th international conference on miniaturized systems for chemistry and life sciences, San Diego, USA, pp 1473–1475

Chapter 4

Method for the Quantification of Aquatic Primary Production and Net Ecosystem Metabolism Using In Situ Dissolved Oxygen Sensors

Joseph A. Needoba, Tawnya D. Peterson, and Kenneth S. Johnson

Abstract

We present an example-based methodology for the "open-water" technique to determine net ecosystem metabolism (or net community metabolism) of coastal and oceanic environments using dissolved oxygen measurements collected with in situ sensors. We briefly discuss strategies for instrument deployment, data collection, and performance evaluation. The main focus is a presentation of the analytical steps necessary to convert raw dissolved oxygen measurements into daily estimates of primary production, aerobic respiration, and the resulting net metabolic sum. The data manipulation is based on a compilation of approaches from the literature that span multiple decades of research. We provide a summary of the foundational concepts, a brief discussion of the necessary assumptions, and a list of additional variables required for the parameterization of the individual components, such as the air–water diffusion term. The methods are presented in a step-by-step example format using representative data sets from two contrasting environments, the coastal ocean (Monterey Bay, California) and an estuary (Columbia River Estuary, Oregon/ Washington).

Key words: Dissolved oxygen, Net ecosystem metabolism, Primary production, In situ sensors, Water quality, Primary production, Ocean observatories, Net ecosystem production, Respiration rates, Air–water diffusion flux

1. Introduction

1.1. Overview

This chapter provides the methodology for the collection and analysis of high-resolution in situ dissolved O_2 (DO) measurements for determining net ecosystem metabolism (NEM), net ecosystem production (NEP), gross primary production (GPP), and ecosystem respiration (ER) of coastal and estuarine aquatic ecosystems. We focus on the "open-water" or "diel oxygen" techniques

Sonia M. Tiquia-Arashiro (ed.), *Molecular Biological Technologies for Ocean Sensing*, Springer Protocols Handbooks, DOI 10.1007/978-1-61779-915-0_4, © Springer Science+Business Media New York 2012

employed to measure the daily rates of organic matter production and aerobic respiration, the sum of which determines the daily NEM (1). An accurate determination of the changes in oxygen attributable to biological activity relies on having corresponding data to (i) correct for temperature and salinity effects on oxygen solubility (2), and (ii) characterize the potential importance of physical inputs or losses to the system (3–5). Our approach is similar to that of Staehr et al. (6), which presents the methodology for calculating NEM in lake ecosystems, and thus should be consulted for studies pertaining to lakes. In addition, for a more detailed overview of the assumptions and uncertainties associated with NEM estimates based on the use of high-resolution sensors, the reader is encouraged to consult Staehr et al. (7).

1.2. Dissolved Oxygen as a Measure of Ecosystem Metabolism

The determination of NEM, broadly defined as the difference between primary production and aerobic respiration within an ecosystem (or community of organisms), is increasingly being employed as a means to assess the health of aquatic environments and to quantify the biogeochemical fluxes of organic carbon, oxygen, and other important nutrients at regional and global scales. For example, in many estuaries an external nutrient input can stimulate primary productivity that results in an initial increase of inorganic carbon biomass, but it is typically followed by elevated organic matter respiration rates that can lead to hypoxia and shifts in community structure and function. Observations and measurements that accurately quantify rates of primary production and respiration are very useful for characterizing ecosystem-scale processes and determining the metabolic balance of a system. Therefore, NEM is closely tied to habitat quality and indicates the nature of carbon flux within a system, which can be used to inform stewardship and remediation activities.

A realistic representation of biogeochemical variability within aquatic environments (including processes associated with NEM) requires measurements performed at appropriate temporal and spatial scales. Yet, in practice, the effort required to collect water samples over long duration and large spatial areas is often impractical or cost prohibitive. For example, in many coastal environments the variations in water quality parameters such as algal biomass or nutrient concentration occur over timescales from minutes (tidal variability) to years (seasonal and climate variability), and can be altered by episodic weather events at unpredictable times. Periodic sampling, especially in coastal ecosystems, is likely to under-sample this variability. Even when data are collected at high enough frequency to capture variability appropriately, the intense personnel effort required for these efforts renders the observational periods short and unsustainable for multiple seasons or years. Thus, methods other than traditional ship-based field sampling are needed to capture high-resolution variability while also providing sufficient temporal coverage to allow the identification of seasonal and inter-annual changes in biogeochemical parameters. For these reasons, the development of robust, relatively inexpensive instruments capable of continuous deployment in the aquatic environment has dramatically improved the quantification of biogeochemical cycles and characterization of ecological function (8–11). As evidenced by other papers in this volume, the development of biological and chemical sensors is an emerging field in environmental sciences and a variety of technologies seek to adopt biological methods typically used in a laboratory setting and apply them to a range of new applications designed for in situ and autonomous operation.

Many aquatic monitoring programs now incorporate some kind of autonomous platform observational approach to provide a routine measure of water quality parameters that can be used as a complement to more advanced investigations of habitat structure or function and to support modeling studies. Dissolved oxygen (DO) is one such routinely measured parameter; it is relatively easy to measure autonomously and yields rich information regarding the health of an ecosystem. Oxygen gas (O_2) is released by primary producers as a product of the light reactions of photosynthesis and is consumed during cellular aerobic respiration with the following approximate stoichiometry (9):

$$106CO_2 + 16HNO_3 + H_3PO_4 + 78H_2O \rightleftharpoons C_{106}H_{175}O_{42}N_{16}P + 138O_2 \quad (4.1)$$

Therefore, changes in O_2 can be used to calculate rates of organic matter production and respiration if measured at the appropriate resolution and considering other processes that may simultaneously occur. The utility of the diel DO approach has increased considerably with the development of robust sensor technologies capable of high accuracy, long-term deployments in a variety of platforms, including moorings (12), profiling floats (13, 14), and autonomous vehicle surveys (15, 16).

1.3. Overview of Primary Production and Net Ecosystem Metabolism (NEM)

The magnitude of organic carbon production within an ecosystem is often used to classify aquatic environments (or ecoregions) at a global scale. Longhurst (17) defines ecological regions in the ocean based on rates of primary production derived primarily from satellite observations of surface chlorophyll concentrations. Nixon (18) identifies the trophic status of marine systems based solely on organic carbon production. For example, oligotrophic systems are defined as having organic carbon production rates of <100 g C m^{-2} y^{-1} and eutrophic systems range between 301 and 500 gC m^{-2} y^{-1}. Traditional methods of measuring primary productivity require sample collection (usually on a ship or small boat) followed by incubation of water with a tracer such as ^{14}C-labelled HCO_3^- (19). Alternatively, changes in DO or CO_2 concentration can be monitored over the course of an incubation to yield a net or GPP rate. Depending on the technique, bottle incubations do not provide a measure of carbon respiration (i.e., net productivity) and they often suffer from so-called bottle effects that lead to biases in rate estimates (20–22).

Odum (23, 24) first suggested the use of changes in open-water dissolved oxygen (DO) to provide integrated estimates of metabolic processes in aquatic systems. In this holistic view, the "metabolic status" of a given environment reflects the balance of autochthonous organic matter production (i.e., primary production), allochthonous delivery of organic matter (advection), and respiration of organic matter (from whichever source) at an ecosystem scale. He coined the term "net ecosystem metabolism" (NEM), alternatively called "net community production" in oceanographic literature, to describe the difference between oxygen production (photosynthesis) and community respiration. Ecosystem, or community, metabolism can be calculated based on oxygen, carbon dioxide, or macronutrient measurements. The former has been most widely adopted because oxygen sensors are more practical: they are cheaper to produce, require less power to operate, and tend to be more robust compared to carbon dioxide or macronutrient sensors (25). Before the wide availability of robust in situ oxygen sensors (26, 27), it was necessary to make measurements periodically or over short time spans; however, in recent years there have been many examples where NEM has been determined over extended periods. Examples of NEM measurements are common in lake (7, 28) and estuarine (29–31) systems, but are also possible in open ocean environments (9).

At face value, daily measurements of NEM are simply the difference between O_2 produced through photosynthesis minus that used by aerobic respiration (of the community) in a 24 h period, corrected for nonbiological gas exchange. At the ecosystem scale, and over sufficiently long periods, positive NEM suggests that the system is a sink for inorganic carbon (and ultimately atmospheric CO_2), while negative NEM values implies an autochthonous source of organic matter is available in addition to the supply from primary production from within the ecosystem. In addition, many systems likely operate relatively close to a "balanced" NEM (no net change in organic carbon); however, the daily variability is an important indicator of ecosystem function, i.e., high productivity and high respiration versus low productivity and low respiration. For example, in low-productivity regions such as the oligotrophic ocean, the small signals of oxygen production have sparked debate about whether NEM is positive or negative, with some studies suggesting that respiration (mainly by bacteria) exceeds production by phytoplankton (32–34). Others, citing systematic methodological biases in the literature (35) and geochemical estimates of excess seasonal oxygen production (36) argue the opposite. Recently, long-term records of DO from autonomous floats have improved the temporal resolution of direct measurements of DO that suggest that these regions have positive NEM (14).

1.4. Physical Processes Affecting Oxygen Concentration

The utility of DO measurements for calculating NEM is hindered by physical processes that are sometimes difficult to validate. These include changes associated with air–water diffusion of oxygen and the consistency of the water mass over the measurement period. These are not trivial issues and should be carefully considered and tailored to the system of interest. The propensity for dissolution of oxygen in natural waters depends strongly on temperature, since—according to Henry's Law—the concentration of a gas is a product of its partial pressure and its solubility in the aqueous medium, as well as salinity (2). In order to separate the physical and biological fluxes of dissolved oxygen, the atmospheric flux (air–water exchange) must be measured or estimated (24).This is facilitated by knowledge of the temperature and salinity of the water mass for each oxygen measurement.

In addition to oxygen solubility in water, both wind speed and water velocity affect air–water gas exchange (11, 37). The magnitude of effect of each of these processes, however, may differ from system to system. In a fast-flowing system, such as a large river, water velocity is more likely to play an important role in air–water exchange when wind speeds are less than 7 $m s^{-1}$ (37). Conversely, in a low-flow system and in the open ocean, wind speed is more likely to be important (38). Thus, it is necessary to identify and measure the factors that most strongly influence air–water gas exchange for a given environment.

1.5. Assumptions Made When Calculating Net Ecosystem Metabolism (NEM)

The fundamental assumption in NEM calculations is that all measurements come from a water mass that has the same recent history, which allows for point measurements in flowing waters to be compared over time. This assumption is difficult to assess in many systems and is often not satisfied, especially in estuaries where significant horizontal exchange occurs due to tidal forcing. However, some basic rules can be applied to determine if the system of interest is appropriate for the calculation of NEM. Ideally, water residence times should be sufficiently long that the same water mass is likely measured over a 24 h period. Ancillary measurements, such as temperature and salinity, are useful for determining if mixing or water movement is introducing new water masses over short periods of time. Ideally, before beginning a monitoring program where NEM criteria are to be employed, a thorough analysis of water mass variability of the system should be undertaken.

A second important assumption is that respiration rates are constant over a diel cycle. It is likely that daytime respiration exceeds nighttime respiration (39, 40), although it is nontrivial to determine the true respiration rates (41). It is therefore commonly assumed that photosynthesis is zero at night, and therefore the diffusion-corrected decrease in oxygen is a direct measure of respiration. The hourly rate determined at night can therefore be extrapolated to a daily rate.

Another point to note is that, in practice, dissolved oxygen sensors are not always deployed at multiple depths, and single point measurements may not be representative of photosynthetic rates throughout the water column. In order to calculate water column primary productivity accurately, the simplest approach is to assume that sufficient mixing occurs throughout the water column such that concentrations at a fixed depth are representative of the entire vertical section—in this case, rates are typically reported as a real rates (m^{-2}) per unit time. If this condition is not met, rates can be reported as volume rates (m^{-3}). Unless water depth is accounted for during the calculation, the latter method will overestimate the importance of air–water diffusion if the measurement is then extrapolated to more than 1 m depth, and therefore should be avoided in most cases.

Finally, respiration rates measured by changes in dissolved oxygen are aerobic respiration rates only, thus in environments where significant anaerobic organic matter decomposition occurs this method will underestimate the true NEM.

1.6. General Considerations of Working with In Situ Sensors and Observatory Data

There are many choices for sensor and platform operations, and the methods presented here are mainly concerned with post-data acquisition steps to the extent possible. It is assumed that the reader has some knowledge about sensor calibration and necessary time series configurations when purchasing or building instruments and implementing observatory platforms in the field. However, given the myriad ways in which environmental observatory science is evolving, it is increasingly possible for anyone to access data from a public database to make the calculations presented here, and, depending on the data source, may never directly communicate with the people who collected the original data. Thus, it is important to understand the practical considerations for data collection even if it is external to the research project.

It is important to underscore that in order to determine primary productivity, NEM, or metabolic changes in any aquatic system, a suite of sensors must be deployed together as a unified platform with similar sampling frequencies to provide the necessary physical context needed to correct for solubility and air–water gas transfer. At a minimum, a conductivity-temperature-depth (CTD) package must be co-deployed with an oxygen sensor. A meteorological (Met) station that provides wind speed data (speed and direction) and a current meter that provides water velocity will improve the estimates of air–water diffusion. Additional biological and chemical sensors are not needed but can often add synergistic information that helps to interpret the changes in DO. For example, sensors that determine inorganic carbon concentrations or pH (42, 43) are useful for further constraining the biological changes in DO, a chlorophyll fluorometer will provide context regarding algal biomass, and in situ optical nitrate sensors (44) or other "wet chemistry sensors" (45) can capture variability in nutrient biogeochemistry.

The frequency and length of sensor deployment depends on the variability of the biological DO signal, which varies according to the environment in which the sensors are to be deployed. A further constraint is the practical limitation of sensor maintenance schedules, which can be labor intensive, and thus incur substantial costs. The goal is to minimize loss of data due to baseline drift, bio or geo-fouling, degradation of instrument components, data storage limitations, or power interruptions. The Nyquist theorem (46) suggests that sampling should occur at twice the frequency of the signal to be detected. Ideally, in estuaries and tidal systems, the sensors are programmed to collect data at hourly time scales (or shorter) to capture sub-tidal variability and to allow for long-term deployments. Collecting data at too high a frequency shortens the length of deployments due to limitations on sensor function or data storage capabilities. To avoid potential biases introduced by small-scale processes (e.g., internal waves (3)), it is recommended that the data be band-pass filtered (low-pass filtered) or smoothed, if such physical processes are likely to influence the observations. Similarly, if low-frequency variability (>33 h) is expected, then the observations should be high-pass filtered (47).

Many oxygen sensors are subject to irregular changes in calibration coefficients (48), requiring frequent and careful inspection of the data for baseline drift. The calibration coefficients should therefore be determined frequently (on a monthly basis) or the data should be compared with recently calibrated instruments in the field, side by side. In high-productivity waters where biofouling may be a significant problem, it is best to choose a deployment length of approximately a week for oxygen sensors to avoid problems with fouling or instrument drift. On the other hand, deployment of oxygen sensors can be challenging in oligotrophic environments since biomass is low and therefore O_2 fluxes are expected to be small, yielding low signal-to-noise ratios that make it difficult to detect hourly changes in O_2. In all cases, quality control is best achieved by strictly following manufacturer specified operation and calibration procedures.

1.7. Example Datasets

To facilitate the description below of the basic steps to calculate NEM parameters, we refer to sample data sets from sensor platforms in two contrasting environments, a marine, near-shore coastal environment (~21 m depth) in Monterey Bay (Fig. 1a–e) and an estuarine environment (~7 m depth) in the Lower Columbia River Estuary in the US Pacific Northwest (Fig. 1f–j). In both locations, variability in temperature and salinity of the water mass remain relatively constant over 2–3 day periods or longer (Fig. 1a, f) and thus meet the assumption that the same community is represented throughout the diel cycle. The additional parameters nitrate and chlorophyll are complementary to the NEM measurements and are included for contextual purposes (Fig. 1b, g).

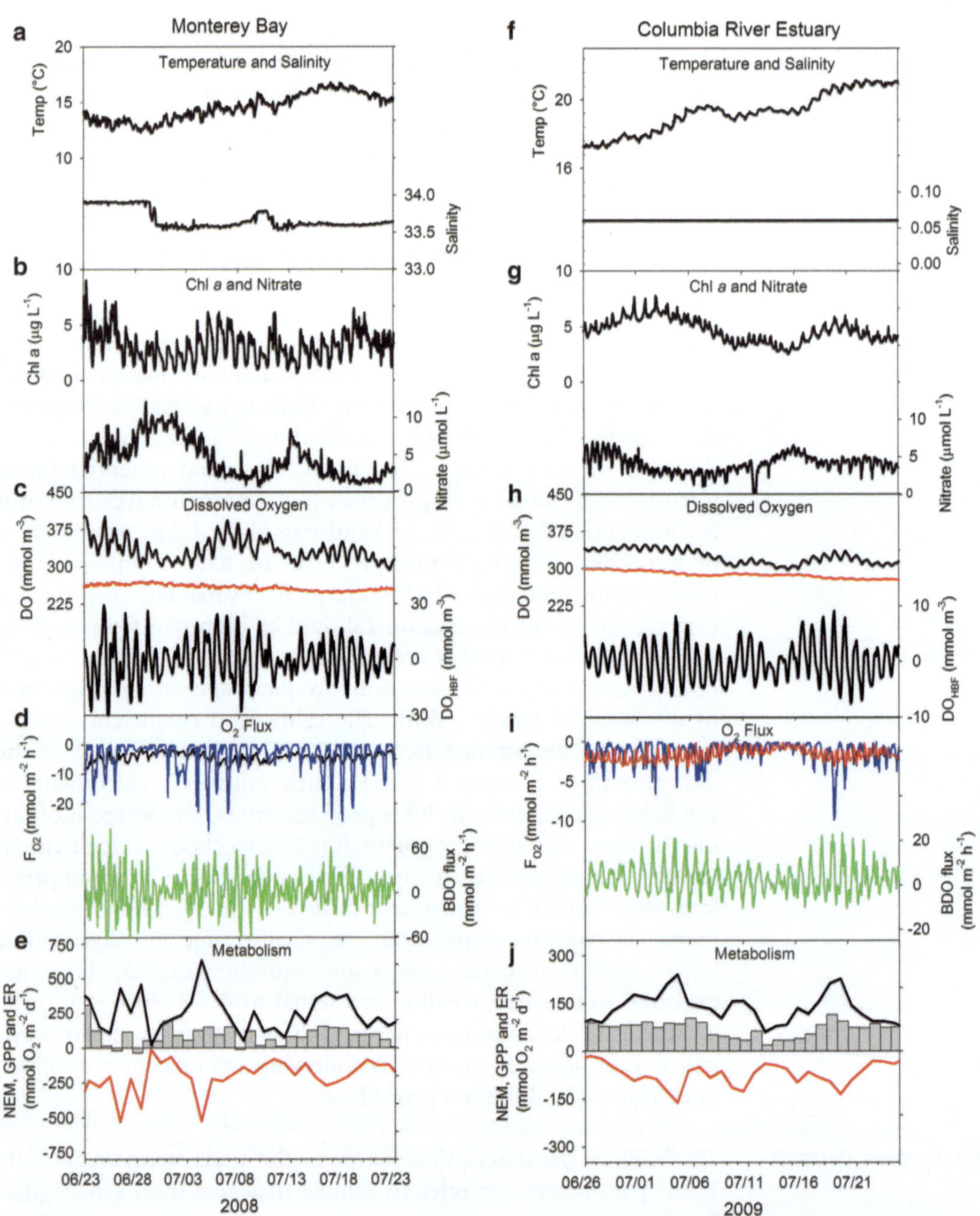

Fig. 1. Example datasets from two contrasting environments, Monterey Bay, CA (**a–e**) and the Columbia River Estuary, OR (**f–j**). (**a, f**) Hourly measurements of temperature and salinity. (**b, g**) Hourly measurements of chlorophyll (Chl *a*) and nitrate concentration. (**c, h**) Hourly measurements of dissolved oxygen concentration (DO) and high band-pass filtered DO (DO_{HBF}). (**d**) Hourly calculated air–water diffusion flux (F_{O_2}) using wind speed correction (*blue line*) or a constant correction factor (*black line*) and the biological oxygen change (BDO flux). (**i**) Hourly calculated air–water diffusion flux (F_{O_2}) using wind speed correction (*blue line*) or water velocity correction (*red line*) and the biological oxygen change (BDO). (**e, j**) Daily calculated gross primary production (GPP—*black line*), ecosystem respiration (ER—*red line*), and net ecosystem metabolism (NEM—*histogram*).

2. Materials (Required Measurements)

(A) Dissolved Oxygen

- Instrumentation: Clark-type polarographic electrode or oxygen optode (Note 1)
- Units: µmol L^{-1}

(B) Temperature

- Instrumentation: Thermister integrated with oxygen sensor, or external sensor (Note 2)
- Units: °C

(C) Salinity

- Instrumentation: CTD sensor (Note 3)
- Units: Practical Salinity Units (psu)

(D) Wind speed (m s^{-1})

- Instrumentation: Anemometer (Note 4)
- Units: m s^{-1}

(E) Water velocity

- Instrumentation: Current meter or Acoustic Doppler instrumentation
- Units: m s^{-1}

(F) Photoperiod

- Instrumentation: PAR (photosynthetically active radiation) sensor or sunrise and sunset predictions (49)
- Units: h

Example

Both example datasets described here were collected using Land/Ocean Biogeochemical Observatory (LOBO) platforms. LOBO is a commercially available multisensor platform developed in collaboration with the Monterey Bay Aquarium Research Institute, Moss Landing, CA (12). The instrumentation platform measures a variety of water quality parameters including dissolved oxygen, salinity, temperature, chlorophyll *a* fluorescence, turbidity, and nitrate concentration. The first example dataset is from the L20 mooring located 1 km offshore in Monterey Bay and operated by the Monterey Bay Aquarium Research Institute, and the second example dataset is from the SATURN-05 platform located at river mile 53 of the Columbia

(continued)

River Estuary on a USGS water monitoring structure and operated by the Center for Coastal Margin Observation and Prediction at Oregon Health & Science University. Both stations collected measurements at an hourly frequency. Wind speed data for the Monterey Bay example was acquired from the nearby MBARI M0/CIMT mooring (approximately10 km from the L20 mooring). For the Columbia River Estuary example, wind speed was acquired from the NOAA NOS/CO-OPS station 94470422 and water velocity was acquired from the USGS National Streamflow Information Program Station 14246900, which is co-located with the SATURN-05 instrument platform in the Columbia River Estuary.

3. Methods

The step-by-step procedure presented below is intended to provide a complete description of the calculations needed to convert time series observations into rates of primary production, aerobic respiration, and NEM for a given data set. For each step, the example data sets are shown for a deployment period (Fig. 1) and a detailed calculation over 24 h (Tables 1 and 2). The basic steps are listed here and described in greater detail below:

1. Collect high-resolution in situ measurements from one or more locations in the system for at least 24 h, starting at sunrise. Convert oxygen measurements to appropriate units.
2. Calculate the air–water diffusion flux (F_{O_2}) for each time interval.
3. If necessary, filter the time series data to remove variability associated with processes occurring at frequencies longer than the diel cycle [high band-pass filter (HBF)).
4. Calculate the biological oxygen change for each time interval.
5. Sum the biological oxygen change over 24 h to calculate NEP and night respiration.
6. Calculate the aerobic respiration rate from the night respiration and extrapolate to 24 h to calculate daily ecosystem respiration.
7. Calculate the daily GPP from the NEP plus the correction for daytime respiration, assuming a constant respiration rate.
8. Convert NEM measurements to units of organic carbon

Table 1

Example dataset for Monterey Bay, July 4, 2008 to demonstrate measurements and calculations for determining DO_{sat}, ΔDO_{HBF}, F_{O_2}, and BDO as described in the text

Date (PST)	Temperature (°C)	Salinity (psu)	(DO) (mmol m^{-3})	DO_{sat} (mmol m^{-3})	Wind speed (m s^{-1})	Water velocity (m s^{-1})	ΔDO_{HBF} (mmol m^{-2})	F_{O_2} (mmol m^{-2})	BDO_d (mmol m^{-2})	BDO_n (mmol m^{-2})
7/4/2008 5:28	13.5	33.6	310	264	2.8	n.d.	8.1	-1.0	9.1	
7/4/2008 6:28	13.5	33.6	314	265	3.0	n.d.	33.2	-0.9	34.1	
7/4/2008 7:28	13.2	33.6	320	266	0.1	n.d.	39.9	0.0	39.9	
7/4/2008 8:28	13.9	33.6	325	262	1.7	n.d.	27.4	-0.6	28.0	
7/4/2008 9:28	14.1	33.6	333	261	0.1	n.d.	37.7	0.0	37.7	
7/4/2008 10:28	13.6	33.6	339	264	1.8	n.d.	20.8	-0.7	21.4	
7/4/2008 11:28	14.0	33.6	341	262	3.7	n.d.	-6.7	-3.1	-3.6	
7/4/2008 12:28	13.8	33.6	341	263	6.0	n.d.	-12.0	-8.1	-3.9	
7/4/2008 13:28	13.9	33.6	342	262	8.1	n.d.	-5.1	-15.2	10.1	
7/4/2008 14:28	13.9	33.6	343	262	8.0	n.d.	-7.2	-14.2	7.0	
7/4/2008 15:28	13.9	33.6	344	262	9.4	n.d.	-3.6	-19.5	15.9	
7/4/2008 16:28	13.7	33.6	346	263	10.6	n.d.	-1.8	-25.5	23.6	
7/4/2008 17:28	14.1	33.6	350	261	9.4	n.d.	14.5	-22.0	36.5	
7/4/2008 18:28	14.0	33.6	355	261	10.7	n.d.	11.7	-29.1	40.7	
7/4/2008 19:28	14.1	33.6	356	261	5.9	n.d.	-3.7	-10.0	6.3	
7/4/2008 20:28	14.6	33.6	357	259	5.1	n.d.	-7.0	-7.7		0.6
7/4/2008 21:28	14.9	33.6	357	257	3.8	n.d.	-15.5	-4.0		-11.4
7/4/2008 22:28	14.6	33.6	354	259	0.5	n.d.	-27.5	-0.1		-27.4

(continued)

Table 1 (continued)

Date (PST)	Temperature (°C)	Salinity (psu)	(DO) (mmol m^{-3})	DO_{sat} (mmol m^{-3})	Wind speed (m s^{-1})	Water velocity (m s^{-1})	ΔDO_{HBF} (mmol m^{-2})	F_{O_2} (mmol m^{-2})	BDO_d (mmol m^{-2})	BDO_n (mmol m^{-2})
7/4/2008 23:28	14.6	33.6	350	258	0.1	n.d.	–26.1	0.0		–26.1
7/5/2008 0:28	14.3	33.6	344	260	0.2	n.d.	–28.7	0.0		–28.7
7/5/2008 1:28	14.0	33.6	338	261	0.1	n.d.	–27.9	0.0		–27.8
7/5/2008 2:28	14.2	33.6	332	260	1.5	n.d.	–26.5	–0.4		–26.1
7/5/2008 3:28	14.0	33.6	324	261	3.0	n.d.	–31.2	–1.6		–29.6
7/5/2008 4:28	13.9	33.6	316	262	2.3	n.d.	–19.4	–0.8		–18.6

Table 2
Example dataset for the Columbia River Estuary July 19, 2009 to demonstrate measurements and calculations for determining DO_{sat}, ΔDO_{HBF}, F_{O_2}, and BDO as described in the text

Date (PST)	Temperature (°C)	Salinity (psu)	(DO) (mmol m^{-3})	DO_{sat} (mmol m^{-3})	Wind speed (m s^{-1})	Water velocity (m s^{-1})	ΔDO_{HBF} (mmol m^{-2})	F_{O_2} (mmol m^{-2})	BDO_d (mmol m^{-2})	BDO_n (mmol m^{-2})
7/19/2009 6:00	20.5	<0.1	320	281	0.5	0.97	−5.0	−2.3	−2.7	
7/19/2009 7:00	20.4	<0.1	318	281	0.6	1.00	−3.0	−2.2	−0.8	
7/19/2009 8:00	20.4	<0.1	317	281	2.2	1.00	−0.8	−2.7	1.9	
7/19/2009 9:00	20.4	<0.1	316	281	2	1.01	5.1	−2.6	7.7	
7/19/2009 10:00	20.4	<0.1	317	282	1.7	0.90	7.6	−2.3	9.9	
7/19/2009 11:00	20.4	<0.1	318	282	3.1	0.75	14.9	−3.0	17.8	
7/19/2009 12:00	20.4	<0.1	321	281	3	0.44	20.9	−2.6	23.5	
7/19/2009 13:00	20.6	<0.1	323	281	3.6	0.15	12.2	−3.0	15.1	
7/19/2009 14:00	20.7	<0.1	324	280	3.5	0.11	5.5	−3.2	8.7	
7/19/2009 15:00	20.6	<0.1	327	281	3.9	0.21	14.8	−3.6	18.4	
7/19/2009 16:00	20.6	<0.1	329	280	4.8	0.42	10.0	−5.0	15.0	
7/19/2009 17:00	20.8	<0.1	330	279	4.8	0.60	−2.0	−6.8	4.8	
7/19/2009 18:00	20.9	<0.1	332	279	5.4	0.73	5.8	−8.9	14.7	
7/19/2009 19:00	20.9	<0.1	335	279	4	0.79	5.9	−6.2	12.1	
7/19/2009 20:00	20.9	<0.1	335	279	4.9	0.77	−9.4	−7.8	−1.5	
7/19/2009 21:00	20.9	<0.1	333	279	4.2	0.69	−14.6	−6.2		−8.4
7/19/2009 22:00	20.9	<0.1	332	279	2.5	0.27	−12.7	−2.9		−9.8

(continued)

Table 2 (continued)

Date (PST)	Temperature (°C)	Salinity (psu)	(DO) (mmol m^{-3})	DO_{sat} (mmol m^{-3})	Wind speed (m s^{-1})	Water velocity (m s^{-1})	ΔDO_{HBF} (mmol m^{-2})	F_{O_2} (mmol m^{-2})	BDO_d (mmol m^{-2})	BDO_n (mmol m^{-2})
7/19/2009 23:00	20.9	<0.1	331	279	3.6	0.14	−13.9	−3.6		−10.4
7/20/2009 0:00	20.8	<0.1	329	279	1.5	0.46	−10.4	−2.4		−8.0
7/20/2009 1:00	20.8	<0.1	328	279	3	0.46	−6.3	−3.5		−2.9
7/20/2009 2:00	20.8	<0.1	327	279	2	0.37	−5.2	−2.4		−2.8
7/20/2009 3:00	20.8	<0.1	326	279	1.5	0.09	−3.2	−1.2		−2.0
7/20/2009 4:00	20.8	<0.1	324	279	1.4	0.53	−4.9	−2.3		−2.6
7/20/2009 5:00	20.8	<0.1	323	279	2.1	0.74	−5.8	−3.0		−2.8

3.1. Collect High-Resolution In Situ Measurements from One or More Locations in the System for at Least 24 h, Starting at Sunrise. Convert Measurements to Appropriate Units

Observations should be collected and reported in molar mass, as this allows for conversion and comparison with other measures of productivity such as nutrient drawdown, and also facilitates the conversion to units of carbon (see below). However, dissolved oxygen is measured and reported in a variety of common units. For clarity, we provide the routine unit conversion for DO at sea level here.

DO measured in μmol L^{-1} is converted to mg L^{-1} by multiplying by the molar mass of O_2 and dividing by 1,000:

$$O_2(\text{mg L}^{-1}) = O_2(\mu\text{mol L}^{-1}) \times 0.0319 \tag{4.2}$$

DO measured in mg L^{-1} can be converted to a molar volume of O_2 (mL L^{-1}) at standard temperature and pressure (STP; 0 °C, 1 atm) by:

$$O_2(\text{mL L}^{-1}) = O_2(\text{mg L}^{-1}) \times \frac{22.4}{31.9} \tag{4.3}$$

The numerator in (4.3) is the standard volume of 1 mol of an ideal gas and the denominator is the molar mass as in (4.2). DO solubility is also reported in relation to atmospheric equilibrium:

$$O_{2\%\text{Sat}} = \frac{O_{2\text{meas}}}{O_{2\text{sat}}} \times 100 \tag{4.4}$$

where $O_{2\text{meas}}$ is the measured value, corrected for salinity and temperature, and $O_{2\text{sat}}$ is the computed equilibrium concentration of oxygen with respect to the atmosphere at sea level as a function of water temperature and salinity (2, 50). For routine computation of $O_{2\text{sat}}$, we follow the recommended procedure of Garcia and Gordon (50):

$$\begin{aligned}\ln O_{2\text{sat}}(\text{mL L}^{-1}) = {} & A_0 + A_1T_s + A_2T_s^2 + A_3T_s^2 + A_3T_s^3 + A_4T_s^4 \\ & + A_5T_s^5 + S(B_0 + B_1T_s + B_2T_s^2 + B_3T_s^3) + C_0S^2,\end{aligned} \tag{4.5}$$

where T_s is a scaled temperature function, S is salinity (psu), and A, B, and C are coefficients. T_s is calculated by:

$$T_s = \ln[(298.15 - t) \times (273.15 + t)^{-1}], \tag{4.6}$$

where t is the measured temperature (°C). The coefficients A, B, and C are originally from (51): $A_0 = 2.00907$, $A_1 = 3.22014$, $A_2 = 4.05010$, $A_3 = 4.94457$, $A_4 = -2.56847 \times 10^{-1}$, $A_5 = 3.88767$, $B_0 = -6.24523 \times 10^{-3}$, $B_1 = -7.37614 \times 10^{-3}$, $B_2 = -1.03410 \times 10^{-2}$, $B_3 = -8.17083 \times 10^{-3}$, $C_0 = -4.88682 \times 10^{-7}$. For example, solving (4.5) for 10 °C and 35 psu yields 6.315 mL L^{-1} (i.e., 281.9 μmol L^{-1}).

Example

DO for the two example datasets are shown for approximately 1 month during mid-summer (Fig. 1c, h) and at high resolution in Tables 1 and 2. The DO sensors were automatically corrected for the effects of temperature on the raw measurement using an internal temperature sensor. For the Monterey Bay dataset, salinity was measured externally using a CTD (Fig. 1a) and salinity corrections for the DO measurement were applied using manufacturer specified protocols. For the Columbia River Estuary dataset, salinity was always below 0.1 (Fig. 1f), therefore no correction was required. DO concentrations are reported in mmol m^{-3} (i.e., μmol L^{-1}). O_2 solubility for each time step was calculated from (4.5) and concurrent temperature and salinity measurements.

3.2. Calculate the Air–Water Diffusion Flux (F_{O_2}) for Each Time Interval

Several factors influence the diffusion and transfer rates of DO (52–54), including the concentration gradient between the water mass and the atmosphere, temperature- and salinity-dependent oxygen solubility, wind speed (38, 55), and water velocity (37, 56). The last two factors will vary in their relative importance depending on the environment; therefore we consider four different parameterization schemes for oxygen diffusion: (i) the open ocean deep-water environment, (ii) the flowing shallow-water environment, (iii) environments where both water velocity and wind must be taken into account, and (iv) a constant air–water correction factor. In all cases, oxygen solubility for each time interval must be calculated to determine the diffusion gradient. For example, when the oxygen concentration is above the calculated solubility, O_2 will diffuse from the water to the air, producing a negative flux. Except for the case of the constant correction factor, the flux is described by calculating the piston velocity (νO_2):

$$\nu O_2 = k_{\mathrm{wind}} + k_{\mathrm{flow}}, \tag{4.7}$$

where k_{wind} is the flux caused by wind-induced turbulence at the surface, and k_{flow} is the flux caused by turbulence generated by water moving over the benthos. If the environment is dominated by water velocity (e.g., in many rivers and shallow estuaries), k_{wind} may be ignored and $\nu O_2 \approx k_{\mathrm{flow}}$, if the water column is deep and velocities are low (often observed in lakes and ocean waters), then $\nu O_2 \approx k_{\mathrm{wind}}$ (56).

Once piston velocity and the oxygen gradient are known, the concentration-dependent flux is determined by:

$$F_{O_2} = -\nu O_2 \times (O_{2\mathrm{meas}} - O_{2\mathrm{sat}}). \tag{4.8}$$

F_{O_2} is expressed as a mass transferred per unit surface area (i.e., integrated over the entire water column) and unit time (e.g., mmol $O_2\,m^{-2}\,h^{-1}$).

3.2.1. Air–Water Diffusion: Open-Water Environment

The wind-generated O_2 diffusion, k_{wind} (cm h^{-1}) can be approximated for each time interval based on wind speed and the Schmidt number (ratio of viscous diffusion rate to mass diffusion rate of a fluid) (38):

$$k_{wind} = 0.31 \times u_{10}^2 \left(\frac{Sc}{660}\right)^{-0.5}, \quad (4.9)$$

where u_{10} is wind speed (m s^{-1}) at 10 m above the water surface and Sc is the Schmidt number for O_2. In fresh water (i.e., 0 salinity), Sc_0 can be determined from the empirical relationship:

$$Sc_0 = 1800.6 - 120.10T + 3.7818T^2 - 0.047608T^3, \quad (4.10)$$

where T is temperature (°C). Sc is also dependent on salinity and can be determined as follows:

$$Sc = Sc_0(1 + 3.14 \times 10^{-3} S), \quad (4.11)$$

where Sc_0 is calculated from (4.10) and S is salinity.

Example

The Monterey Bay LOBO platform was deployed at a site where the seafloor was 21 m deep, and the euphotic zone (i.e., mixed layer) was approximated to be 6 m during the time period of the example dataset. Therefore this dataset meets the criteria for using k_{wind} as the best estimate of piston velocity (vO_2) since the bottom has no measurable influence on the turbulence at the surface of the mixed layer. Therefore, (4.9)–(4.11) was used to determine air–water flux. Prior to calculating F_{O_2}, k_{wind} was converted from units of cm h^{-1} to m h^{-1}. Wind measurements and the resulting air–water flux are provided for a 24 h period in Table 1. Note that both wind speed and the DO concentration both changed throughout the day that result in variability in the hourly estimate of F_{O_2}. The hourly measurements of F_{O_2} for the month long time series are shown in Fig. 1d (left axis, blue line).

3.2.2. Air–Water Diffusion: Flowing Water

The gas transfer velocity in flowing waters depends mainly on the kinematic viscosity of the liquid (v), the molecular diffusion of oxygen (D), and the rate of water velocity (U). Gas exchange is highly dependent on turbulence and the transfer of large eddies from the bottom to the surface (56, 58), therefore water depth (h) is also an important parameter, and k_{flow} (cm h^{-1}) is calculated by:

$$k_{flow} = U\left(\frac{v}{D}\right)^{-\frac{1}{2}}\left(\frac{Uh}{v}\right)^{-\frac{1}{2}} = \sqrt{\frac{UD}{h}}. \quad (4.12)$$

For (4.12), U (cm s^{-1}), and h (cm) are directly measured and D (cm^2 s^{-1}) is calculated from dynamic viscosity (η), molar weight, and molar volume according to Wilke and Chang (57):

$$D = 7.4 \times 10^{-8} \times \frac{(xM)^{0.5} T}{\eta V^{0.6}}, \qquad (4.13)$$

where x is the association factor of water (=2.26), M is the molar weight of water (=18), T is temperature in Kelvin (=°C+273.15), η is the dynamic viscosity of water (in centipoise), and V is the molar volume at the boiling point (=25.6). A practical approximation of η as a function of temperature is

$$\eta = (2.414 \times 10^{-2}) \times \left(10^{\frac{247.8}{(T-140)}} \right), \qquad (4.14)$$

where T is in Kelvin (=°C+273.15).

3.2.3. Air–Water Diffusion: Wind and Flowing Water

The effects of wind speed and water velocity must both be accounted for in shallow systems (37, 58). These are combined in an additive fashion to yield the overall piston velocity as in (4.7). See Note 5 for further discussion of piston velocity considerations.

Example

A LOBO platform was deployed in the main-stem of the Columbia River Estuary in a vertically well-mixed, freshwater region with an average depth of approximately 7 m. The river velocity is a result of both discharge volume and the tidal cycle (even though ocean water does not reach the site, the river flow responds to the changing sea level and can even flow "backwards" during very high tides). In addition, the river is sufficiently wide (>500 m) that wind speed must also be considered. Therefore, this example dataset meets the criteria for using k_{wind} and k_{flow} as the best estimate of piston velocity. Equations (4.12)–(4.14) were used to determine k_{flow} and (4.9) and (4.10) were used to determine k_{wind}. The flux (F_{O_2}) for flow and wind calculated separately are shown in Fig. 1i (left axis) to demonstrate the relative importance of each factor over time. The combined F_{O_2} calculated from (4.8) is shown for hourly measurements in Table 2, and used for the NEM calculations below.

3.2.4. Constant Correction Factor, K

Due to uncertainties or measurement errors in some of the terms used to calculate gas diffusion, Caffrey (1, 29) found it useful to employ a constant K value to describe air–water exchange in estuaries. This approach is the simplest of the presented scenarios and is most accurate when wind speeds are >2 and <5 m s^{-1} and water velocity is ignored. Since air–water diffusion is nonlinearly dependent on wind speed, assuming a constant of 0.5 for K may significantly underestimate exchange when wind speeds are >5 $\mathrm{m\,s^{-1}}$ and overestimate exchange when wind speeds are <2 $\mathrm{m\,s^{-1}}$. Assuming the constant diffusion coefficient of $K=0.5$ (in g $O_2\,m^{-2}\,h^{-1}$), then air–water diffusion can be calculated by:

$$F_{O_2} = \left(1 - \frac{O_{2\%\mathrm{Sat},t1} + O_{2\%\mathrm{Sat},t_2}}{200}\right) \times K \times \mathrm{d}t, \tag{4.15}$$

where dt is the time interval between two consecutive measurements, and $O_{2\%\mathrm{Sat}}$ is the percent oxygen saturation calculated from (4) for the two time intervals t_1 and time t_2.

Example

A comparison of F_{O_2} calculated with the constant diffusion coefficient (4.15) converted to molar concentration, and the wind speed diffusion (from Sect. 3.2.1) is shown in Fig. 1d (left axis). As expected, the constant correction leads to a higher F_{O_2} when the flux approaches zero, and to a lower F_{O_2} when the flux becomes large as a result of high wind speeds.

3.3. If Necessary, Filter the Time Series Data to Remove Variability Associated with Processes Occurring at Frequencies Longer Than the Diel Cycle (HBF)

Changes in dissolved oxygen occurring at longer frequencies than the day/night cycle are inherent to most natural systems and can be mathematically removed from time series data while still preserving the short-term variability associated with diurnal oxygen production and consumption. This can be accomplished with Fourier transformation by using a HBF (59).

Example

Figure 1c, h (left axis) show oxygen data for a 1 month period where spring/neap tidal and other forcing may be changing the background oxygen concentrations. A HBF for frequencies greater than 33 h (47) was applied to both datasets (Fig. 1c, h, right axis) to yield [DO] with frequencies longer than the diurnal cycle removed (DO_{HBF}).

3.4. Calculate the Biological Oxygen Change for Each Time Interval

The production and consumption of DO caused by biological processes (BDO) for each time interval (t) is calculated from the change in DO_{HBF} for each time interval (t), the air–water diffusion (F_{O_2}) for time interval (t) using the appropriate approach from Sect. 3.2, and the depth of the mixed layer (h in m):

$$\mathrm{BDO}_t = (\mathrm{DO}_t - \mathrm{DO}_{t-1}) \times h - F_{O_2}. \tag{4.16}$$

Example

BDO was determined for each example dataset by calculating the hourly change from the DO_{HBF} data shown in Fig. 1c, h (right axis) and integrating over a depth of 6 and 7 m for the Monterey Bay and Columbia River Estuary sites respectively. Tables 1 (Monterey Bay) and 2 (Columbia River Estuary) show the hourly measurements for a 24 h period. Note that the change in DO_{HBF}, i.e., ΔDO_{HBF} is depth-integrated and therefore assumes that the single point measurement is representative of the entire mixed layer (see Note 6). From Tables 1 and 2 it is evident that the BDO calculation is significantly influenced by the F_{O_2} calculation, although the term becomes less important as depth (i.e., h in (4.16)) increases.

3.5. Sum the Biological Oxygen Change Over 24 h to Calculate NEM, NEP, and Night Respiration

The total daily NEM is calculated by adding the hourly values calculated in Sect. 3.4 over a 24 h period. The choice of the start time will cause small differences in the NEM calculation. Depending on the site and the time series duration, it is more appropriate to begin at sunrise rather than the calendar day, as this will integrate one complete photoperiod and one complete dark period, rather than combining the dark periods before and after the photoperiod in a calendar day. By differentiating the day into a photoperiod and a dark period, the NEM sum can be divided into a "net ecosystem production" term and a "night respiration" term that together sum to the "daily net ecosystem metabolism" term.

Example

Tables 1 and 2 show the hourly BDO during the day (BDO_d) and night (BDO_n) for one 24 h period starting at sunrise. The sum of BDO_d is the NEP and the sum of BDO_n is the night respiration. Table 3 shows these values along with the overall sum, i.e., NEM. Note that the data columns of Table 3 are different representations of the initial calculation in mmol $O_2\,m^{-2}$. The Daily NEM was 107.6 mmol $O_2\,m^{-2}\,d^{-1}$ for Monterey Bay (July 4, 2008) and 95.0 mmol $O_2\,m^{-2}\,d^{-1}$ for the Columbia River Estuary example (July 19, 2009). From Fig. 1e, j (histogram), the variability in NEM is such that the Monterey Bay site has both positive and negative daily NEM values over the month long time series and the Columbia River site had always positive (but variable) daily NEM values.

Table 3
Summary of NEM calculations and unit conversions for Monterey Bay and the Columbia River (CR) Estuary as described in the text

	mmol O_2 m^{-2}	g O_2 m^{-2}	mmol O_2 m^{-3}	g O_2 m^{-3}	mmol OrgC m^{-2}	g OrgC m^{-2}	mmol OrgC m^{-3}	g OrgC m^{-3}
Monterey Bay July 4, 2008								
Net ecosystem production (d^{-1})	302.8	9.69	50.5	1.61	232.9	2.79	38.8	0.47
Night respiration (d^{-1})	-195.1	-6.24	-32.5	-1.04	-150.1	-1.80	-25.0	-0.30
Net ecosystem metabolism (d^{-1})	107.6	3.44	17.9	0.57	82.8	0.99	13.8	0.17
Average respiration rate (h^{-1})	-21.7	-0.69	-3.6	-0.12	-16.7	-0.20	-2.8	-0.03
Ecosystem respiration (d^{-1})	-520.4	-16.65	-86.7	-2.78	-400.3	-4.80	-66.7	-0.80
Gross primary production (d^{-1})	628.0	20.10	104.7	3.35	483.1	5.80	80.5	0.97
CR Estuary July 19, 2009								
Net ecosystem production (d^{-1})	144.6	4.63	20.7	0.66	111.3	1.34	15.9	0.19
Night respiration (d^{-1})	-49.6	-1.59	-7.1	-0.23	-38.2	-0.46	-5.5	-0.07
Net ecosystem metabolism (d^{-1})	95.0	3.04	13.6	0.43	73.1	0.88	10.4	0.13
Average respiration rate (h^{-1})	-5.5	-0.18	-0.8	-0.03	-4.2	-0.05	-0.6	-0.01
Ecosystem respiration (d^{-1})	-132.3	-4.23	-18.9	-0.60	-101.8	-1.22	-14.5	-0.17
Gross primary production (d^{-1})	227.3	7.27	32.5	1.04	174.9	2.10	25.0	0.30

3.6. Calculate the Aerobic Respiration Rate from the Night Respiration and Extrapolate to 24 h to Calculate Daily Ecosystem Respiration

For the period when there is no daylight, it is assumed that the only process altering the BDO flux is aerobic respiration. Therefore, the hourly respiration rate (mmol $O_2\,m^{-2}\,h^{-1}$) is

$$\text{Hourly respiration rate} = \frac{\text{Sum of BDOn}}{\text{Hours of night}} \tag{4.17}$$

And the daily oxygen respiration is determined by:

$$\text{Daily respiration rate} = 24 \times \text{Hourly respiration rate.} \tag{4.18}$$

> **Example**
>
> Table 3 shows the calculated hourly respiration rates from the BDO_n values in Tables 1 and 2. The night respiration rate is 21.7 and 5.5 mmol $O_2\,m^{-2}\,h^{-1}$ for Monterey Bay and Columbia River Estuary, respectively. It is clear from the hourly measurements in Tables 1 and 2 that the respiration rate is not the same each hour, and therefore the average rate calculation likely introduces error into the daily respiration rate calculation. From the hourly rate, a daily respiration rate is determined from (4.18) (520.4 mmol $O_2\,m^{-2}\,d^{-1}$ for Monterey Bay and 132.3 mmol $O_2\,m^{-2}\,d^{-1}$ for the Columbia River Estuary). Figure 1e, j (red line) show the ecosystem respiration rate during the time series. Although significant variability exists in both regions, overall, the rates are higher in Monterey Bay than the Columbia River estuary.

3.7. Calculate the Daily GPP from the NEP Plus the Correction for Daytime Respiration, Assuming a Constant Respiration Rate

The daily GPP is the sum of the NEP (from Sect. 3.5), plus a correction for the DO removed by respiration during the day:

$$\begin{aligned}\text{GPP} = \text{NEP} + (&\text{Hours of daylight} \\ &\times \text{Absolute value of hourly res. rate}).\end{aligned} \tag{4.19}$$

> **Example**
>
> Table 3 shows the calculations of GPP for the example datasets. For the dates shown, GPP is 628.0 and 227.3 mmol $O_2\,m^{-2}\,d^{-1}$ for Monterey Bay and Columbia River Estuary, respectively. GPP varies over time with a temporal relationship to nitrate and chlorophyll *a* (Fig. 1b, g) which is an indicator of primary productivity, however, a statistical examination of these relationships is beyond the scope of this methodology paper (see Note 7).

3.8. Convert NEM Measurements to Units of Organic Carbon

Changes in BDO are a direct measure of metabolic processes; however, changes in productivity and respiration are often expressed in units of organic carbon (OrgC). In general, the stoichiometric relationship for production and consumption can be modeled after the Redfield ratio of OrgC:O_2 of 106:138 (see Note 8). Therefore:

$$\mathrm{OrgC}(\mathrm{mmol\ m^{-2}\,d^{-1}}) = \frac{O_2(\mathrm{mmol\ m^{-2}\,d^{-1}})}{1.3}. \tag{4.20}$$

Example

Table 3 provides a summary of the NEM calculations for the two example datasets. The first column of Table 3 provides the NEM calculations for the example datasets in units of mmol $O_2\,m^{-2}\,d^{-1}$. Column 2 was determined by applying (4.2) to the values from Column 1. Columns 3 and 4 are the calculated unit volume rates, determined by dividing Columns 1 and 2, respectively, by water depth. Columns 5–8 are the rates calculated in Columns 1–4 converted to units of carbon using (4.20). For example, Column 5 shows that the daily GPP rate for the two example datasets are 5.80 and 2.10 $g\,C\,m^{-2}\,d^{-1}$ for Monterey Bay and the Columbia River Estuary, respectively. The average values for metabolic rates over the course of the time series (Fig. 1e, j) in units of $g\,C\,m^{-2}\,d^{-1}$ for Monterey Bay are 2.3 ± 1.3 (GPP), 1.9 ± 1.3 (ER), and 0.84 ± 0.6 (NEM). For the Columbia River Estuary the results are 1.3 ± 0.4 (GPP), 0.8 ± 0.4 (ER) and 0.67 ± 0.2 (NEM). Therefore, despite higher rates of productivity and respiration between the two ecosystems, the NEM are relatively similar during this time series comparison.

4. Notes

1. Oxygen sensors are available from many oceanographic and environmental sensor manufactures. Clark electrodes have traditionally demonstrated a fast response time; however, they suffer from calibration drift and a propensity for biofouling. Optodes tend to be more stable in long-term deployments and recent models have a faster response time (27, 60).
2. Temperature is required to accurately calculate DO concentration and can be included as a component of in situ oxygen sensors. However, temperature can also be measured separately and oxygen measurements can be post-corrected following manufacturer protocols.

3. Salinity is required to accurately calculate the DO concentration and is usually not included as a component of in situ oxygen sensors. Salinity must be measured separately and used to correct the DO measurement. Salinity is calculated from the temperature-corrected conductivity of the water mass (therefore referred to as Practical Salinity Units). The measurement is normalized to a standard solution and therefore salinity is a unit-less measurement.
4. Wind speeds should be collected at 10 m above the water level or extrapolated to 10 m assuming a linear change in wind speed with altitude (61, 62). For example, NDBC wind speeds are calculated from an algorithm that uses wind speeds at 10 m, 20 m and a constant relative humidity of 85 %, a constant sea-level pressure of 1,013.25, and the air and water temperatures (63).
5. The two most common causes of turbulence that lead to enhanced air–water gas transfer are wind speed and water velocity. Here we assume that the two components can be determined separately and added together as in (58). The approach described here is based on empirical relationships described in (38) for wind-generated turbulence, and O'Connor and Dobbins (56) for water velocity-generated turbulence. This later approach method was originally developed for streams, but is applicable to rivers and estuaries (64). However, multiple recent empirical determinations of piston velocity for estuaries are also available (see (65) for a summary). If possible, it is best to calculate piston velocity for the range of conditions experienced in a desired location, e.g., (66). On the other hand, estimates related to wind speed and water velocity, or more direct measures of near surface turbulence, e.g., (67, 68) are available, and there are multiple options for choosing an empirical expression that is suited to a particular environment, e.g., shallow-water estuaries (30, 31) or turbid estuaries (69).
6. In the case where diel oxygen measurements are collected from only one depth, it must be assumed that the rate of change is representative of the entire water column or mixed layer. In water column greater than a few meters deep, this requires rapid vertical mixing since photosynthesis is more likely to be light limited at deeper depths, and therefore the assumption will not hold true unless the oxygen signal is quickly integrated throughout the mixed layer. A better approach is to deploy oxygen sensors at multiple depths to provide an average mixed layer concentration (e.g., (30)).
7. In addition to O_2, inorganic carbon, nitrogen, and phosphorus are all necessary components of primary production (see (4.1)) and therefore their daytime disappearance is a potential measure of GPP. In addition, carbon dioxide is both the substrate for photosynthetic carbon fixation and the direct product of cellular respiration, thus its application to NEM is analogous

to the oxygen methodology described above, and it provides a more comprehensive measure of ecosystem respiration since it is also the product of anaerobic organic matter respiration. Several components of ecosystem metabolism have been successfully determined from in situ measurements of pCO_2 (9, 70) and nitrate (9, 15, 47, 71). In addition, recent sensors designed to measure pH (42), ammonium (45), and inorganic phosphorus (72) in seawater can also be adapted for one or more aspects of NEM calculations. Here, we briefly summarize the advantages and disadvantages to adapting these measurements to open-water techniques.

Carbon dioxide: pCO_2 can be determined in situ using wet chemistry techniques (73) and by infrared spectrometry (74). As stated above, changes in CO_2 provide a comprehensive measurement of ecosystem respiration that accounts for all respiration processes including those occurring in anoxic sediments or hypoxic zones, which may be common in estuarine environments.

Nitrate: NO_3^- can be accurately determined in situ using a variety of instrumentation, including electrochemical, colorimetric, and UV absorption techniques. Of these, UV absorption has proven reliable for high resolution, long-term deployments. Nitrate disappearance can be used as a measure of phytoplankton growth and has the major advantage of the absence of the need to correct for air–water gas diffusion. The disadvantage is the complicating factors caused by other processes in the aquatic nitrogen cycle, such as nitrification and denitrification, which also alter the nitrate concentration over short time scales. In addition, other nitrogen substrates, such as ammonium, may also be accessible to primary producers.

Ammonium and phosphate: The low concentration of NH_4^+ and PO_4^{3-} that are typical of most oceanic environments, coupled with the low relative changes compared to either O_2 or CO_2 make these analytes less useful for direct diel productivity measurements, except perhaps in highly eutrophic or polluted systems. However, measurements of either analyte can provide useful information about organic matter remineralization and potential limiting growth conditions, and therefore high resolution measurements are desirable to complement oxygen NEM methodology.

8. The Redfield ratio of 106:138 is dependent on a number of intrinsic factors, most notably the photosynthetic quotient and respiration quotient. Anderson (75) suggests that a ratio of 106:150 is more representative of the oxygen content of algal cells. The actual ratio is dependent on a variety of environmental and physiological variables (41), thus introducing additional uncertainty when calculating ecosystem metabolism in units of organic carbon.

5. Future Directions

NEM (or, net community metabolism) is a useful metric for assessing the organic carbon balance in a given ecosystem, whether it be at large (ocean basin) or small (estuary or bay) scales. Intense debate over the balance of in situ oxygen and carbon production and consumption in vast oceanic regions based on relatively low-resolution data are being resolved through analysis of high-resolution datasets arising from the deployment of in situ oxygen sensors, particularly when they are coupled with in situ nutrient sensors (15, 47, 76) and mounted on profiling floats (13) or autonomous vehicles (15, 16) that provide improved spatial resolution over fixed moorings.

For regional applications, the emergence of in situ monitoring programs and increased data accessibility is likely to lead to more opportunities for employing calculations of NEM as a practical and cost-effective metric to assess the status of a given environment for water quality applications.

References

1. Caffrey JM (2003) Production, respiration and net ecosystem metabolism in U.S. estuaries. Environ Monit Assess 81:207–219
2. Weiss RF (1970) The solubility of nitrogen, oxygen and argon in water and seawater. Deep-Sea Res A 17:721–735
3. Hanson PC, Carpenter SR, Kimura N, Wu C, Cornelius SP, Kratz TK (2008) Evaluation of metabolism models for free-water dissolved oxygen methods in lakes. Limnol Oceanogr Methods 6:454–465
4. Kemp WM, Boynton WR (1980) Influence of biological and physical processes on dissolved oxygen dynamics in an estuarine system: implications for measurement of community metabolism. Estur Coast Mar Sci 11:407–431
5. McCutchan JH, Lewis WM, Saunders JF (1998) Uncertainty in the estimation of stream metabolism from open-channel oxygen concentrations. J N Am Benthol Soc 17:155–164
6. Staehr PA, Bade D, Van de Bogert MC, Koch GR, Williamson C, Hanson P, Cole JJ, Kratz T (2010) Lake metabolism and the diel oxygen technique: state of the science. Limnol Oceanogr Methods 8:628–644
7. Staehr PA, Testa JM, Kemp WM, Cole JJ, Sand-Jensen K, Smith SV (2011) The metabolism of aquatic ecosystems: history, applications, and future challenges. Aquat Sci 74:15–29. doi:10.1007/s00027-011-0199-2
8. Chapin TP, Caffrey JM, Jannasch HW, Coletti LJ, Haskins JC, Johnson KS (2004) Nitrate sources and sinks in Elkhorn Slough, California: results from long-term continuous in situ nitrate analyzers. Estuaries 27(5):882–894
9. Johnson KS (2010) Simultaneous measurements of nitrate, oxygen, and carbon dioxide on oceanogrpahic moorings: observing the Redfield ratio in real time. Limnol Oceanogr 55(2):615–627
10. Johnson KS, Needoba JA, Riser SC, Showers WJ (2007) Chemical sensor networks for the aquatic environment. Chem Rev 107:623–640
11. Van de Bogert MC, Carpenter SR, Cole JJ, Pace ML (2007) Assessing pelagic benthic metabolism using free water measurements. Limnol Oceanogr Methods 5:145–155
12. Jannasch HW, Coletti LJ, Johnson KS, Fitzwater SE, Needoba JA, Plant JN (2008) The Land/Ocean Biogeochemical Observatory: a robust networked mooring system for continuously monitoring complex biogeochemical cycles in estuaries. Limnol Oceanogr Methods 6:263–276
13. Martz TR, Johnson KS, Riser SC (2008) Ocean metabolism observed with oxygen sensors on profiling floats in the Pacific. Limnol Oceanogr 53:2094–2111
14. Riser S, Johnson KS (2008) Net production of oxygen in the subtropical ocean. Nature 451:323–325

15. Johnson KS, Needoba JA (2008) Mapping the spatial variability of plankton metabolism using nitrate and oxygen sensors on an autonomous underwater vehicle. Limnol Oceanogr 53(5): 2237–2250
16. Nicholson D, Emerson S, Eriksen C (2008) Net community production in the deep euphotic zone of the subtropical North Pacific gyre from glider surveys. Limnol Oceanogr 53:2226–2236
17. Longhurst AR (2007) Ecological geography of the sea, 2nd edn. Elsevier, New York, NY, 560 p
18. Nixon SW (1995) Coastal marine eutrophication: a definition, social causes, and future concerns. Ophelia 41:199–219
19. Steemann-Nielsen E (1952) The use of radioactive carbon (C14) for measuring organic production in the sea. J Cons Int Explor Mer 18:117–140
20. Bender M, Grande K, Johnson K, Marra J, Williams PJL, Sieburth J, Pilson M, Langdon C, Hitchcock G, Orchardo J, Hunt C, Donaghay P (1987) A comparison of four methods for determining planktonic community production. Limnol Oceanogr 32:1085–1098
21. Chen CC, Petersen JE, Kemp WM (2000) Nutrient uptake in experimental estuarine eco systems: scaling and partitioning rates. Mar Ecol Prog Ser 200:103–116
22. Pratt DM, Berkson H (1959) Two sources of error in the oxygen light and dark bottle method. Limnol Oceanogr 4(3):328–334
23. Odum HT (1956) Primary production in flowing waters. Limnol Oceanogr 1:102–117
24. Odum HT, Wilson RF (1962) Further studies on reaeration and metabolism of Texas Bays, 1958–1960. Publication of the Institute of Marine Science, Texas, pp 23–55
25. Hanson PC, Blade DL, Carpenter SR, Kratz TK, (2003) Lake metabolism: relationships with dissolved organic carbon and phosphorus. Limnol Oceanogr 48:1112–1119
26. Martini M, Butman B, Michelson MJ (2007) Long term performance of Aanderaa optodes and Sea-Bird SBE-43 dissolved-oxygen sensors bottom mounted at 32 m in Massachusetts Bay. J Atmos Oceanic Technol 24:1924–1935
27. Tengberg A, Hovdenes J, Andersson HJ, Brocandel O, Diaz R, Hebert D, Arnerich T, Huber C, Körtzinger A, Khripounoff A, Rey F, Rönning C, Schimanski J, Sommer S, Stangelmayer A (2006) Evaluation of a lifetime-based optode to measure oxygen in aquatic systems. Limnol Oceanogr Methods 4:7–17
28. Staehr PA, Sand-Jensen K (2007) Temporal dynamics and regulation of lake metabolism. Limnol Oceanogr 52:108–120
29. Caffrey JM (2004) Factors controlling net ecosystem metabolism in U.S. estuaries. Estuaries 27(1):90–101
30. D'Avanzo C, Kremer JN, Wainright SC (1996) Ecosystem production and respiration in response to eutrophication in shallow temperate estuaries. Mar Ecol Prog Ser 141:263–274
31. Russell MJ, Montagna PA, Kalke RD (2006) The effect of freshwater inflow on net ecosystem metabolism in Lavaca Bay, Texas. Estuar Coast Shelf Sci 68:231–244
32. del Giorgio PA, Cole JJ, Cimbleris A (1997) Respiration rates in bacteria exceed phytoplankton in unproductive aquatic systems. Nature 385:148–151
33. Duarte CM, Agusti S (1998) The CO_2 balance of unproductive aquatic ecosystems. Science 281:234–236
34. Williams PJL (1998) The balance of plankton respiration and photosynthesis in the open oceans. Nature 394:55–57
35. Geider RJ (1997) Photosynthesis or planktonic respiration? Nature 388:132
36. Karl DM, Laws EA, Morris P, Williams PJL, Emerson S (2003) Metabolic balance in the open sea. Nature 426:32
37. Chu CR, Jirka GH (2003) Wind and stream flow induced reaeration. J Environ Eng 129(12):1129–1136
38. Wanninkhof R (1992) Relationship between wind speed and gas exchange over the ocean. J Geophys Res 97(C5):7373–7382
39. Pace ML, Prairie YT (2005) Respiration in lakes. In: del Giorgio PA, Williams PJB (eds) Respiration in aquatic ecosystems. Oxford University Press, New York, NY, pp 103–121
40. Tobias CR, Bolke JK, Harvey JW (2007) The oxygen-18 isotope approach for measuring aquatic metabolism in high-productivity waters. Limnol Oceanogr 52:1439–1453
41. Falkowski PG, Raven JA (2007) Aquatic photosynthesis. Princeton University Press, Princeton, NJ, 484 p
42. Martz TR, Connery JG, Johnson KS (2010) Testing the Honeywell Durafet for seawater pH applications. Limnol Oceanogr Methods 8:172–184
43. Cullison Gray SE, DeGrandpre MD, Moore TS, Martz TR, Friederich GE, Johnson KS (2011) Applications of in situ pH measurements for inorganic carbon calculations. Mar Chem 125(1–4):82–90
44. Johnson KS, Coletti LJ (2002) In situ ultraviolet spectrophotometry for high resolution and long-term monitoring of nitrate, bromide and bisulfide in the ocean. Deep-Sea Res Pt I 49:1291–1305

45. Plant JN, Johnson KS, Needoba JA, Coletti LJ (2009) NH4-Digiscan: an in situ and laboratory ammonium analyzer for estuarine, coastal, and shelf waters. Limnol Oceanogr Methods 7:144–156
46. Nyquist H (1932) Regeneration theory. Bell Syst Tech J 11:126–147
47. Johnson KS, Coletti LJ, Chanvez FP (2006) Diel nitrate cycles observed with in situ sensors predict monthly and annual new production. Deep-Sea Res Pt I 53:561–573
48. Wainwright SC, Kremer JN, D'Avanzo C (1995) Evaluation of ENDECO 1184 C dissolved oxygen recorders for use in temperate estuaries. Water Res 29:2035–2042
49. Iqbal M (1983) An introduction to solar radiation. Academic Press, New York, NY, 390 pp
50. Garcia HE, Gordon LI (1992) Oxygen solubility in seawater: better fitting equations. Limnol Oceanogr 37(6):1307–1312
51. Benson BB, Krause JD (1984) The concentration and isotopic fractionation of oxygen dissolved in freshwater and seawater in equilibrium with the atmosphere. Limnol Oceanogr 29(3):620–632
52. Holley ER (1977) Oxygen transfer at the air-water interface. In: Gibbs RJ, Shaw RP (eds) Transport processes in lakes and oceans. Plenum Press, New York, NY, pp 117–150
53. Jahne B, Haubecker H (1998) Air-water gas exchange. Annu Rev Fluid Mech 30:443–468
54. Jahne B, Munnich KO, Bosinger R, Dutzi A, Huber W, Libner P (1987) On the parameters influencing air-water gas exchange. J Geophys Res 92:1937–1949
55. Crusius J, Wanninkhof R (2003) Gas transfer velocities measured at low wind speed over a lake. Limnol Oceanogr 48:1010–1017
56. O'Connor DJ, Dobbins WE (1958) Mechanisms of reaeration in natural streams. Trans Am Soc Civ Eng 123:641–684
57. Wilke CR, Chang P (1955) Correlation of diffusion coefficients in dilute solutions. AIChE J 1:264–270
58. Vanderborght JP, Wollast R, Loijens M, Regnier P (2002) Application of a transport-reaction model to the estimation of biogas fluxes in the Scheldt Estuary. Biogeochemistry 59:207–237
59. Press WH, Teukolsky SA, Vetterling WT, Flannery BP, (1986) Numerical recipes: the art of scientific computing. Cambridge University Press, Cambridge, 848 p
60. Uchida H, Kawano T, Kaneko I, Fukasawa M (2008) In-situ calibration of optode-based oxygen sensors. J Atmos Oceanic Technol 25:2271–2281
61. Hartman B, Hammond DE (1985) Gas exchange in San Francisco Bay. Hydrobiologia 129(1):59–68
62. Smith SV (1985) Physical, chemical and biological characteristics of CO_2 gas flux across the air-water interface. Plant Cell Environ 8:387–398
63. Liu WT, Katsaros KB, Businger JA (1979) Bulk parameterization of air-sea exchange in heat and water vapor including the molecular constraints at the interface. J Atmos Sci 36:1722–1735
64. Borges AV, Delille B, Schiettecatte L, Gazeau F, Abril G, Frankignoulle M (2004) Gas transfer velocities of CO_2 in three European estuaries (Randers Fjord, Scheldt, and Thames). Limnol Oceanogr 49(5):1630–1641
65. Raymond PA, Cole JJ (2001) Gas exchange in rivers and estuaries: choosing a gas transfer velocity. Estuaries 24(2):312–317
66. Moore TS, Nuzzio DB, Di Toro DM, Luther GW III (2009) Oxygen dynamics in a well mixed estuary, the lower Delaware Bay, USA. Mar Chem 117:11–20
67. Zappa CJ, Raymond PA, Terray EA, McGillis WR (2003) Variation in surface turbulence and the gas transfer velocity over a tidal cycle in a macro-tidal estuary. Estuar Coast 26(6):1401–1415
68. Zappa CJ, McGillis WR, Raymond PA, Edson JB, Hintsa EJ, Zemmelink HJ, Dacey JWH, Ho DT (2007) Environmental turbulent mixing controls on air-water gas exchange in marine and aquatic systems. Geophys Res Lett 34(L10601)
69. Abril G, Commarieu M, Sottolichio A, Bretel P, Guérin F (2009) Turbidity limits gas exchange in a large macrotidal estuary. Estuar Coast Shelf Sci 83(3):342–348
70. Wesslander K, Hall P, Hjalmarsson S, Lefevre D, Omstedt A, Rutgersson A, Sahlée E, Tengberg A (2011) Observed carbon dioxide and oxygen dynamics in a Baltic Sea coastal region. J Mar Syst 86(1–2):1–9
71. Heffernan JB, Cohen MJ (2010) Direct and indirect coupling of primary production and diel nitrate dynamics in a subtropical spring-fed river. Limnol Oceanogr 55(2):677–688
72. Gilbert ML (2011) Nutrient loading and transformations in the Columbia River Estuary determined by high resolution in situ sensors. M.S. thesis, Division of Environmental & Biomolecular Systems, Oregon Health & Science University 166 p.

73. DeGrandpre MD, Hammar TR, Smith SP, Sayles FL (1995) In situ measurements of seawater pCO_2. Limnol Oceanogr 40(5):969–975
74. Friederich GE, Brewer PG, Herlien R, Chavez FP (1995) Measurements of sea surface partial pressure of CO_2 from a moored buoy. Deep-Sea Res Pt I 42(1175–1186)
75. Anderson LA (1995) On the hydrogen and oxgyen content of marine phytoplankton. Deep-Sea Res Pt II 42(9):1675–1680
76. Emerson S, Stump C, Johnson B, Marl DM (2002) In situ determination of oxygen and nitrogen dynamics in the upper ocean. Deep-Sea Res Pt I 49:941–952

Chapter 5

DNAzyme-Based Sensing for Metal Ions in Ocean Platform

Jung Heon Lee, Zidong Wang, and Yi Lu

Abstract

The ocean contains a number of metal ions that are either beneficial or detrimental to marine lives or ecology. Developing sensors for on-site and real-time detection of these metal ions plays an important role in our understanding the ocean as well as its protection. DNAzymes, DNA molecules with enzymatic functions, have emerged as a new class of sensing molecules for metal ions, because DNAzymes with high affinity and specificity for almost any metal ions at a specific oxidation state can be obtained through in vitro selection. By integrating the DNAzyme with different signal transduction molecules, such as fluorophores or nanoparticles, DNAzyme-based sensors for a broad range of metal ions with high sensitivity (with limit of detection down to ppt) and selectivity (with over a million fold) have been reported. In this chapter, we summarize recent progress in DNAzyme-based sensors for metal ions and describe detailed protocols in designing fluorescent and colorimetric sensors for uranium and mercury. The diverse range of metal ions it can detect as well as its excellent sensing properties makes DNAzyme an excellent choice for ocean sensing.

Key words: DNAzyme, Sensing, Seawater, Uranium, Mercury, Colorimetric sensor, Fluorescent sensor, Fluorophores, Nanoparticles, Ocean

1. Introduction

The ocean contains many metal ions, such as copper, molybdenum, and vanadium, that are important for marine organisms and plants (1). At the same time, heavy metal ion contaminations from sewage run-offs and air pollution (such as mercury from power plants) are becoming an increasingly serious problem in ocean. Monitoring these metal ions in seawater is very important to protect the ocean from environmental contamination which can threaten the safety of the ecosystem including human being. The recent damage of Fukushima Daiichi Nuclear Power Plant due to tsunami that hit Japan on March 11, 2011 and the resulting release of several toxic radioactive metals into the ocean highlights the importance of

Sonia M. Tiquia-Arashiro (ed.), *Molecular Biological Technologies for Ocean Sensing*, Springer Protocols Handbooks,
DOI 10.1007/978-1-61779-915-0_5,

developing effective metal sensors for on-site and real-time detection. However, accurate and fast detection of metals in seawater is quite challenging because of their low concentration (nM to pM range) and high interferences from the competing species (2). Thus, seawater generally has to go through several processes, including preconcentration and separation, before complicated instrumental analysis can be performed (3, 4). Furthermore, most detection methods rely on expensive and sophisticated instruments, such as inductively coupled plasma mass spectrometry and atomic absorption spectrometry, which are not suitable for on-site real-time monitoring (2, 5, 6).

DNA has been regarded as a passive molecule, containing genetic information of a living organism. However, in 1994, the existence of short DNA strands with active catalytic functions has been reported and they were called catalytic DNA, deoxyribozyme, or DNAzyme (7). These DNAzymes perform catalytic function, such as cleavage of the complementary substrate strand, in the presence of metal ion cofactors. Since this reaction occurs in response to certain metal ion cofactors, DNAzymes can be used as a universal platform for their detection. Furthermore, because these DNAzymes can be obtained through in vitro selection from a large DNA library in the presence of metal cofactors (7), DNAzymes specific for a variety of metal ions, including Pb^{2+} (7, 8), Cu^{2+} (9–11), Zn^{2+} (12), Co^{2+} (13, 14), Mn^{2+} (15), Hg^{2+} (16), and UO_2^{2+} (17), have been selected. Recognizing this new capability, we and others have converted these DNAzymes as a new class of metal ion sensors (18–32). In this chapter, we summarize recent advances in and provide detailed protocols for DNAzyme-based fluorescent and colorimetric sensors for detection of metal ions, using uranium and mercury sensing as examples.

2. DNAzyme-Based Metal Detection

2.1. Uranium-Specific DNAzyme

Uranium is one of the radioactive metals which exist ubiquitously in our environment, and it is also a core component of nuclear energy generation and nuclear weapon. Therefore, precise monitoring of uranium in our environment is highly demanded to protect human health and our environment.

Recently, our group has used in vitro selection to obtain a DNAzyme with high specificity for uranyl (UO_2^{2+}), which is the most stable form of uranium in water (see Fig. 1) (17). UO_2^{2+}-specific DNAzyme is composed of two strands of DNA: a substrate strand (39S) and an enzyme strand (39E). The DNAzyme remains hybridized in the absence of UO_2^{2+}. However, the substrate strand which contains hydrolytically susceptible ribo-adenosine (rA) is

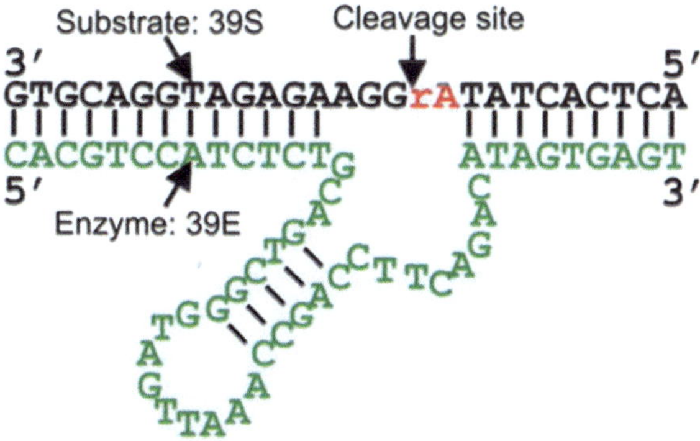

Fig. 1. The secondary structure of an UO_2^{2+}-specific DNAzyme. Reproduced by permission of National Academy of Sciences of the United States of America (17).

cleaved by the enzyme strand in the presence of UO_2^{2+} and breaks into two fragments. Uranyl-specific cleavage is a highly efficient multiple turnover reaction, with dissociation constant of 469 nM for uranyl, making it an ideal choice as UO_2^{2+} sensor (33).

2.2. Fluorescence-Based Sensors

In order to design a sensor, at least two components are needed: target recognition and signal transduction. The high sensitivity and selectivity of DNAzyme in response to specific analytes allow them to serve as an ideal target recognition element for sensing. However, we also need a part which transforms the recognition event into detectable signal.

Since DNA can be chemically modified with a variety of organic fluorescent dyes of choice, DNAzymes have been converted into fluorescent sensors (8, 24). Recently, our group developed uranyl-specific fluorescent sensor by introducing a fluorophore on the substrate strand of the DNAzyme and a quencher on its enzyme strand (see Fig. 2a) (17). With this simple design, the catalytic activity of the DNAzyme remains intact and the fluorescent uranyl sensor with a detection limit of 11 parts per trillion (45 pM) and selectivity of more than one million fold over other metal ions has been observed. This sensor rivals one of most sensitive analytical instruments for uranium detection.

This strategy can also be applied to detection of other metal ions, such as mercury. Hg^{2+} can bind specifically to two DNA thymine (T) mismatches and form stable thymine–Hg^{2+}–thymine complex. We introduced several thymine mismatches to the catalytic site of the uranyl DNAzyme, which is only activated in the presence of Hg^{2+} (See Fig. 2b). By attaching fluorophore/quencher pairs to the DNAzyme complex, the DNAzyme could function as highly sensitive and selective Hg^{2+} sensor in aqueous solution with detection limit of 2.4 nM and no interference from competing metal ions (34).

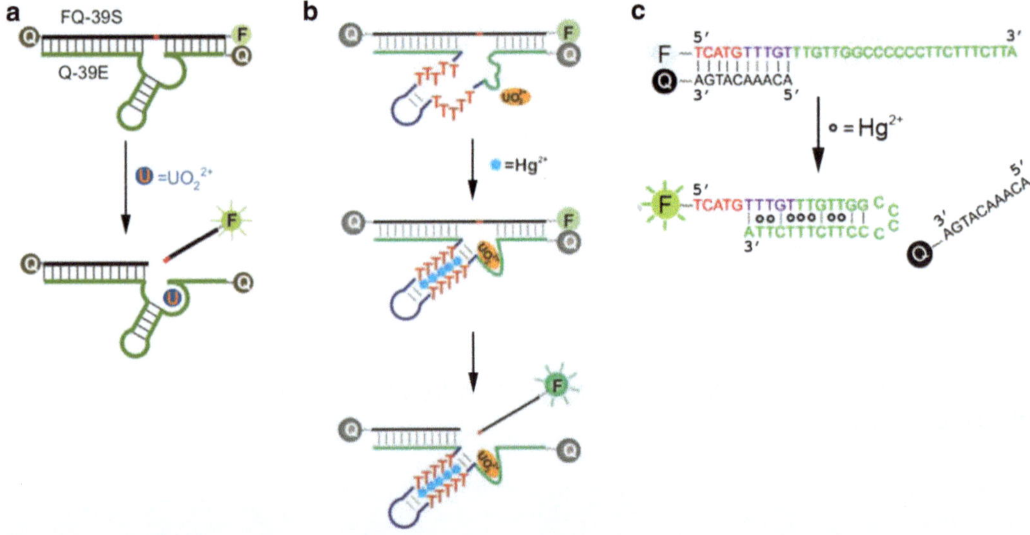

Fig. 2. DNAzyme-based fluorescent metal ion sensors: (**a**) uranyl sensor, (**b**, **c**) mercury sensors. Sensor (**b**) needs uranyl for detection of mercury while sensor (**c**) does not. Reproduced by permission of National Academy of Sciences of the United States of America (17), Wiley-VCH Verlag GmbH & Co. KGaA (34) and the Royal Society of Chemistry (35).

While this modified uranyl DNAzyme sensor shows excellent performance for detection of Hg^{2+}, it functions only in the presence of micromolar concentration of UO_2^{2+}. A fluorescence sensor which is also based on the thymine (T) mismatches but does not need UO_2^{2+} has also been developed from our group (see Fig. 2c) (35). This "turn-on" sensor shows excellent performance in water with 3.2 nM (0.6 ppb) detection limit and no response from other metal ions. The practical application of this sensor has also been demonstrated for detection of mercury in pond water. These few examples suggest that DNA can be an ideal tool for detection of broad range of metal ions in seawater.

2.3. Interaction Between Gold Nanoparticles and DNA: The Labeled and Label-Free Systems

Even though fluorophores are commonly used as signal reporter molecules for the development of high performance sensors, the requirement of analytical instruments such as fluorimeters makes them less convenient for on-site and remote monitoring. On the other hand, noble metal nanoparticles such as gold nanoparticles (AuNPs) have several unique optical properties (36). For instance, AuNPs have extremely high extinction coefficients; so the color of a few nanomolar AuNP can be distinguished with naked eye without the need of any instrument. In addition, the extinction wavelength of AuNPs can be significantly altered by changing the distances between the nanoparticles. For example, 13 nm AuNPs have reddish color in dispersed phase while their color changes bluish after aggregation. These interesting optical properties of AuNPs make them an ideal reporter for colorimetric sensing.

While DNA has the capability to identify a broad range of target molecules, they can also interact with AuNPs in a certain way and control their aggregation and disassembly status. This essential role of DNA will construct direct correlation between the presence of analyte and the color change of the sensor. The DNA-based AuNP colorimetric sensors can be designed in either labeled or label-free methods based on the way DNA interacts with AuNPs (26, 27, 37).

2.3.1. Labeled U Colorimetric Sensing

In the labeled method, AuNPs are chemically functionalized with DNA and their aggregation and disassembly status is controlled by the hybridization of the DNA (37, 38). Two batches of AuNP–DNA are first prepared by conjugating AuNPs with two different DNA arm strands (Arm 5′ and 3′) through Au-thiol chemistry. When both batches of AuNP–DNA are mixed together, they stay dispersed with reddish color due to the strong negative charge of the DNA. In the presence of the bridging DNA, however, which is complementary to both arm strands, the AuNPs aggregate and the color of the AuNP becomes blue because the distance between the AuNPs becomes close (see Note 1).

Uranyl colorimetric sensor uses uranyl-specific DNAzyme complex (39S and 39E) as the bridging DNA to crosslink AuNPs (see Fig. 3) (26). While the AuNPs of the sensor stay assembled in the absence of uranyl, they start to disassemble in its presence due to the cleavage of the substrate strand (39S). The disassembly of the AuNPs can be significantly accelerated by introducing two invasive DNA strands which are complementary to both ends of the substrate strands (39). The addition of this invasive DNA can help to shorten the detection time.

Figure 4 summarizes various properties of the optimal-labeled uranyl sensor. As shown in Fig. 4a, the blue shift of the plasmonic resonance of the AuNPs could be observed by UV-vis in the presence of uranyl within 30 min and the detection limit was found to be 50 nM. This is lower than the maximum contamination level of uranium (130 nM) defined by the US Environmental Protection Agency (EPA). Furthermore, no color change of the sensor occurred on competing metal ions, which indicates that the sensor only responses to uranyl and not to other metal ions (See Fig. 4b).

2.3.2. Label-Free U Colorimetric Sensing

The label-free method is based on the different adsorption properties of single-stranded (ss) DNA and double-stranded (ds) DNA on the surface of citrate modified AuNP (27, 40). Although citrate-capped AuNPs are negatively charged, they are naturally unstable in the presence of NaCl and can be easily aggregated. Because ssDNA is flexible and can uncoil its structure, positively charged bases of the ssDNA can be easily exposed and adsorbed on negatively charged AuNP surface, resulting in enhanced stability of AuNPs even in the presence of NaCl. On the other hand, because dsDNA is relatively rigid and surrounded by negatively charged phosphate backbone, its binding with AuNP is negligible.

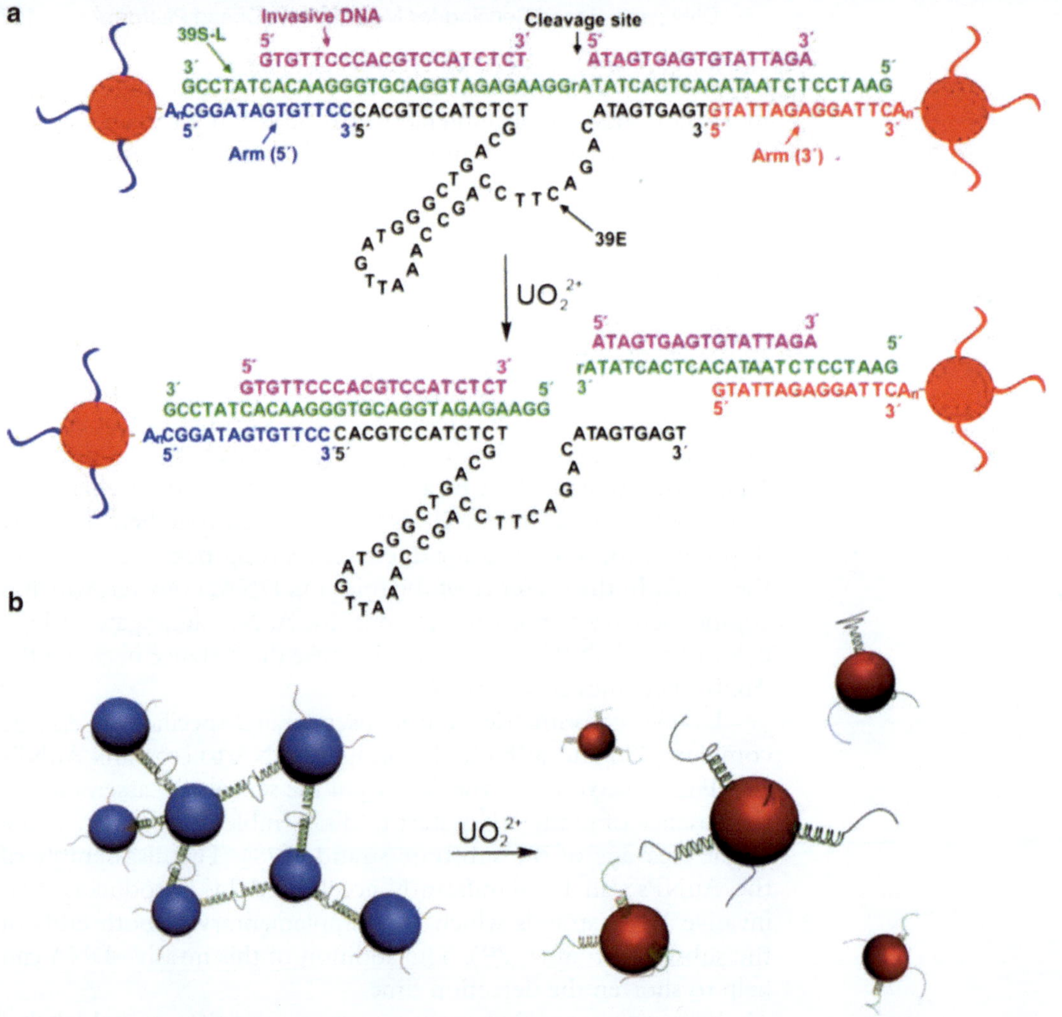

Fig. 3. Scheme of labeled colorimetric UO_2^{2+} sensor: (**a**) Labeled sensor need chemical modification between AuNP and DNA. (**b**) The blue-colored AuNP aggregates disassemble in the presence of UO_2^{2+}, resulting in color change of the sensor. Reproduced with permission of ref. 26, copyright of 2008 American Chemical Society.

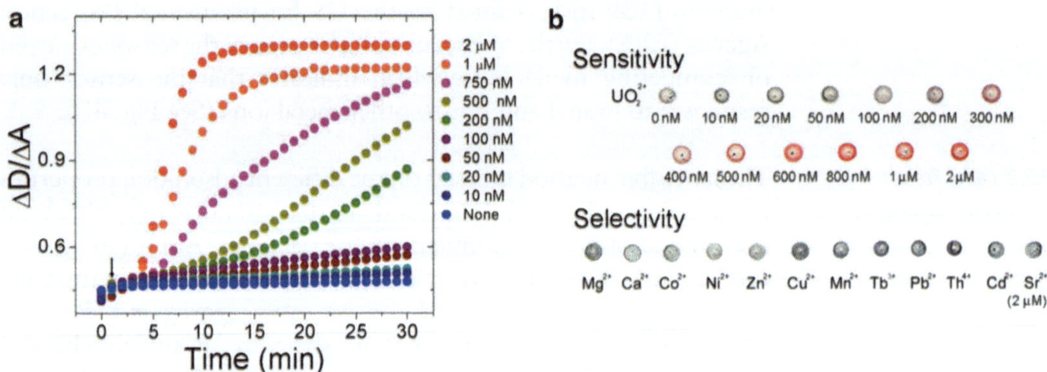

Fig. 4. Properties of labeled uranyl colorimetric sensor: (**a**) Variation of AuNP disassembly kinetics as a function of uranyl concentration (**b**) Image of the color of labeled sensor in response to different concentrations of UO_2^{2+} and other metal ions. Reproduced with permission of ref. 26, copyright of 2008 American Chemical Society.

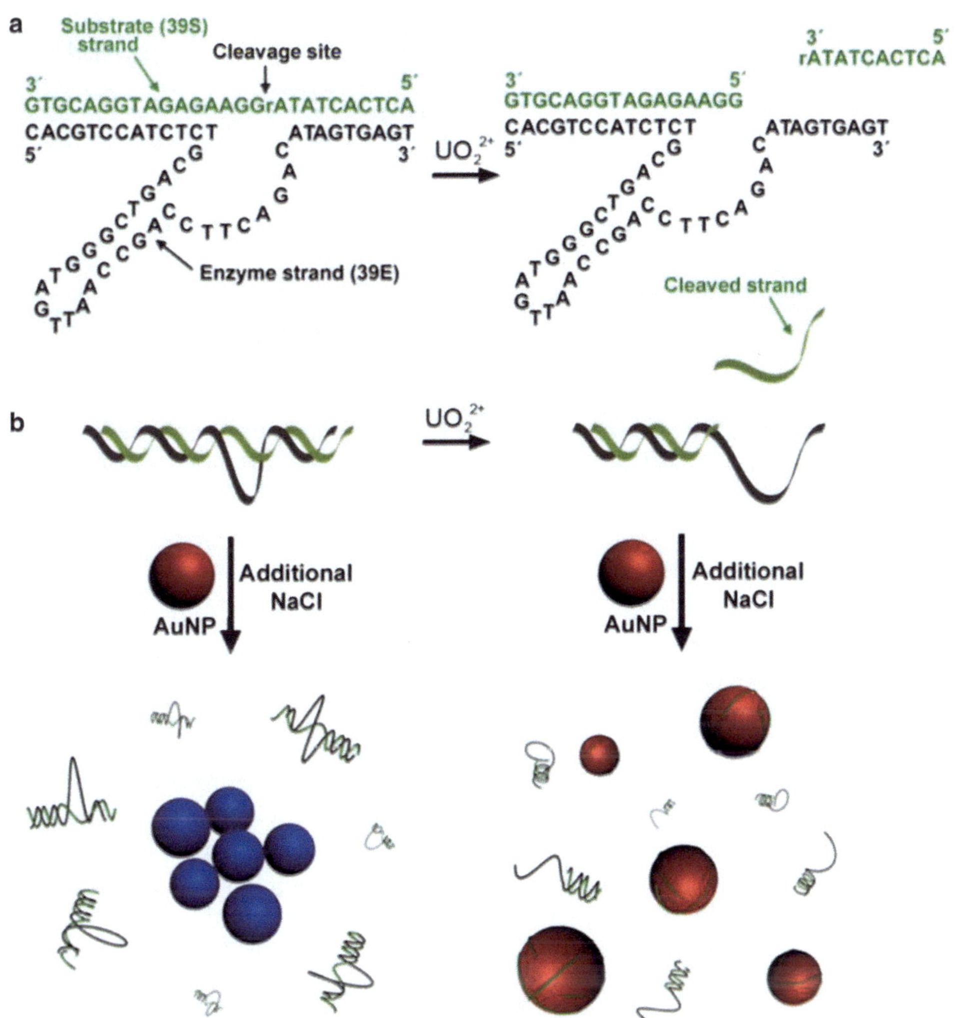

Fig. 5. Scheme of the label-free colorimetric uranyl sensor: (**a**) Label-free sensor does not need chemical modification between DNA and AuNPs. (**b**) The sensor is based on different adsorption properties of ssDNA and dsDNA on citrate modified AuNP. The amount of ssDNA is determined by the concentration of uranyl. Reproduced with permission of ref. 26, copyright of 2008 American Chemical Society.

Therefore, NaCl induced aggregation of AuNPs will occur in the existence of dsDNA.

The label-free uranyl sensor can be prepared by, first hybridizing unmodified 39E and 39S strands to form DNAzyme complex (see Fig. 5). The DNAzyme complex is then treated with uranyl for 6 min, resulting in the release of the short 10-mer DNA strand from the substrate that can interact with AuNPs. The uranyl-dependent cleavage reaction can be quenched by quickly mixing concentrated TRIS base solution with the DNAzyme solution, which results in a shift of the pH from 5.5 to 8. Since UO_2^{2+} DNAzyme is most active at pH 5.5 but nonactive at pH 8 (33), the

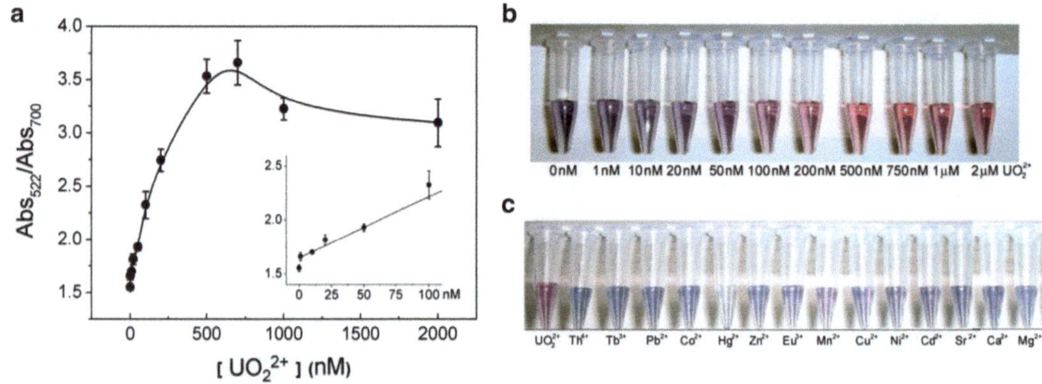

Fig. 6. Properties of label-free uranyl colorimetric sensor: (**a**) Variation of the aggregation status of the label-free sensor at different concentration of uranyl. (**b**) Color change of label-free sensor in response to different amount of uranyl. (**c**) Image of the color of label-free sensor treated with competing metal ions. Reproduced with permission of ref. 26, copyright of 2008 American Chemical Society.

cleavage reaction is quenched after the addition of TRIS. Once 13 nm AuNPs are added to the solution containing DNA, the color of the AuNP will change in correlation to the amount of uranyl in the solution.

The color change of the AuNPs treated with various uranyl concentrations has been monitored by comparing the extinction ratios between 522 and 700 nm by UV-vis (Fig. 6). The sensitivity of sensor was found to be as low as 1 nM, which is even lower than that of the labeled sensor. The color change can also be easily observed by naked eye. No color change of the sensor was observed when it was treated with other metal ions, indicating that the label-free sensor has excellent selectivity over competing metal ions.

2.3.3. Experimental Protocols of U Sensors

Fluorescence Sensor

In order to form DNAzyme complex, 60 nM 39S with both fluorescein and quencher labeled on each end (FQ-39S) and 60 nM quencher modified 39E (Q-39E) were used. The two strands were annealed in a buffer containing 300 mM $NaNO_3$ and 50 mM MES (pH 5.5) by heating the solution up to 70°C and cooling down to room temperature in 1 h (see Fig. 2a).

The detection can be accomplished using a fluorimeter. Five hundred microliters of the prepared sensor solution was transferred to a quartz cuvette with a 0.5 cm path length. The cuvette was placed in a fluorimeter (FluoroMax-P; Horiba Jobin Yvon, Edison, NJ). The excitation and emission wavelengths were set at 490 and 520 nm, respectively. After the initial measurement, a small volume of concentrated metal solution was added to initiate the cleavage reaction.

Labeled Colorimetric Sensor

Thirteen nanometer AuNPs were synthesized by reducing $HAuCl_4$ with sodium citrate based on previous literature (see Note 2). AuNP–

DNA conjugates were formed by reacting thiol modified DNA with AuNPs. First, thiol modification placed on Arm 5′ strand was activated by mixing 9 μL of 1 mM Arm 5′ strand, 1 μL of 500 mM pH 5.5 MES buffer, and 1.5 μL of 10 mM tris (2-carboxyethyl) phosphine hydrochloride (TCEP) solution in a microcentrifuge tube and keeping them for an hour. A parallel procedure was performed with Arm (3′) strand to activate thiol modification on Arm (3′) strand.

In order to react Arm (5′) strand with AuNPs, 3 mL of 13 nm AuNP solution was mixed with 9 μL Arm (5′) strand in a scintillation vial and left overnight. The process was repeated with Arm (3′) strand to react the Arm (3′) strand with AuNPs. The next day, 300 μL of 1 M NaCl and 15 μL of 500 mM TRIS acetate buffer (pH 7.6) were added to each scintillation vial and were stored for another day.

To prepare the labeled sensor, AuNPs functionalized with Arm strands were purified to remove excess DNA. Both 500 μL of Arm (5′) functionalized AuNPs and Arm (3′) functionalized AuNPs were, respectively, placed in 1.5 mL microcentrifuge tube and were centrifuged at 16,110 × *g* for 15 min. The supernatant was replaced with 500 μL of 100 mM NaCl, 50 mM MES (pH 5.5) solution. After additional purification process, the supernatant was replaced with 250 μL of 300 mM NaCl, 50 mM MES (pH 5.5) buffer. In order to form AuNP aggregates, both AuNP–DNA solutions were mixed with 10 μL of 10 μM elongated substrate strand (39S-L), and 20 μL of 10 μM enzyme strand (39E) and the combined solution was heated up to 55 °C and cooled down to room temperature for about 1 h. DNA directed AuNP aggregation should happen, which can be notified by the color change of AuNP solution. The microcentrifuge tube containing the AuNP aggregates should be centrifuged for about 1 min followed by supernatant replacement with 120 μL of 300 mM NaCl, 50 mM MES (pH 5.5) solution to remove excess DNA (39S-L and 39E) in the solution (see Note 3).

Before detection of uranyl, a pair of invasive DNA strands were added to the sensor solution to expedite sensing process. Three hundred and eighty-one microliters of 50 mM MES buffer (pH 5.5), 4.5 μL of 0.2 mM Inva (5), and 1.5 μL of 0.2 mM Inva (3) were mixed with 60 μL sensor solution. Uranyl was added 1 min after the addition of invasive DNA strands and the detection typically took 30 min. Color change can be monitored by either UV-vis or naked eye.

Label-Free Colorimetric Sensor

In order to form DNAzyme complex, 4 μL of 100 μM 39S strand were mixed with identical amount of 39E strand in 70 μL 300 mM NaCl, 10 mM MES buffer (pH 5.5) in a microcentrifuge tube and mixture was annealed by heating it up to 80 °C and cooling it down to room temperature (see Note 4). For the reaction with uranyl, 77 mL of the DNAzyme complex were transferred to a new tube and treated with uranyl for 6 min. To quench UO_2^{2+}-induced cleavage reaction, 1.19 μL of 500 mM TRIS base solution and 56 μL Millipore water were added to the tube containing DNAzyme and uranyl with quick vortexing. In order to observe color change of the sensor, 76 μL of 10 nM AuNPs were added to the tube containing DNAzyme and uranyl. The color change of the solution, which can be monitored by either UV-vis or naked eye, will be determined by the amount of the uranyl in the solution.

2.3.4. Comparison of the Labeled and Label-Free Colorimetric Sensors

DNA–AuNP-based colorimetric sensors have recently received significant attention because the detection can be made for a broad range of analytes at a low cost without the need of any complicated instrument. But it is noteworthy to emphasize that in most cases, DNA–AuNP-based sensors are operated by either labeled or label-free method.

Although both methods take advantage of the altered optical property of the AuNP induced by the introduction of the analytes, the difference in the basic principle of both sensors make their general properties significantly different. Therefore, it is worthwhile to compare the two sensing systems.

In terms of sensor performance, the label-free sensor is much more sensitive than the labeled sensor. Because the labeled sensor relies on uranyl dependent disassembly of AuNPs from aggregated states, certain amount of uranyl concentration and reaction time is necessary for the detection to be recognized. These factors result in relatively lower detection limit (50 nM) and slower kinetics (30 min) of the labeled sensor. On the other hand, because the label-free sensor allows UO_2^{2+}-induced cleavage reaction to happen in the absence of AuNPs and AuNPs are added to the DNA solution afterwards, just to monitor the amount of the released short DNA strands, the reaction can be completed much quickly (6 min) with high sensitivity (with 1 nM detection limit).

In terms of sensor preparation and handling, both sensors have distinct properties as well. The labeled sensor requires more time and effort to be prepared as DNA–AuNP conjugation and AuNP aggregation processes are needed. However, once the labeled sensors are prepared, it could be more convenient to be used because UO_2^{2+} detection can be completed in a single step, by adding it to the sensor solution. On the other hand, although the label-free sensor can be relatively easily prepared, multiple handling steps such as UO_2^{2+} treatment, quenching, and AuNP addition, are required, for detection.

Finally, for practical applications, the labeled sensor has higher flexibility, due to its high stability of AuNP–DNA conjugation and simpler handling process. In contrast, the label-free sensor is normally restricted by the narrow operational condition, such as ionic strength, and multiple handling steps needed for detection.

3. Application of DNAzyme for Ocean Sensing

Because the DNAzyme has been demonstrated to target a broad range of metal ions with excellent sensitivity and selectivity in buffer solution, it holds great promise to serve as an ideal platform for detection of metals in ocean. In comparison with buffer solutions, the ocean water contains much higher level of various metal ions, including Na^{+}, Mg^{2+}, Ca^{2+}, and K^{+}. Although these metal ions might not affect the selectivity of a DNAzyme, they can still have considerable effect on its property. Therefore, the DNAzyme-based sensor will have to be optimized so that it can have maximal performance in the ocean. To obtain a DNAzyme with the best performance in the seawater, it would also be possible to carry out in vitro selection to obtain new DNAzymes directly in seawater.

While it is likely that fluorescence-based sensors could be adapted for ocean sensing without much alteration on the sensor designs, colorimetric sensors may have to be reengineered to adapt the high salt environment in seawater due to the vulnerability of AuNPs to salt. Among the two colorimetric sensors, the labeled sensor has advantages on ocean sensing application as DNA conjugated AuNPs are more resistant to salt-induced aggregation. On the other hand, since the label-free sensor uses naturally unstable citrate-capped AuNPs, the samples have to be diluted to appropriate concentration before optimal sensing performance could be obtained. In both cases, the sensors need to be recalibrated in ocean water with a standard addition method.

Although the DNAzyme sensors can be used for detection of metal ions in solution phase, it can also be integrated into various devices, such as nanocapillary array membranes, microfluidic devices, and dip sticks (41–44). The possibility of adapting DNAzymes onto diverse devices gives us the opportunity to expand the capability of DNAzyme for ocean sensing in other detection platforms.

4. Notes

1. Because the AuNPs are linked by DNA base-pairing interactions, the general property of the labeled sensor is highly dependent on the NaCl concentration.
2. In order to prevent AuNPs from sticking on the glass surface, the scintillation vials were first incubated in 10 M NaOH solution for an hour, and then rinsed with distilled water and subsequently with Millipore water copious times.

3. Even though DNA–AuNP conjugates stored at room temperatures for months can still form aggregates, the degradation of DNA can cause change of the sensing property and performance. To obtain consistent results, it is desirable to use freshly prepared conjugates for sensor preparation.
4. It should be emphasized that exact same concentration of 39E and 39S strands should be used for the hybridization of DNAzyme complex to minimize the background signal.

5. Future Directions

In this study, we have discussed the possibility of using DNAzyme based sensors for ocean sensing, using UO_2^{2+} and Hg^{2+} as examples. Although detection of both UO_2^{2+} and Hg^{2+} is meaningful, there are a number of other metal ions that can be important targets. For example, as a result of the tragic nuclear plant meltdown in Fukushima, Japan, several radioactive metals, such as cesium, plutonium, and strontium, have been released to the ocean that are highly harmful to not only human beings, but also the whole ecosystem (45, 46). Therefore, the development of facile method for detecting these metals becomes crucial. Furthermore, the ocean contains numerous metals that can be important resources. For example, seawater contains 250 billion tons of lithium which exceeds the amount of lithium stored on earth (47). Considering the highly increasing demand of these metals, it will be very useful to detect these metals with high sensitivity in the ocean. The DNAzyme can serve as one of the best candidates for these purposes as DNAzymes specific for almost any target metal ion can be obtained via the in vitro selection process. Furthermore, as the selectivity of the DNAzyme originates from a specific sequence, and each DNAzyme can be labeled with a different reporter group, it is possible use different DNAzymes in one system for multiplex sensing of various metal ions in the ocean.

Acknowledgements

The authors acknowledge the financial support from the US National Institutes of Health (ES016865), Department of Energy (DE-FG02-08-ER64568), National Science Foundation (CTS-0120978, DMR-0117792, and DMI-0328162).

References

1. Butler A (1998) Acquisition and utilization of transition metal ions by marine organisms. Science 281(5374):207–210
2. Ardini F, Magi E, Grotti M (2011) Determination of ultratrace levels of dissolved metals in seawater by reaction cell inductively coupled plasma mass spectrometry after ammonia induced magnesium hydroxide coprecipitation. Anal Chim Acta 706(1):84–88
3. Moore TS, Mullaugh KM, Holyoke RR, Madison ANS, Yucel M, Luther GW (2009) Marine chemical technology and sensors for marine waters: potentials and limits. Ann Rev Mar Sci 1:91–115
4. Achterberg EP, Holland TW, Bowie AR, Fauzi R, Mantoura C, Worsfold PJ (2001) Determination of iron in seawater. Anal Chim Acta 442(1):1–14
5. Takata H, Zheng J, Tagami K, Aono T, Uchida S (2011) Determination of 232 Th in seawater by ICP-MS after preconcentration and separation using a chelating resin. Talanta 85(4):1772–1777
6. Colbert D, Johnson KS, Coale KH (1998) Determination of cadmium in seawater using automated on-line preconcentration and direct injection graphite furnace atomic absorption spectrometry. Anal Chim Acta 377(2–3): 255–262
7. Breaker RR, Joyce GF (1994) A DNA enzyme that cleaves RNA. ChemBiol 1(4):223–229
8. Li J, Lu Y (2000) A highly sensitive and selective catalytic DNA biosensor for lead ions. J Am Chem Soc 122(42):10466–10467
9. Cuenoud B, Szostak JW (1995) A DNA metalloenzyme with DNA ligase activity. Nature 375(6532):611–614
10. Carmi N, Shultz LA, Breaker RR (1996) In vitro selection of self-cleaving DNAs. Chem Biol 3(12):1039–1046
11. Wang W, Billen LP, Li Y (2002) Sequence diversity, metal specificity, and catalytic proficiency of metal-dependent phosphorylating DNA enzymes. Chem Biol 9(4):507–517
12. Santoro SW, Joyce GF, Sakthivel K, Gramatikova S, Barbas CF III (2000) RNA cleavage by a DNA enzyme with extended chemical functionality. J Am Chem Soc 122(11):2433–2439
13. Mei SHJ, Liu Z, Brennan JD, Li Y (2003) An efficient RNA-cleaving DNA enzyme that synchronizes catalysis with fluorescence signaling. J Am Chem Soc 125(2):412–420
14. BruesehoffPJ, Li J, Augustine AJ, Lu Y (2002) Improving metal ion specificity during in vitro selection of catalytic DNA. Comb Chem High Throughput Screen 5(4):327–335
15. Wang Y, Silverman SK (2003) Deoxyribozymes that synthesize branched and lariat RNA. J Am Chem Soc 125(23):6880–6881
16. Hollenstein M, Hipolito C, Lam C, Dietrich D, Perrin DM (2008) A highly selective DNAzyme sensor for mercuric ions. Angew Chem-Int Ed 47(23):4346–4350
17. Liu J, Brown AK, Meng X, Cropek DM, Istok JD, Watson DB et al (2007) A catalytic beacon sensor for uranium with parts-per-trillion sensitivity and millionfold selectivity. Proc Natl Acad Sci U S A 104(7):2056–2061
18. He Q-C, Zhou J-M, Zhou D-M, Nakamatsu Y, Baba T, Taira K (2002) Comparison of metal-ion-dependent cleavages of RNA by a DNA enzyme and a hammerhead ribozyme. Biomacromolecules 3(1):69–83
19. Liu J, Lu Y (2004) Accelerated color change of gold nanoparticles assembled by DNAzymes for simple and fast colorimetric Pb^{2+} detection. J Am Chem Soc 126(39):12298–12305
20. Khachigian LM (2005) DNAzymes targeting immediate-early genes as inhibitors of angiogenesis and restenosis. In: Khachigian LM (ed) Synthetic nucleic acids as inhibitors of gene expression. CRC, New York, pp 153–159
21. Liu J, Lu Y (2006) Fluorescent DNAzyme biosensors for metal ions based on catalytic molecular beacons. Methods in molecular biology, Totowa, NJ, p 335 (Fluorescent energy transfer nucleic acid probes, pp 275–288)
22. Cheglakov Z, Weizmann Y, Beissenhirtz MK, Willner I (2006) Ultrasensitive detection of DNA by the PCR-induced generation of DNAzymes: the DNAzyme primer approach. Chem Commun 30:3205–3207
23. Cheglakov Z, Weizmann Y, Basnar B, Willner I (2007) Diagnosing viruses by the rolling circle amplified synthesis of DNAzymes. Org Biomol Chem 5(2):223–225
24. Liu J, Lu Y (2007) A DNAzyme catalytic beacon sensor for paramagnetic Cu^{2+} ions in aqueous solution with high sensitivity and selectivity. J Am Chem Soc 129(32):9838–9839
25. Liu J, Lu Y (2007) Colorimetric Cu^{2+} detection with a ligation DNAzyme and nanoparticles. Chem Commun 46:4872–4874
26. Lee JH, Wang ZD, Liu JW, Lu Y (2008) Highly sensitive and selective colorimetric sensors for Uranyl (UO_2^{2+}): development and comparison of labeled and label-free DNAzyme gold nano particle systems. J Am Chem Soc 130(43): 14217–14226
27. Wang Z, Lee JH, Lu Y (2008) Label free colorimetric detection of metal ions using gold

nanoparticles and DNAzyme with 3 nM detection limit and tunable dynamic range. Adv Mater 20:3263–3267

28. Lan T, Furuya K, Lu Y (2010) A highly selective lead sensor based on a classic lead DNAzyme. Chem Commun 46(22):3896–3898
29. Hung YL, Hsiung TM, Chen YY, Huang YF, Huang CC (2010) Colorimetric detection of heavy metal ions using label-free gold nanoparticles and alkanethiols. J Phys Chem C 114(39):16329–16334
30. Kong RM, Zhang XB, Chen Z, Meng HM, Song ZL, Tan WH et al (2011) Unimolecular catalytic DNA biosensor for amplified detection of l-histidine via an enzymatic recycling cleavage strategy. Anal Chem 83(20): 7603–7607
31. Zhang XB, Kong RM, Lu Y (2011) Metal ion sensors based on DNAzymes and related DNA molecules. In: Cooks RG, Yeung ES (eds) Annual review of analytical chemistry, vol 4. Annual Reviews, Palo Alto, CA, pp 105–128
32. Liu JW, Cao ZH, Lu Y (2009) Functional nucleic acid sensors. Chem Rev 109(5): 1948–1998
33. Brown AK, Liu J, He Y, Lu Y (2009) Biochemical characterization of a uranyl ion-specific DNAzyme. ChemBioChem 10(3): 486–492
34. Liu J, Lu Y (2007) Rational design of "turn-on" allosteric DNAzyme catalytic beacons for aqueous mercury ions with ultrahigh sensitivity and selectivity. Angew Chem-Int Ed 46(40): 7587–7590
35. Wang ZD, Lee JH, Lu Y (2008) Highly sensitive "turn-on" fluorescent sensor for Hg^{2+} in aqueous solution based on structure-switching DNA. Chem Commun 45:6005–6007
36. Lee JH, Yigit MV, Mazumdar D, Lu Y (2010) Molecular diagnostic and drug delivery agents based on aptamer-nanomaterial conjugates. Adv Drug Deliv Rev 62(6):592–605
37. Liu J, Lu Y (2003) A colorimetric lead biosensor using DNAzyme-directed assembly of gold nanoparticles. J Am Chem Soc 125(22): 6642–6643
38. Mirkin CA, Letsinger RL, Mucic RC, Storhoff JJ (1996) A DNA-based method for rationally assembling nanoparticles into macroscopic materials. Nature 382(6592):607–609
39. Liu J, Lu Y (2005) Stimuli-responsive disassembly of nanoparticle aggregates for light-up colorimetric sensing. J Am Chem Soc 127(36): 12677–12683
40. Li H, Rothberg L (2004) Colorimetric detection of DNA sequences based on electrostatic interactions with unmodified gold nanoparticles. Proc Natl Acad Sci U S A 101(39): 14036–14039
41. Wernette DP, Swearingen CB, Cropek DM, Lu Y, Sweedler JV, Bohn PW (2006) Incorporation of a DNAzyme into Au-coated nanocapillary array membranes with an internal standard for Pb(II) sensing. Analyst 131(1):41
42. Dalavoy TS, Wernette DP, Gong M, Sweedler JV, Lu Y, Flachsbart BR et al (2008) Immobilization of DNAzyme catalytic beacons on PMMA for Pb^{2+} detection. Lab Chip 8(5): 786–793
43. Mazumdar D, Liu JW, Lu G, Zhou JZ, Lu Y (2010) Easy-to-use dipstick tests for detection of lead in paints using non-cross-linked gold nanoparticle-DNAzyme conjugates. Chem Commun 46(9):1416–1418
44. Liu J, Mazumdar D, Lu Y (2006) A simple and sensitive "dipstick" test in serum based on lateral flow separation of aptamer-linked nanostructures. Angew Chem Int Ed 45(47):7955–7959
45. Prants SV, Uleysky MY, Budyansky MV (2011) Numerical simulation of propagation of radioactive pollution in the ocean from the Fukushima Dai-ichi nuclear power plant. Doklady Earth Sci 439(2):1179–1182
46. Kinoshita N, Sueki K, Sasa K, Kitagawa J, Ikarashi S, Nishimura T et al (2011) Assessment of individual radionuclide distributions from the Fukushima nuclear accident covering central-east Japan. Proc Natl Acad Sci U S A 108(49):19526–19529
47. Takeuchi T (1980) Extraction of lithium from sea-water with metallic aluminum. J Nucl Sci Technol 17(12):922–928

Chapter 6

Ultrasensitive Visual Fluorescence Detection of Heavy Metal Ions in Water Based on DNA-Functionalized Hydrogels

Juewen Liu, Neeshma Dave, and Po-Jung Jimmy Huang

Abstract

Heavy metal contamination of oceans, lakes, and other water resources can occur by both natural and human-related processes. Human exposure to heavy metals such as mercury is known to cause a number of serious health problems. Due to its high toxicity and bioaccumulative properties, the maximum toxic level of mercury in drinking water is set to be 10 nM or 2 parts-per-billion by the US EPA. Therefore, detection of mercury at such a low concentration poses an analytical challenge. While analytical instruments such as ICP-MS are still very widely used for heavy metal analysis, biosensors, are emerging as a cost-effective alternative allowing on-site and real-time detection. We herein describe a protocol for preparing polyacrylamide hydrogel-based biosensors functionalized with a thymine-rich DNA that can effectively detect mercury in water. Detection is achieved by the selective binding of Hg^{2+} between two thymine bases inducing a hairpin structure where upon the addition of SYBR Green I dye, green fluorescence is observed. In the absence of Hg^{2+}, the addition of the dye results in yellow fluorescence. This hydrogel-based sensor can easily detect 10 nM Hg^{2+} using the naked eye, can be regenerated using a simple acid treatment, and can be dried for storage and easily rehydrated. This sensor is also used to detect Hg^{2+} from Lake Ontario water samples spiked with mercury. In the case where a cationic gel formulation is used, the background fluorescence can be effectively suppressed to increase sensitivity. The future research directions of using such gels to detect other metal ions and to detect metal ions in ocean water are also discussed.

Key words: Mercury, Hydrogels, DNA, Biosensors, Fluorescence, DNA-functionalized Hydrogels, Real-time detection, Hydrogel-based sensor, Heavy metal detection, Ocean sensing

1. Introduction

Heavy metal ions such as Hg^{2+}, Cu^{2+}, Cd^{2+}, and Pb^{2+} are known to cause various adverse health effects including immune system dysfunction, damage to the brain, nerve system, kidney, and other organs (1, 2). The adverse effects are not only limited to human beings, but also to other species in the environment, such as those in the ocean (3). For example, the accumulated Hg^{2+} in fish and

Sonia M. Tiquia-Arashiro (ed.), *Molecular Biological Technologies for Ocean Sensing*, Springer Protocols Handbooks,
DOI 10.1007/978-1-61779-915-0_6,

other marine species can effectively pass through the food chain to reach human beings. The release of heavy metals into water sources is a result of both natural processes and human activities. The largest body of water is sea water. The high saline content of the oceans makes the analytical chemistry even more challenging since the metal speciation under such conditions is very different from that in other water sources (4, 5).

Traditional methods for metal analysis in water rely heavily on analytical instruments, such as atomic absorption and atomic emission spectroscopy. While high sensitivity and simultaneous detection of multiple metal ions can be achieved, such instruments are only available in centralized laboratories, making on-site and real-time detection difficult. In addition, sample pretreatment, skilled operators, and high operation costs are also associated with these detection methods. As a result, the development of chemical and biological sensors for heavy metal detection becomes highly attractive (6, 7).

1.1. DNA for Heavy Metal Detection

Given the similar charge and size of metal ions, the rational design of highly selective metal sensor is difficult. While several successful examples have been reported, the generality still remains to be demonstrated (2, 8, 9). An interesting recent advancement is the use of DNA for metal recognition (10, 11). For a long time, DNA has been only known for its genetic function. With the development of chemical and biological tools, it becomes possible to explore new functions of DNA that does not exist in nature. For example, catalytic DNA or DNAzymes are DNA sequences that act as catalysts (12). All known DNAzymes are isolated using a combinatorial technique known as in vitro selection. For these DNAzymes to work, divalent metal ions are usually required. As a result, it is now possible to obtain DNAzymes that work only in the presence of specific metal ions. Using this method, many highly selective and sensitive sensors for metal ions such as Pb^{2+}, Hg^{2+}, Cu^{2+}, and UO_2^{2+} have been reported (13–16).

In addition to being a catalyst, certain DNA sequences can also selectively bind to other molecules without exerting catalytic reactions. These binding DNAs are called aptamers (17). Using a similar combinatorial selection method, DNA aptamers for Zn^{2+} have been recently isolated (18). Several highly specific metal/DNA interactions have also been recently discovered. For example, Hg^{2+} can be selectively chelated by two thymine bases to turn a T-T mismatch into T-Hg^{2+}-T base pair (see Fig. 1c), whose stability is even slightly higher than that of T-A base pairs (19, 20). The high selectivity of this reaction stems from the fact that no other metal ion can perform the same recognition. Similarly, Ag^+ can form a C-Ag^+-C base pair (21), and K^+ or Pb^{2+} can be recognized by the formation of G-quadruplex (22, 23). Therefore, the rich but still under-explored bioinorganic chemistry of nucleic acids provides us a unique opportunity to obtain highly effective metal sensors.

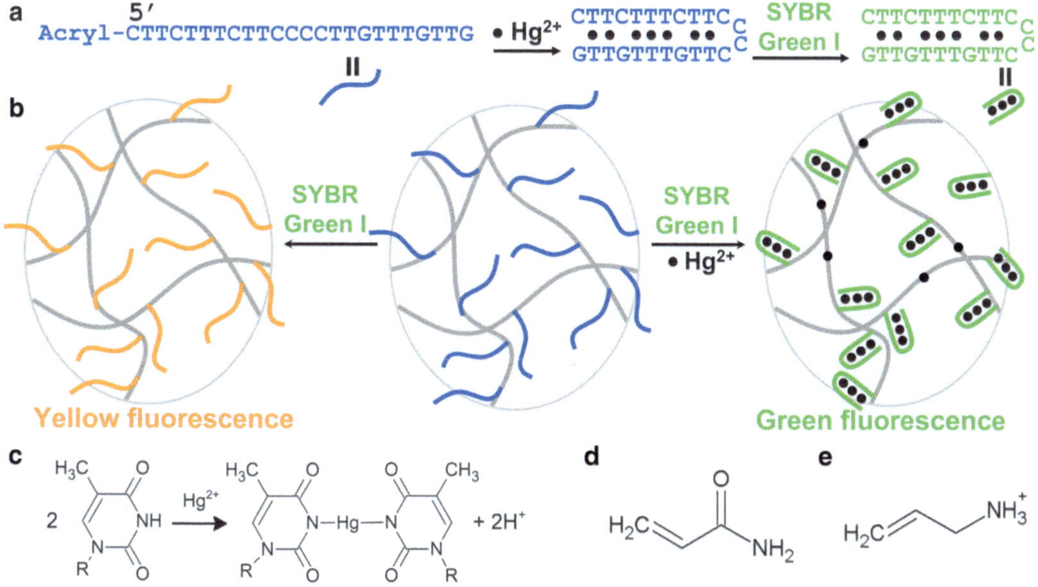

Fig. 1. The DNA sequence of acrydite-Hg-DNA and the fluorescence signal generation in the presence of Hg^{2+} after the addition of SYBR Green I. The 5′-end is modified with an acrydite group for hydrogel attachment (**a**). Covalent DNA immobilization in a polyacrylamide hydrogel and the interaction with Hg^{2+} and SYBR Green I produces a visual fluorescence signal (**b**). The chemical reaction scheme of Hg^{2+} binding with thymine base pairs (**c**), as well as, structures of acrylamide, (**d**) and allylamine (**e**) monomers. Copyright 2010 American Chemical Society. Reproduced from Dave et al., 2010 (34) with permission from American Chemical Society.

1.2. Detection of Hg^{2+}

There are many sensors that can effectively detect Hg^{2+} with either a fluorescence or color change (2, 24); most are based on small molecule chelators. Recently, the number of sensors for Hg^{2+} detection based on the thymine base is growing very rapidly. As mentioned above, this is due to the highly specific thymine/Hg^{2+} interaction with a high binding affinity. A large number of fluorescent (20, 25–27), colorimetric (28, 29), and electrochemical sensors (30) have been designed. In some cases, detection limits in the lower nanomolar have allowed such sensors to be used for Hg^{2+} detection in drinking water (toxic level = 10 nM mercury or 2 parts-per-billion as reported by the US Environmental Protection Agency [US EPA]). However, most of the reported sensors are based on DNA probes freely dispersed in solution; for practical applications, immobilized sensors are preferred.

1.3. Sensor Design

To design a biosensor, a signal transduction mechanism needs to be coupled to the target recognition element. Since DNA is a highly predictable molecule, it has allowed a large degree of versatility in terms of sensor design (10). Our aim is to achieve a cost-effective sensor with minimal use of analytical instruments. At the same time, high sensitivity needs to be maintained. Therefore, a design first reported by the Liu group was employed (27). A rich thymine DNA sequence shown in Fig. 1a contains seven hypothetic

Hg^{2+}-binding sites. In the absence of Hg^{2+}, the DNA sequence adopts a random coil structure in water and binds weakly with SYBR Green I (SG) through electrostatic interactions. SG is a commonly used dye for staining double-stranded DNA and the quantum yield of SG is close to zero in the absence of DNA. At low DNA and SG concentrations (e.g., 15 nM DNA and 90 nM SG), the SG fluorescence is close to zero if no Hg^{2+} is present. The addition of Hg^{2+} quickly folds the DNA into a hairpin structure and SG can strongly bind to the duplex region through minor groove binding and intercalation to emit an intense green fluorescence. This design is attractive because of its label-free nature; there is no covalent DNA modification.

To visually observe the fluorescence, however, much higher DNA and SG concentrations are required. For example, if 1 μM DNA and 6 μM SG are mixed, intense yellow and green fluorescence can be respectively observed in the absence and presence of Hg^{2+} by the eye using a handheld UV lamp under the 365 nm excitation. To covalently attach the DNA to a polyacrylamide hydrogel, the 5′-end of the DNA is modified with an acrydite group that allows the DNA to be copolymerized with acrylamide and bis-acrylamide during gel formation.

1.4. Hydrogel Design

Hydrogels are crosslinked hydrophilic polymer networks. With the majority of the gel volume being water, immobilized biomolecules can retain their native structure and function (31–34). There are a number of advantages of using hydrogels for sensor immobilization. First, a hydrogel has a much larger loading capacity compared to planar surfaces because the immobilization occurs in 3D. This facilitates the generation of a strong signal for visual detection. Second, hydrogels are transparent with a very low optical background, making it easy to observe visual signals. Third, hydrogels can be readily dried and rehydrated, allowing long-term storage. This property is particularly useful for DNA-based sensors since denatured DNA can be readily renatured upon rehydration. Fourth, the gel matrix can protect the immobilized DNA from nuclease degradation. Fifth, immobilization also allows for sensor regeneration. Finally, the gel backbone property can be engineered to improve sensor performance.

We choose to use polyacrylamide gels since they can be conveniently prepared in aqueous solutions and acrydite-modified DNA can be copolymerized (monomer structure in Fig. 1d). In addition, a wide range of monomers are available to control the property of the gel. For example, by using allylamine, the gel backbone carries a positive charge. In this chapter, we describe the protocols for preparing DNA-functionalized hydrogels for visual fluorescence detection of Hg^{2+}. The various features of this system have been explored including the active adsorption of mercury for sensitive detection, hydrogel drying, rehydration, and regeneration. The effect of hydrogel charge on the detection is also discussed.

2. Reagents and Equipment

2.1. Hydrogel-Related Reagents

1. 40 % Acrylamide/*N*,*N'*-Methylenebisacrylamide 29:1 gel stock solution (VWR, CA-99500-938)
2. Ammonium persulphate (APS) (VWR, CA99501-026)
3. *N*,*N*,*N'*,*N'*-Tetramethylethylenediamine (TEMED) (VWR, CA99501-596)

2.2. DNA Samples

1. Acrydite-Hg-DNA (Acrydite-5′-CTTCTTTCTTCCCCTTG TTTGTTG-3′)

2.3. Metal Ions

1. Mercury(II) perchloride hydrate (99.998 %, Sigma-Aldrich, 529656-5G)
2. Copper(II) sulfate dihydrate (99.99 + %, Sigma-Aldrich, 467847-50G)
3. Zinc chloride (99.999 %, Sigma-Aldrich, 229997-10G)
4. Manganese(II) chloride tetrahydrate (99.99 %, Sigma-Aldrich, 203734-5G)
5. Cobalt(II) chloride hexahydrate (Sigma-Aldrich, C8661-25G)
6. Lead(II) acetate trihydrate (99.999 %, Sigma-Aldrich, 316512-5G)
7. Magnesium chloride hexahydrate (Mandel Scientific, Canada, AMR-0288-500G)
8. Calcium chloride dihydrate (Mandel Scientific, Canada, AMR-0556-500G)

2.4. Buffer and Salts

1. Sodium nitrate (>99 %, Mandel Scientific, Canada, AMR-0598-500G)
2. Tris(hydroxymethyl)aminomethane (Tris) (Mandel Scientific, Canada, AMR-0497-500G)
3. 4-(2-Hydroxyethyl)-1-piperazineethanesulfonic acid (Mandel Scientific, Canada, AMR-0511-1KG)
4. 4-(2-Hydroxyethyl)piperazine-1-ethanesulfonic acid sodium salt (Mandel Scientific, Canada, AMR-0485-500G)

2.5. Water

1. MilliQ water (18.2 MΩ Millipore Milli-Q system, Billerica, MA)
2. Lake Ontario water (Colonel Samuel Smith Park in Toronto, Ontario, Canada)

2.6. DNA Staining dye and Solvent

1. SYBR Green I (10,000× in dimethyl sulfoxide, Invitrogen, S-7585)
2. Dimethyl sulfoxide (DMSO) (VWR, CA99501-202)

2.7. Plastic Tubes

1. Microcentrifuge tubes (VWR, 10011-722)
2. 15 mL conical tubes (VWR, 89039-664)
3. 50 mL conical tubes (VWR, 89039-656)

2.8. Handheld UV Lamp

1. Dual wavelength 254/365 nm 4 W compact handheld UV lamp (Fisher Scientific, 95002112)

2.9. Digital Camera

1. Canon PowerShot SD 1200 IS

2.10. Gel Documentation System

1. Alpha Innotech FluorChem FC2

2.11. Fluorometer

1. PTI spectrofluorometer

2.12. Fluorometer Cuvette

1. Micro-cuvette for fluorometer (Starna, 3-4.45/SOG/4)
2. Ada ptor for micro-cuvette (Starna, FCA4)

2.13. UV Protection Goggle

1. Goggle for UV radiation protection (VWR, 33002-078)

2.14. Tube Shaker

1. Thermolyne Labquake tube shaker (Fisher Scientific, 13-687-10Q)

2.15. pH Meter

1. Denver Instrument UltraBasic benchtop pH meters (Fisher Scientific, 02-226-213)

2.16. Analytical Balance

1. Mettler Toledo (AL204) analytical balance

2.17. Weighing Boat

1. Antistatic polystyrene weigh boat (VWR, 89106-764)

2.18. Syringes and Syringe Filters

1. Nalgene PTFE syringe filters (0.2 μm, VWR, 28195-868)
2. BD 10 mL syringe (10 mL, VWR, WLBBD309604)

2.19. Mini-Centrifuge

1. Mini-centrifuge (Fisher Scientific, 05-090-100)

2.20. 96-Well Plate

1. 96-Well polypropylene microplates (VWR, 82050-636)

3. Protocol

3.1. Preparing Stock Solutions

3.1.1. Preparing 500 mM Tris Nitrate Buffer, pH 8.0

1. Weigh 2.42 g Tris base in a weighing boat transfer it to a 50 mL conical tube. Add ~30 mL water to dissolve it.
2. Add 12 M HNO_3 drop wise using a glass Pasteur pipette and monitoring the pH using a pH meter.

3. Stop the addition when pH reaches 8.0.
4. Fill with water to 40 mL.
5. Filter the buffer through a 0.22 μm syringe filter and store at 4 °C.

3.1.2. Preparing 2 M $NaNO_3$ Solution

1. Weigh 6.8 g $NaNO_3$ and dissolve in 40 mL water.
2. Filter the solution through 0.22 μm syringe filter and store at 4 °C.

3.1.3. Preparing Buffer A (8 mM Tris Nitrate, pH 8.0, 20 mM $NaNO_3$)

1. Transfer 720 μL of 500 mM Tris nitrate buffer, pH 8.0 to a 50 mL conical tube.
2. Transfer 450 μL of 2 M $NaNO_3$ to the conical tube.
3. Fill with MilliQ water up to 45 mL.

3.1.4. Prepare 10 mM HEPES Buffer, pH 7.6

1. Weigh 5.96 g HEPES-free acid and dissolve in 50 mL water to make 500 mM solution.
2. Weigh 6.51 g HEPES sodium salt and dissolve in 50 mL water to make 500 mM solution.
3. Mix 20 mL of each above solution in a 50 mL conical tube to achieve a pH 7.6 buffer.
4. Filter using a syringe filter.
5. Transfer 100 μL of the above buffer to a 15 mL conical tube and add MilliQ water to 10 mL to obtain 5 mM HEPES solution.

3.1.5. Preparing Hg^{2+} and Other Salt Solutions

1. Weigh ~100 mg $Hg(ClO_4)_2 \cdot xH_2O$ in a plastic weighing boat. Since this mercury salt is in the crystalline form, it is difficult to weigh the exact designated amount. Therefore, the final volume needs to be adjusted to make the appropriate concentration. Add 2 mL water and an orange color can be observed.
2. Add 12 M HNO_3 drop wise until the salt is completely dissolved and the solution becomes clear. Fill with water to a final volume of 5 mL to obtain an Hg^{2+} concentration of 50 mM.
3. Make subsequent tenfold dilutions in 10 mM HNO_3.
4. Other metal salt solutions are prepared in a similar way.

3.1.6. Preparing SG Solution

1. Mix 195 μL MilliQ water and 195 μL DMSO solution.
2. Add 10 μL 10,000× SG to prepare 250× SG (500 μM) and store it at −20 °C.

3.1.7. Preparing DNA Solution

1. Centrifuge the received lyophilized DNA powder using a mini centrifuge.
2. Add 5 mM HEPES buffer, pH 7.6 to dissolve the DNA so that the final DNA concentration 500 μM. Store the DNA at −20 °C until use.

3.2. Preparing Hydrogels

3.2.1. Preparing Initiator Solution

1. Weigh 50 mg APS in a plastic weighing boat using an analytical balance.
2. Add 0.5 mL MilliQ water to dissolve the APS. To this solution, add 25 μL TEMED in a fume hood. Transfer the mixed solution to a microcentrifuge tube (see Note 1).

3.2.2. Preparing Hydrogel Sol

For 4 % Polyacrylamide Gel, Mix the Following (75 μL for Each gel). Read the Following Steps in Sect. 3.2 Before Starting This Step

MilliQ water	104 μL
$NaNO_3$ (2 M)	15 μL
Tris nitrate buffer (pH 8.0, 0.5 M)	30 μL
29:1 Acrylamide:bisacrylamide (40 %)	30 μL
DNA (500 μM)	6 μL
Initiator solution (to be added the last)	15 μL

For 6 % Polyacrylamide/Allylamine Gel (50 μL for Each Gel)

MilliQ water	107.8 μL
$NaNO_3$ (2 M)	10 μL
Tris nitrate buffer (pH 8.0, 0.5 M)	20 μL
Allylamine (35 %)	9.6 μL
29:1 Acrylamide:bisacrylamide (40 %)	33.6 μL
Bisacrylamide (2.5 %)	4.8 μL
DNA (500 μM)	4.2 μL
Initiator solution (to be added the last)	10 μL

3.2.3. Gel Casting

1. After mixing all of the gel components, add the initiator solution last followed by immediate vortexing of the solution.
2. Transfer 75 μL aliquots of the acrylamide gel solution or 50 μL for the allylamine containing solution using a pipette to a wellt of the round bottom 96-well plate (see Note 2).

3.2.4. Gel Harvesting

1. Add 50 μL MilliQ water for lubrication to each of the wells containing the polymerized gels.
2. Wait for 2 min and use two 200 μL pipette tips to push each gel from the side into a 50 mL conical tube (see Note 3).

3.2.5. Gel Washing

1. Soak the gels in 50 mL conical tubes with 45 mL Millipore water on a rotating tube shaker for 6 h at room temperature.
2. Remove all of the water carefully using a pipette. Pay particular attention not to damage the gel in this process.
3. Fill the tube containing the gels with 45 mL of fresh MilliQ water again and soak the gels overnight (see Note 4).

4. The next day, remove the water and place the gels in 1.7, 15, or 50 mL microcentrifuge tubes.

3.3. Hg^{2+} Detection in Gel

Hg^{2+} detection can be achieved in one of the following formats. For example, Hg^{2+} and SG can be added at the same time to soak the gel. Alternatively, Hg^{2+} can be added first in a large volume to soak the gel for a long time (e.g., overnight). This is to enrich Hg^{2+} inside the gel since the DNA has a very high binding affinity. Therefore, even very low concentration of Hg^{2+} can be detected.

3.3.1. Adding Hg^{2+} and SG at the Same Time

1. Place each gel in a 1.7 mL microcentrifuge tube.
2. Add 1 mL of buffer A.
3. Add 1 μL of 1 mM $Hg(ClO_4)_2$ solution to achieve a final 1 μM Hg^{2+} concentration.
4. Add 2 μL of 500 μM SG and soak the gel for 1 h on the tube shaker.

3.3.2. Adding Hg^{2+} First

1. Fill a 50 mL conical tube with buffer A. Add $Hg(ClO_4)_2$ solution to the desired final Hg^{2+} concentration. Soak the gel on a shaker in the dark overnight to allow Hg^{2+} to diffuse into the gel and bind with the DNA.
2. Recover the gel and transfer it into a 1.7 mL microcentrifuge tube. Add 1 mL fresh buffer A and 2 μL 500 μM SG.
3. Soak the gel for 1 h at room temperature on a tube shaker in the dark.
4. Observe the gel using the handheld UV lamp in a dark room under the excitation of 365 nm. Place the lamp at a distance of ~10 cm from the gel (see Note 5).
5. Record the gel fluorescence using a digital camera.
6. Transfer the gel into a gel documentation system for quantification. Use the 365 nm excitation and the SG filter for collecting the emission light into the CCD camera.

3.4. Hg^{2+} Detection in Solution

1. Add final concentration of 15 nM DNA and 90 nM SG in 500 μL buffer A.
2. Transfer the solution into a quartz micro-cuvette.
3. Scan fluorescence emission from 500 to 650 nm by exciting at 485 nm (see Note 6).
4. Add 2.25 μL of 200 μM Hg^{2+} to make the final Hg^{2+} concentration 90 nM. Vortex the cuvette immediate after the addition of Hg^{2+}.
5. Scan the fluorescence spectrum again using the same settings.
6. Alternatively, add 1 μL of 500 μM DNA and 6 μL of 500 μM SG to 493 μL buffer A. Measure fluorescence in the absence of Hg^{2+}. Add 2 μL 1 mM Hg^{2+} to make a final of 4 μM Hg^{2+} and

measure the fluorescence again. Alternatively, take 100 μL aliquots of this mixture into 0.6 mL microcentrifuge tubes and add varying concentrations of Hg^{2+}. Observe the fluorescence using the handheld UV lamp in a dark room with the 365 nm excitation.

3.5. Hydrogel Drying and Rehydration

1. Soak the gels in 1 mL MilliQ water for 1 h. Replace with fresh water and soak for another hour.
2. Transfer the gels onto a plastic weighing boat. Place the round side of the gel in contact with the weighing boat surface. Otherwise, the gel dries to form a film sticking to the plastic surface.
3. Leave the gels overnight at room temperature to allow complete drying. This is confirmed by weighing the dried gels. The mass of each gel before drying is ~80 mg. After drying, the mass is reduced to 3–4 mg.
4. To rehydrate the gel, soak the dried gels in buffer A for 3 h at room temperature. The gel mass can be recovered to the original value and the gels are then ready for Hg^{2+} detection.

3.6. Hydrogel Regeneration

After incubation with SYBR Green I and Hg^{2+}, the hydrogels show green fluorescence. To regenerate hydrogel, both Hg^{2+} and SG need to be removed.

1. Soak the gels in 1 mL of 1 % HCl for 3 min.
2. Discard the HCl solution using a pipette and wash the gels with 10 mL of MilliQ water for 1 min and then soak in 10 mL buffer A for 20 min.
3. Soak the gels again in 1 mL of 1 % HCl followed by washing. This process is repeated five times. After the last soaking in buffer A, soak an additional hour in 10 mL buffer A. After that, the gels can be observed using a handheld UV lamp to ensure no fluorescence is observed, and these gels are used for Hg^{2+} detection.

3.7. Detection of Hg^{2+} in Lake Ontario Water Samples

ICP-MS analysis showed no detectable mercury in the collected Lake Ontario water samples. Therefore, $Hg(ClO_4)_2$ was added to simulate contaminated water.

1. Transfer 15 mL of the collected water samples into 15 mL conical tubes.
2. Add $Hg(ClO_4)_2$ solution to make appropriate Hg^{2+} concentration in the lake water.
3. Add one hydrogel to each of the tubes and soak for 1 day to allow Hg^{2+} binding.
4. Transfer the gels into 1.5 mL microcentrifuge tubes and add 1 mL buffer A containing 1 μM SYBR. After 1 h, image the gels using a digital camera.

4. Typical Protocol Results

4.1. Detecting Hg^{2+} Using the Non-Immobilized Sensor

Before attaching DNA to the hydrogel, Hg^{2+} detection in solution is tested. This is to ensure that all of the reagents and buffers work. With 1 μM DNA and 6 μM SG, a yellow fluorescence can be observed using a handheld UV lamp under the 365 nm excitation as shown in the inset of Fig. 2b. In the presence of 4 μM Hg^{2+}, a bright green fluorescence can also be observed. This is a quick test of our system. To quantitatively understand the fluorescence change, a fluorometer can be used for measuring the spectra of these samples. By exciting the sample at 485 nm, the emission peaks at 521 and 526 nm appear for the samples with and without Hg^{2+}, respectively (Fig. 2b). Even though the wavelength shift is only 5 nm, the fluorescence color has completely changed. Using a lower DNA and SG concentration, as shown in Fig. 2a, the fold of fluorescence increase is much larger. After confirming that the system works in the solution phase, the next step is to immobilize the DNA in the hydrogel.

4.2. Detecting Hg^{2+} Using Polyacrylamide Hydrogel

The DNA-functionalized polyacrylamide hydrogels are transparent. Since the gels are prepared in a round-bottom 96-well plate, they are hemispherical. After adding 1 μM Hg^{2+} and 1 μM SG to hydrogels soaked in 1 mL buffer A, a gradual increase in green fluorescence can be observed with increasing Hg^{2+} concentration. As shown in Fig. 3a, at least 200 nM Hg^{2+} is required for visual detection. For quantitative analysis, a gel documentation system is used. The gels

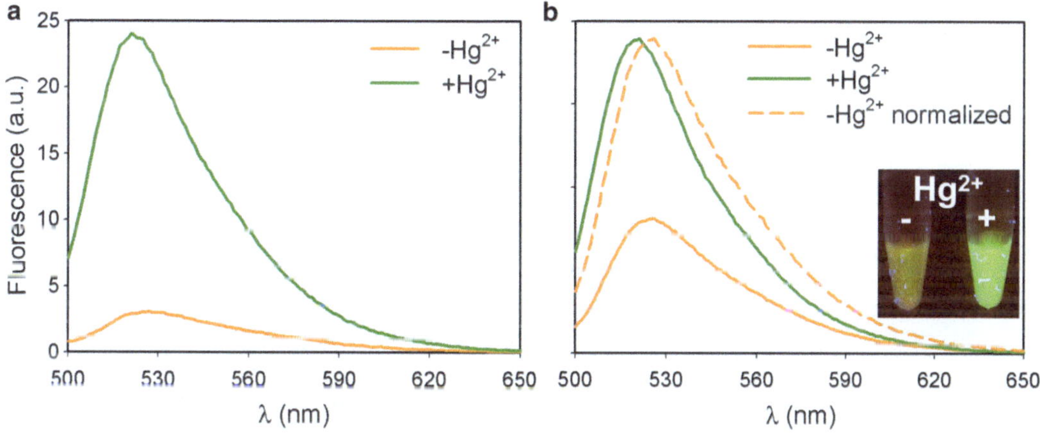

Fig. 2. Fluorescence spectra of SYBR green I and Hg-DNA free in solution where the Hg-DNA concentration is 15 nM (**a**) and 1 μM (**b**). SYBR green I dye and DNA ratio is maintained at 6:1 for both cases. The Hg^{2+} concentration is 90 nM in (**a**) and 4 μM in (**b**) and the inset shows a photograph for 1 μM DNA with and without 4 μM Hg^{2+} excited at 365 nm using a handheld UV lamp. The normalized curve in (**b**) is achieved by multiplying the yellow curve by a factor such that the same peak intensity as the green curve is obtained. Copyright 2010 American Chemical Society. Reproduced from Dave et al., 2010 (34) with permission from American Chemical Society.

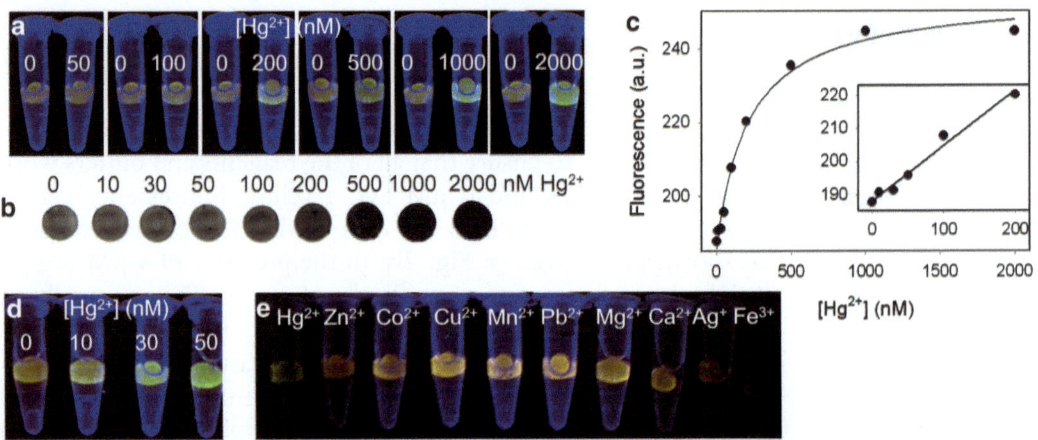

Fig. 3. Sensor sensitivity documented using a digital camera (**a**) and a fluorescence gel documentation system (**b**) and its quantification (**c**) using 1 mL of water sample. (**d**) Sensitivity test using 50 mL water sample. (**e**) Shows the selectivity test of the hydrogels with 1 μM of various metal ions in 1 mL samples. Copyright 2010 American Chemical Society. Reproduced from Dave et al 2010 (34) with permission from American Chemical Society.

are excited at 365 nm and the emission is collected using a CCD camera through a green filter (Fig. 3b). The plot of fluorescence intensity versus Hg^{2+} concentration is shown in Fig. 3c. The intensity initially increased linearly with Hg^{2+} concentration (inset) and saturated at ~1 μM Hg^{2+}. The detection limit is determined to be 75 nM based on the Hg^{2+} concentration required to generate a signal greater than three times the standard deviation of the background noise. Since the wavelength shift is only ~5 nm, it is difficult to use a filter to separate the green and yellow fluorescence spectra. Therefore, only the intensity of fluorescence is measured. Due to the intense yellow background fluorescence in the absence of Hg^{2+}, the final fluorescence intensity increase is only ~30 % with a saturating concentration of Hg^{2+}.

An important property of this acrylamide hydrogel is its ability to actively adsorb Hg^{2+}. The amide nitrogen in the gel backbone has a binding affinity for Hg^{2+} and the gel acts as a selective sponge to concentrate Hg^{2+} inside the gel. Therefore, unlike most sensors whose response is limited by the target concentration, the hydrogel-based sensor should have higher sensitivity by simply increasing the sample volume. For example, after soaking the gels in 50 mL buffer A (previously in 1 mL) containing varying concentrations of Hg^{2+}, as shown in Fig. 3d, even 10 nM Hg^{2+} (the toxic level in drinking water) shows a visible green fluorescence and the sample containing 30 nM Hg^{2+} is highly fluorescent green. This sensitivity is among the highest one reported for Hg^{2+} sensors where no analytical instruments or signal amplification methods are used for detection purposes. If the gel is incubated with 1 μM of other metal ions, only yellow fluorescence can be observed (Fig. 3e), suggesting that high selectivity of the DNA is still maintained within the hydrogel matrix.

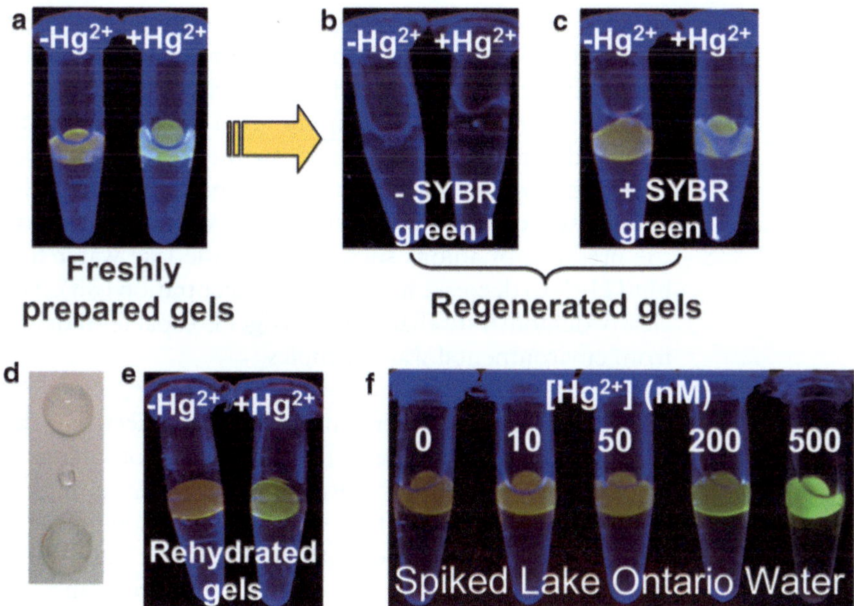

Fig. 4. The test of freshly prepared (**a**) and regenerated (**b**, **c**) hydrogels. (**d**) A photo of freshly prepared (*top*), dried (*middle*), and rehydrated (*bottom*) gels. (**e**) Detection of Hg^{2+} with a rehydrated hydrogel sensor. (**f**) Detection of Hg^{2+} in spiked Lake Ontario water samples. Reproduced from Dave et al., 2010 (34) with permission from American Chemical Society.

4.3. Hydrogel Regeneration, Drying, and Rehydration

Immobilized sensors may allow regeneration by washing away target molecules under denaturing conditions. Since the DNA is immobilized in 3D in a hydrogel, it takes a longer time for the content inside the gel to diffuse out. We employ an acid treatment method for regeneration. After incubating Hg^{2+} and SG-treated gels (Fig. 4a) with 1 % HCl for 3 min, the gels are transferred into buffer A. This process is repeated five times. As shown in Fig. 4b, the hydrogels become nonfluorescent after regeneration, suggesting that both Hg^{2+} and the dye have been removed. The addition of Hg^{2+} and SG to these regenerated gels regains the sensor response (Fig. 4c). This experiment confirms that it is possible to achieve regeneration for the hydrogel immobilized sensor.

The majority of the gel volume is occupied by water. Therefore, another feature of hydrogel-immobilized sensor is the possibility of drying. Drying provides a convenient means for long term gel storage, DNA protection, and ease of handling. The gels can be dried under ambient conditions and the dry mass is ~4 % of the fully hydrated gel mass. The dried gels can be easily rehydrated by soaking in buffer A to the original volume (Fig. 4d). These rehydrated gels can still effectively detect Hg^{2+} (Fig. 4e).

4.4. Detecting Hg^{2+} in Lake Ontario Water

To evaluate whether the hydrogel-based sensor is capable of detecting and removing Hg^{2+} from environmental water samples, samples from Lake Ontario are tested. Since these water samples do not contain Hg^{2+} as determined by ICP-MS, Hg^{2+} is deliberately added to simulate contaminated water. After soaking each gel in a 15 mL conical tube with no additional salt or buffer overnight, the gels are transferred into 1 mL of buffer A with 1 μM SG. After 1 h, the gels are imaged. As shown in Fig. 4f, a weak green fluorescence is observed for 50 nM Hg^{2+} and an intense green fluorescence is observed for 200 nM Hg^{2+}. This sensitivity is slightly lower in comparison to that obtained in buffer A. This may be attributed to the presence of anions such as Cl^- in the lake water that can also bind Hg^{2+} to decrease its effective concentration (35). These results clearly demonstrate that our hydrogel is capable of detecting Hg^{2+} from environmental water samples.

4.5. Detecting Hg^{2+} Using Allylamine/Polyacrylamide Hydrogel

While the above polyacrylamide gel can effectively detect Hg^{2+} based on a yellow-to-green fluorescence color change, the background fluorescence is quite high, disallowing high sensitivity using the gel documentation system. The reason for the high background can be attributed to the electrostatic interaction between DNA and SG. At high DNA concentration, SG can bind to DNA to emit yellow fluorescence. Therefore, by introducing a fraction of cationic monomer (e.g., allylamine shown in Fig. 1e), it may be possible to promote DNA/gel interaction and reduce DNA/SG interaction therefore reducing the background fluorescence (36). The schematics of Hg^{2+} detection in a cationic gel is shown in Fig. 5a. With the hydrogel containing 20 % allylamine/80 % acrylamide and a sample volume of 50 mL, the background fluorescence can be significantly reduced (Fig. 5b). The increase of fluorescence intensity can reach ~300 % (Fig. 5d, black dots). A detection limit of 1.1 nM can be achieved using this gel formulation. Under the same condition but using the neutral polyacrylamide gel, very high background fluorescence is still observed (Fig. 5c) and the fluorescence increase is only 55 % (Fig. 5d, red triangles).

5. Notes

1. APS is a hygroscopic powder and its activity is reduced after absorbing water. Make sure that the container is capped tightly. The weighed APS can be slightly deviated from 50 mg, as long as the amount of water and TEMED are adjusted accordingly. The initiator solution needs to be prepared fresh daily.
2. After adding the initiator, the solution needs to be immediately mixed to avoid high local initiator concentration. The step of transferring the sol into a 96-well plate needs to be performed in less than 2 min to avoid gel formation before completing the transfer step. Leave the plate in a drawer for 1 h at room temperature (~25 °C) for polymerization.

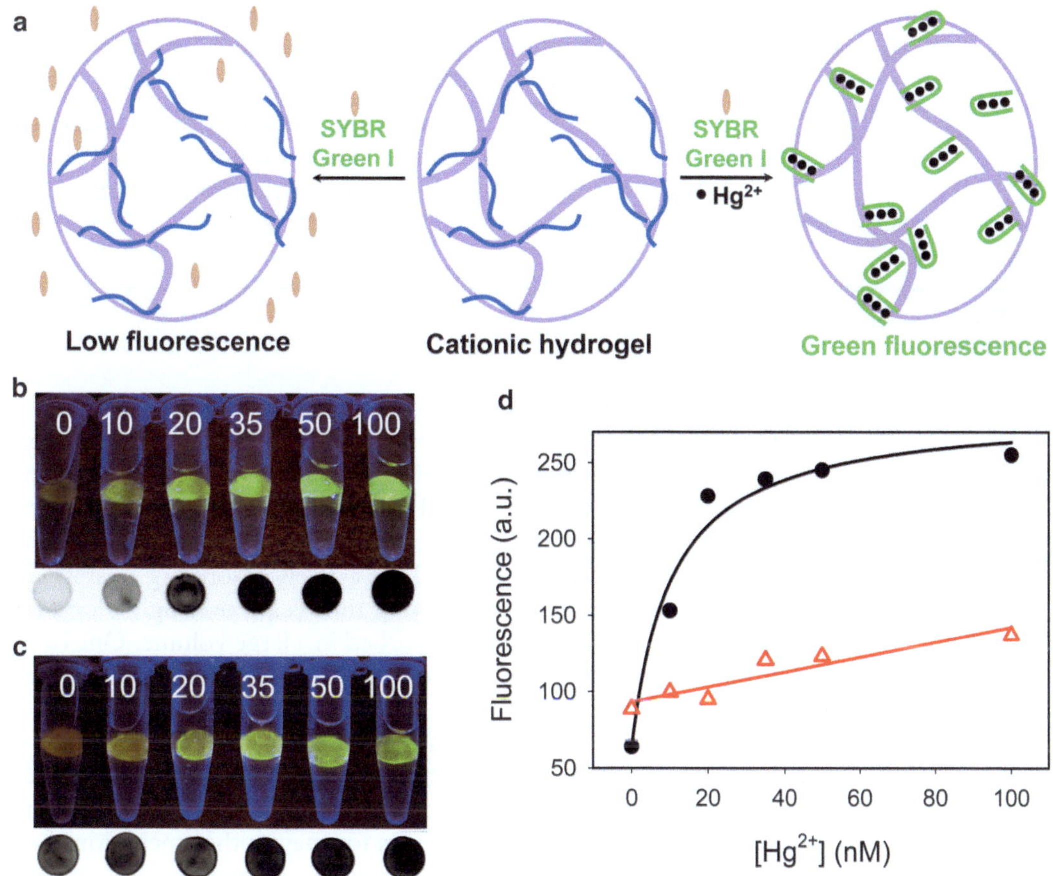

Fig. 5. Schematic presentation of covalent immobilization of DNA in a positively charged hydrogel (containing allylamine) (**a**). In this gel, the DNA interacts more with the gel backbone in the absence of Hg^{2+} and the diffusion of SG into the gel is retarded, giving rise to low background fluorescence. Hydrogel mercury sensor sensitivity obtained using a digital camera (the *top panels*) and a gel documentation system (the *lower panels*) for cationic (20 % allylamine) (**b**) and neutral (**c**) hydrogels. (**d**) Responses of the sensors are quantified using the gel documentation system. Cationic gel: *black dots*; neutral gel: *red triangles*. Reproduced from Joseph et al., 2010 (36) with permission from American Chemical Society.

3. To avoid damaging the gel, do not pinch the gel from the center. It should be quite easy to harvest 4–8 % gels since they are soft and can be reversibly deformed. Higher percentage gels (e.g., 20 %) can break more easily during this process. The polymerization should not take longer than 1 h. Exposure of the gel to air for too long can result in gel drying.

4. Gel soaking is important for Hg^{2+} detection because it removes initiators, free monomers, and DNA. The initiator contains ammonium persulfate and TEMED, which can bind to Hg^{2+} interfering with the fluorescence signal. Free monomers and non-incorporated DNA also need to be removed to obtain consistent results. These chemicals can be removed by repeated soaking the gels in ample amounts of water. The amount of incorporated DNA is estimated to be 50–60 % and the rest is washed away.

5. Hg^{2+} is a toxic metal ion and a waste container should be prepared for discarding solutions containing Hg^{2+}. Wear the UV protection goggles to observe the fluorescence under the UV lamp.
6. It may be necessary to adjust the slit width of the excitation and emission light so that the detector is not saturated at high DNA/SG concentrations when using the fluorometer.

6. Future Directions

The hydrogel-immobilized DNA sensor has showed several important advantages for Hg^{2+} detection in water samples. For example, ultrahigh sensitivity with visual detection can be achieved and the gel can be dried for storage and rehydrated. Sensor regeneration can also be achieved using a simple acid treatment. The high sensitivity comes from the fact that the gel can actively adsorb Hg^{2+}. At its current form, however, it takes a long time (e.g., overnight) for a low concentration of Hg^{2+} to be adsorbed in a large volume. One of the future directions is therefore to make the adsorption kinetics faster. This can be potentially achieved by using a flow through system.

In addition to the detection of Hg^{2+}, hydrogels responsive to other heavy metal ions also need to be demonstrated. The eventual goal is to construct a sensor array. Although DNAzyme and aptamer-based fluorescent sensors for many other metal ions have been demonstrated in solution, covalent immobilization of these sensors in hydrogel is still very challenging. This is because majority of such sensors are designed to contain several DNA pieces annealed together. It is highly desirable to have just a single DNA strand to contain all of the necessary components. Otherwise, the sensor fabrication and response could be adversely affected. With one of the ends being occupied by sensor immobilization, it is difficult to add all the other modifications and still keeping the cost and yield at an acceptable level. Therefore, innovative sensor design methods are required.

In addition, the gels need to be tested more extensively in different water samples. The hydrogel-based Hg^{2+} detection has been tested in a pure buffer and in Lake Ontario water. The sensitivity in lake water is slightly lower compared to that in the pure buffer, which is attributed to the presence of interfering anions such as Cl^-. In the case of samples with extremely high salt, ocean water represents an even greater challenge. Under such conditions, the different speciation of metal ions needs to be considered. For example, it has been reported that a significant fraction of Hg^{2+} in sea water is present as the $HgCl_3^-$ and $HgCl_4^{2-}$ (5). The overall Hg^{2+} concentration is in the sub-nM to low nM. Therefore, it is often necessary to perform pre-concentration steps before determination by spectroscopic techniques. Since the hydrogel can actively adsorb Hg^{2+}, which is effectively a pre-concentrating step, the chemistry and physics of hydrogels may allow researchers to overcome these challenges.

Finally, from the engineering perspective, the hydrogel-based sensor can be integrated into an autonomous platform in the ocean and other water bodies. The gel is optically transparent with immobilized DNA sensors being protected, allowing long-term monitoring after incorporation into a device. A remote signaling mechanism can be designed so that if a certain Hg^{2+} level has reached, an electromagnetic signal will be sent out. If this can be realized, a profound impact on environmental monitoring can be envisioned using this hydrogel-based technology.

References

1. Jarup L (2003) Hazards of heavy metal contamination. Br Med Bull 68:167–182
2. Nolan EM, Lippard SJ (2008) Tools and tactics for the optical detection of mercuric ion. Chem Rev 108:3443–3480
3. Prego R, Cobelo-Garcia A (2004) Cadmium, copper and lead contamination of the seawater column on the Prestige shipwreck (NE Atlantic Ocean). Anal Chim Acta 524:23–26
4. Mason RP, Fitzgerald WF (1990) Alkylmercury species in the equatorial pacific. Nature 347: 457–459
5. Han F, Shan XQ, Zhang SZ, Wen B (2004) Mercury speciation in China's coastal surface seawaters. Int J Environ Anal Chem 84: 583–598
6. Oehme I, Wolfbeis OS (1997) Optical sensors for determination of heavy metal ions. Mikrochim Acta 126:177–192
7. Wernette DP, Liu JW, Bohn PW, Lu Y (2008) Functional-DNA-based nanoscale materials and devices for sensing trace contaminants in water. MRS Bull 33:34–41
8. Que EL, Domaille DW, Chang CJ (2008) Metals in neurobiology: probing their chemistry and biology with molecular imaging. Chem Rev 108:1517–1549
9. Jiang P, Guo Z (2004) Fluorescent detection of zinc in biological systems: recent development on the design of chemosensors and biosensors. Coord Chem Rev 248:205–229
10. Liu J, Cao Z, Lu Y (2009) Functional nucleic acid sensors. Chem Rev 109:1948–1998
11. Zhang X-B, Kong R-M, Lu Y (2011) Metal ion sensors based on DNAzymes and related DNA molecules. Annu Rev Anal Chem 4:105–128
12. Breaker RR (1997) DNA aptamers and DNA enzymes. Curr Opin Chem Biol 1:26–31
13. Li J, Lu Y (2000) A highly sensitive and selective catalytic DNA biosensor for lead ions. J Am Chem Soc 122:10466–10467
14. Hollenstein M, Hipolito C, Lam C, Dietrich D, Perrin DM (2008) A highly selective DNAzyme sensor for mercuric ions. Angew Chem Int Ed 47:4346–4350
15. Liu J, Lu Y (2007) A DNAzyme catalytic beacon sensor for paramagnetic Cu2+ ions in aqueous solution with high sensitivity and selectivity. J Am Chem Soc 129:9838–9839
16. Liu J, Brown AK, Meng X, Cropek DM, Istok JD, Watson DB, Lu Y (2007) A catalytic beacon sensor for uranium with parts-per-trillion sensitivity and millionfold selectivity. Proc Natl Acad Sci U S A 104:2056–2061
17. Wilson DS, Szostak JW (1999) In vitro selection of functional nucleic acids. Annu Rev Biochem 68:611–647
18. Rajendran M, Ellington AD (2008) Selection of fluorescent aptamer beacons that light up in the presence of zinc. Anal Bioanal Chem 390: 1067–1075
19. Miyake Y, Togashi H, Tashiro M, Yamaguchi H, Oda S, Kudo M, Tanaka Y, Kondo Y, Sawa R, Fujimoto T et al (2006) MercuryII mediated formation of thymine-HgII-thymine base pairs in DNA duplexes. J Am Chem Soc 128:2172
20. Ono A, Togashi H (2004) Molecular sensors: highly selective oligonucleotide-based sensor for mercury(II) in aqueous solutions. Angew Chem Int Ed 43:4300–4302
21. Ono A, Cao S, Togashi H, Tashiro M, Fujimoto T, Machinami T, Oda S, Miyake Y, Okamoto I, Tanaka Y (2008) Specific interactions between silver(I) ions and cytosine-cytosine pairs in DNA duplexes. Chem Comm 39:4825–4827
22. Ueyama H, Takagi M, Takenaka S (2002) A novel potassium sensing in aqueous media with

a synthetic oligonucleotide derivative. Fluorescence resonance energy transfer associated with guanine quartet-potassium ion complex formation. J Am Chem Soc 124:14286–14287

23. Li T, Dong S, Wang E (2010) A lead(II)-driven DNA molecular device for turn-on fluorescence detection of lead(II) ion with high selectivity and sensitivity. J Am Chem Soc 132: 13156–13157
24. Nolan EM, Lippard SJ (2003) A "turn-on" fluorescent sensor for the selective detection of mercuric ion in aqueous media. J Am Chem Soc 125:14270–14271
25. Liu J, Lu Y (2007) Rational design of "turn-on" allosteric DNAzyme catalytic beacons for aqueous mercury ions with ultrahigh sensitivity and selectivity. Angew Chem Int Ed 46: 7587–7590
26. Wang Z, Lee JH, Lu Y (2008) Label-free colorimetric detection of lead ions with a nanomolar detection limit and tunable dynamic range by using gold nanoparticles and DNAzyme. Adv Mater 20:3263–3267
27. Wang J, Liu B (2008) Highly sensitive and selective detection of Hg2+ in aqueous solution with mercury-specific DNA and Sybr Green I. Chem Comm:4759–4761
28. Lee J-S, Han MS, Mirkin CA (2007) Colorimetric detection of mercuric ion (Hg2+) in aqueous media by DNA-functionalized gold nanoparticles. Angew Chem Int Ed 46: 4093–4096
29. Li D, Wieckowska A, Willner I (2008) Optical analysis of Hg2+ ions by oligonucleotide-gold-nanoparticle hybrids and DNA-based machines. Angew Chem Int Ed 47:3927–3931
30. Liu S-J, Nie H-G, Jiang J-H, Shen G-L, Yu R-Q (2009) Electrochemical sensor for mercury(II) based on conformational switch mediated by interstrand cooperative coordination. Anal Chem 81:5724–5730
31. Peppas NA, Hilt JZ, Khademhosseini A, Langer R (2006) Hydrogels in biology and medicine: from molecular principles to bionanotechnology. Adv Mater 18:1345–1360
32. Um SH, Lee JB, Park N, Kwon SY, Umbach CC, Luo D (2006) Enzyme-catalysed assembly of DNA hydrogel. Nat Mater 5:797–801
33. Yang HH, Liu HP, Kang HZ, Tan WH (2008) Engineering target-responsive hydrogels based on aptamer: target interactions. J Am Chem Soc 130:6320–6321
34. Dave N, Huang P-JJ, Chan MY, Smith BD, Liu J (2010) Regenerable DNA-functionalized hydrogels for ultrasensitive, instrument-free mercury(II) detection and removal in water. J Am Chem Soc 132:12668–12673
35. Dove A (2009) Long-term trends in major ions and nutrients in Lake Ontario. Aquat Ecosyst Health Manage 12:281–295
36. Joseph KA, Dave N, Liu J (2011) Electrostatically directed visual fluorescence response of DNA-functionalized monolithic hydrogels for highly sensitive Hg2+ detection. ACS Appl Mater Inter 3:733–739

Chapter 7

The Use of Peptide Nucleic Acids in Surface Plasmon Resonance for Detection of Red Tide Algae

Amber R. Bratcher and Laurie B. Connell

Abstract

There is a need for low-cost, rapid, and accurate detection of harmful organisms. Some species of the marine dinoflagellate *Alexandrium* produce paralytic shellfish toxins that can accumulate in shellfish tissue and can cause paralysis and death if ingested by humans. We have developed a method for RNA detection of *Alexandrium* using a portable surface plasmon resonance biosensing instrument and peptide nucleic acid probes. Because our method is user-friendly, cost-effective, and yields quick results, it will greatly improve the abilities of monitoring programs to detect harmful species and take appropriate measures to prevent detrimental human health effects.

Key words: Red tide, Harmful algal blooms, Surface plasmon resonance, Peptide nucleic acids, Biosensors, *Alexandrium*, Paralytic shellfish toxins, Ocean sensing, Bloom detection, Dinoflagellates

1. Introduction

Paralytic shellfish poisoning (PSP) is caused by consumption of shellfish that have fed on toxic algae and is a major health issue worldwide (1). Even at densities as low as 100–200 cells/L, dinoflagellates of the genus *Alexandrium* can produce enough paralytic shellfish toxins (PSTs) to be of public health concern before the water discoloration often associated with harmful algal blooms (HABs) is evident (2, 3). Detection of blooms during early stages is extremely important for human health issues and maintenance of healthy fisheries since HABs can lead to adverse effects on shellfish, zooplankton, coral, seabirds, and marine mammal populations through bioaccumulation of toxin within the food web and ecosystem disruption (2, 4–7). *Alexandrium* species that produce PSTs are difficult to distinguish morphologically from non-PST producing species, and current identification methods are expensive, time-consuming,

Sonia M. Tiquia-Arashiro (ed.), *Molecular Biological Technologies for Ocean Sensing*, Springer Protocols Handbooks,
DOI 10.1007/978-1-61779-915-0_7,

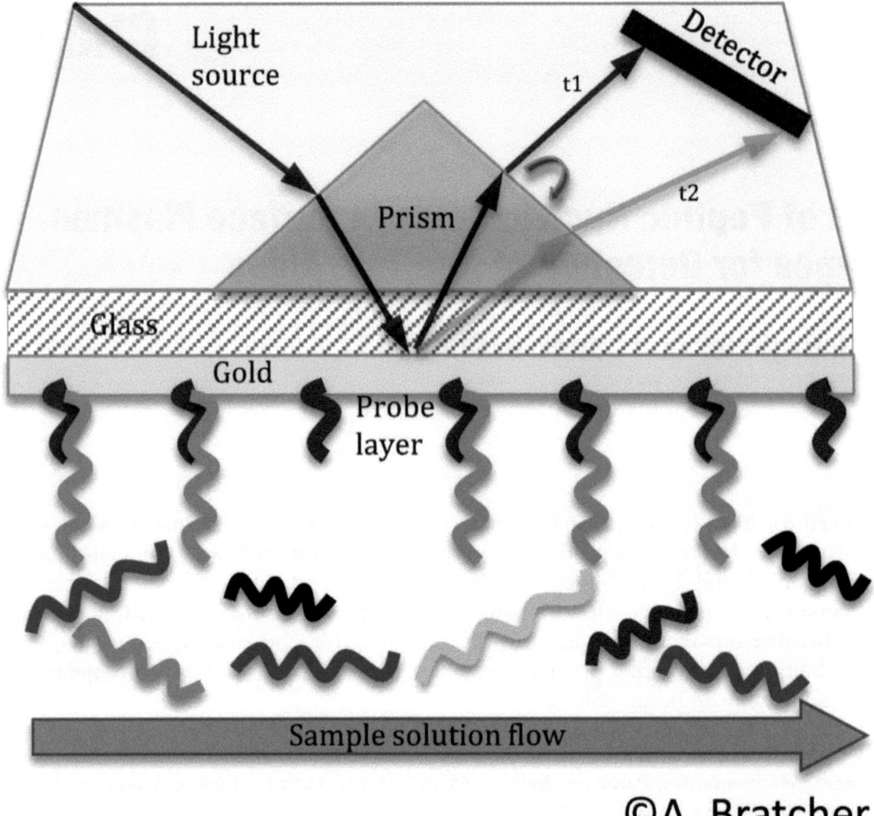

Fig. 1. SPR operation demonstrated with a PNA probe layer on gold sensor surface. Binding of target molecules to the probe layer causes a subsequent change in the refractive index of the surface.

and require special training. HABs vary interannually in location, intensity, and duration, with increasing frequency and geographic distribution over past decades, making detection and prediction challenging areas of current research (8, 9). The development of a rapid, low-cost, and easy-to-use device to detect and monitor *Alexandrium* would be an important advancement in HAB research and management.

Surface plasmon resonance (SPR) is a label-free, optical detection method that measures the change in refractive index after binding (hybridization) of a target to a target-specific probe on a surface (Fig. 1) (10). Polarized light of varying angles is used to illuminate a thin gold surface and is then reflected back to a detector element. At a specific angle, electrons in the gold will resonate and form a surface plasmon wave. This angle is dependent upon the refractive index of the media adjacent to the gold sensor surface opposite the incident light. As the refractive index changes, the angle at which the surface plasmon wave is excited also changes. When target molecules bind to a specific probe immobilized on the gold sensor surface, the subsequent change in refractive index at the sensor surface is detected in real-time and converted to a signal that is expressed in resonance units (RU). One RU is equivalent to an angle shift of 1×10^{-6} degrees (11).

The most well-known SPR instruments are produced by Biacore (GE Health, Piscataway, NJ, USA). These instruments are expensive and designed for bench-top applications only, making them impractical for use in field-based HAB monitoring programs. SPR has been widely used in laboratory detection assays for toxins (12, 13), proteins (14, 15), and other small molecules (16, 17), and is gaining momentum as a tool for detecting nucleic acid sequences (18–20). Most nucleic acid assays for the identification or detection of HAB species to date, however, have made use of DNA probes and fluorescence with whole-cell hybridization (21, 22), sandwich hybridization (21), or utilized a PCR-based approach (23–25). While these methods offer some improvement over traditional microscopic identification, they can still be time-consuming, expensive, require specialized training, and lack portability for use in on-site field detection.

In order to employ a field-based approach to HAB detection, we are using a portable SPR instrument (SPIRIT, Fig. 2) developed by collaborators in Seattle, Washington. SPIRIT employs a reusable, miniature sensor module with a gold-coated surface (Spreeta, by Icx Nomadics, Fig. 3) to which the PNA probes are bound. The Spreeta SPR chips have three distinct channels and are self-contained with an LED light source and array detector that are integrated with the gold sensor surface (26–28). SPIRIT contains eight sensor chips, giving the user the ability to simultaneously detect up to eight different organisms or targets of interest. Liquid samples are flowed over the sensor surface and subsequent binding of target to probe is visible on a connected laptop in real-time (29).

Peptide nucleic acids (PNAs) are short-chain DNA mimics where the negatively charged sugar-phosphate backbone is replaced by a neutral peptide chain (30). Similar to DNA, PNAs bind according to Watson–Crick rules for base pairings (31). PNA probes are superior to the more traditionally used DNA probes due to their high discrimination for mismatches, resistance to protease and nuclease degradation, and the ability to hybridize in low salt concentrations (10, 30). These characteristics make them robust and ideally suited for field applications.

Current detection methods for *Alexandrium* consist of collecting samples from field sites, then transporting them back to a lab for processing, taking up to several days before a positive identification can be made. The standard method for determining bloom toxicity utilizes a mouse bioassay, in which a laboratory mouse is injected with extract from shellfish suspected of containing PSTs, and the occurrence and timing of death corresponds to the

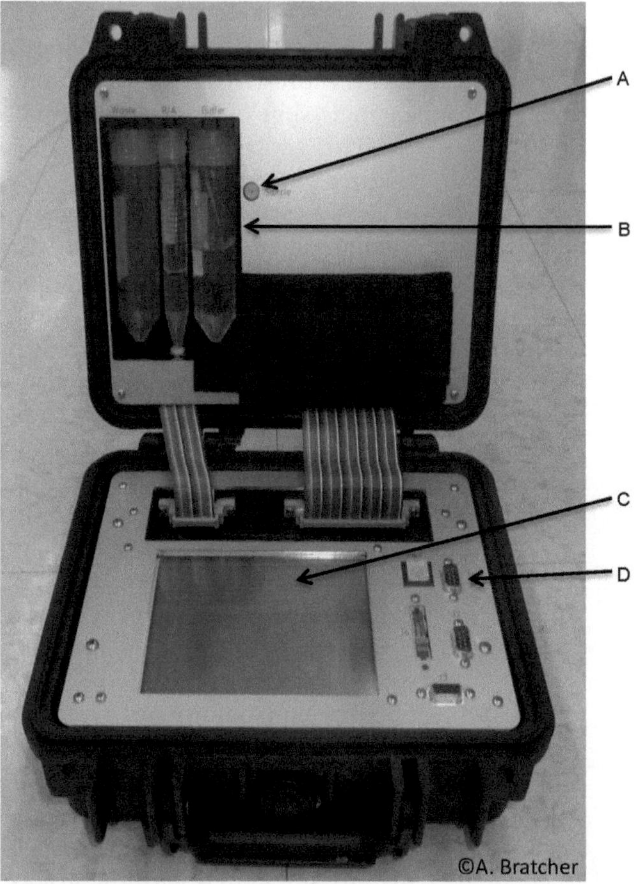

Fig. 2. Custom-built SPR instrument (SPIRIT, Seattle Sensor Systems). A = sample injection port, B = Buffer Tube, C = touch screen controls, D = data output to PC.

Fig. 3. Miniature Spreeta SPR sensor (Icx Nomadics).

level of toxicity present. Due to the high toxicity of blooms of certain species of *Alexandrium*, it would be extremely beneficial to have the ability to rapidly classify blooms as toxic or nontoxic. This would then allow managers to take appropriate action for the closing of affected shellfish harvesting areas before heightened toxicity occurs and human health is adversely affected, and also avoid unnecessary large-scale closures of shellfish fisheries that could lead to huge economic losses. Our custom-built instrument is small and portable, allowing the user the ability to easily transport the instrument to suspected bloom sites and obtain results within minutes.

2. Reagents and Equipment

2.1. Surface Plasmon Resonance

All of our SPR assays used a portable SPR instrument (SPIRIT) developed by Seattle Sensor Systems in Seattle, Washington (26), that uses a miniature sensor module with a modifiable gold-coated surface (Spreeta, by Icx Nomadics) to which the probes were covalently bound via a thiol linkage. The Spreeta SPR chips are designed with three distinct channels, a light source, and an array detector integrated with the gold sensor surface (27–29). SPIRIT is designed so that samples flow through a temperature-controlled (±0.01 °C) flow cell over the sensor surface and subsequent binding of target to probe is visible on a connected laptop in real-time (29).

1. SPIRIT, Seattle Sensor Systems
2. Disposal SPR sensor chips, Icx Nomadics
3. PC laptop computer running Windows operating system.
4. 50 mL conical tubes, BD Falcon
5. 15 mL conical tubes, BD Falcon

2.2. PNA Probes

PNA probes were designed as 16 base long sequences and modified on the C-terminus with carbon spacer groups and either cysteine molecules to provide a reactive thiol group for direct immobilization on the gold sensor surface or biotin to allow for streptavidin–biotin coupling. All sequences were HPLC purified to >95 % and obtained lyophilized from Panagene (Daejeon, Korea) in 100 nmol quantities.

2.3. Synthetic Oligonucleotides

Synthetic sequences of 15, 43, and 60 bases were designed complementary to PNA probe sequences and to mimic the targeted species of interest. Oligonucleotides were purified using standard desalting methods after synthesis and purchased from IDT Technologies (Coralville, Iowa, USA) in quantities of 1 μmol of lyophilized DNA.

2.4. Sample Injection

1. 10 mL luer-lock syringes, BD Falcon
2. 3 mL luer-lock syringes, BD Falcon
3. Blunt-tip, luer-lock needles, Popper and Sons

2.5. SPIRIT Running Buffer and Calibration Components

1. NaCl
2. KCl
3. Na_2HPO_4
4. KH_2PO_4
5. diH_2O
6. Sucrose

2.6. Running Buffer Preparation Equipment

1. Nalgene 500 mL 0.22 μM bottle top filters
2. Pyrex 1,000 mL bottles
3. Vacuum filtration system

2.7. Culture Medium Components

1. Nalgene 500 mL 0.22 μM bottle-top filters
2. Pyrex 1,000 mL bottles
3. L1 Media Kit, Provasoli-Guillard National Center for Culture of Marine Phytoplankton
4. Natural seawater

2.8. Culture Medium Preparation Equipment

1. Nalgene 500 mL 0.22 μM bottle top filters
2. Pyrex 1,000 mL bottles
3. Vacuum filtration system

2.9. Culture Strains

1. *Alexandrium ostenfeldii*, (CCMP1773), Provasoli-Guillard National Center for Culture of Marine Phytoplankton
2. *A. tamarense*, (CCMP1978), Provasoli-Guillard National Center for Culture of Marine Phytoplankton

2.10. Culture Maintenance

Cultures of *A. tamarense* and *A. ostenfeldii* are grown at 15 °C under a photon flux density of 360 μmol/m^2/s, on a 14:10 light/dark cycle in a diurnal illuminated incubator (Shel Lab LI15 Diurnal Plant Growth Chamber; Cornelius, Oregon, USA).

2.11. Cell Concentration and Lysis

1. 5 μM Millipore SVLP 25 mM hydrophilic Durapore filter papers
2. Swinnex 25 mM syringe filter holders
3. 0.45 μM Whatman Puradisc 25 mM Sterile Non-pyrogenic syringe filters

2.12. Cell Lysis Buffer Components

1. 3 M guanidinium thiocyante
2. 50 mM Tris

3. 15 mM EDTA
4. 2 % sarkosyl
5. 0.2 % SDS
6. diH_2O

2.13. Cell Lysis Equipment

1. Heat block
2. Vortex instrument

2.14. RNA Preservation

1. RNAsecure reagent, Ambion

2.15. Miscellaneous Preparatory Items

1. Cylinder of compressed N_2 gas (see Note 1)
2. Pipettes, 20, 200, 1,000 μL, Rainin
3. Disposable pipette tips, MBP Art Low-Retention hydrophobic 20, 200, 1,000 μL sizes
4. 1.5 and 2.0 mL microtubes

3. Protocol

3.1. Preparation of Buffers

3.1.1. SPR Running Buffer (PBS–Tween 0.1 %)

PBS is prepared as 10× stock solutions and filtered by vacuum filtration using the 500 mL 0.22 μM bottle-top filters. Buffer can be diluted to working concentrations by adding 900 mL of diH_2O to 100 mL of 10× stock, adding Tween and allowing to mix completely, then filtering again with a 0.22 μM filter.

Component	For 1 L of 10× stock solution
1. NaCl	80 g
2. KCl	2 g
3. Na_2HPO_4	11.5 g
4. KH_2PO_4	2 g
5. diH_2O	sufficient to bring volume to 1,000 mL

Component	For 1 L of 1× solution
10× PBS	100 mL
Tween	1 mL
diH_2O	889 mL

3.1.2. SPIRIT Calibration Solutions

New solutions must be made every time a new 1× PBST stock is used and filtered to 0.22 μM. Solutions will last at room temperature for 1–2 months; discard and make fresh if solutions become contaminated.

Initialization Solution (PBST–Sucrose 20 %)

Component	For 50 mL of initialization solution
Sucrose	10 g
1× PBST	sufficient to bring solution to 50 mL

Calibration Solution (PBST–Sucrose 0.4 %)

Component	For 50 mL of initialization solution
Initialization solution	1 mL
diH_2O	49 mL

3.1.3. Laboratory Culture Media

Using reagents from L1 media kit (CCMP), mix with 0.22 μM filtered seawater. Media can be stored at 4 °C up to several months.

3.1.4. RNA Extraction Buffer

1. Measure out 200 mL diH_2O in a bottle or beaker on a stir plate.
2. Add sarkosyl.
3. Add Tris base.
4. Add EDTA.
5. Add SDS.
6. Add GuSCN. Let stir several hours or overnight until completely dissolved.
7. Add remaining H_2O to bring to 500 mL.
8. Check pH, adjust to 8.9 using HCl if needed.

Component	For 500 mL
20 % sarkosyl	50 mL
Tris	3.03 g
EDTA	2.79 g
1 M SDS	1 mL
3 M GuSCN	177.24 g
diH_2O	Sufficient to make 500 mL

3.2. Preparation of PNA Probe Stock Solutions

The PNA probes were developed by optimizing previously designed DNA probe sequences ATNA1 for *A. tamarense*, *A. fundyense*, and *A. catenella* species complex (32) and AOST01 for *A. ostenfeldii* (33). A universal probe sequence (UniR) that does not match any eukaryotic sequence found in GenBank was used as a negative control (34). Probe sequences were derived from large-subunit ribosomal RNA sequences for these species and are specific to target 28S ribosomal RNA (32, 33). Probes were modified on the 3′ terminus with either a biotin or cysteine molecule and carbon spacer groups of 1.3 nm long ethylene glycol linkers inserted before the probe sequence. The biotin will bond with streptavidin molecules, while a cysteine molecule provides a reactive thiol group on the probe that allows for direct absorption onto the gold surface of the sensor (35).

1. Dilute PNA to 250 μM concentration in molecular biology grade water.
2. Divide into 5 μL aliquots and store at −20 °C, protected from light.

3.3. Preparation of Synthetic Oligonucleotides

For laboratory assays, 15-mer, 43-mer, and 60-mer target DNA sequences were designed complementary to the probes. The DNA oligonucleotide sequences were synthesized by IDT Technologies (Coralville, Iowa, USA) and were reconstituted to 250 μM stock solutions in molecular biology grade ultrapure H_2O, storing at −20 °C until use. Oligonucleotides were further diluted to the desired concentrations in PBS buffer prior to experimentation.

3.4. RNA Extraction from Alexandrium Cultures

The *Alexandrium* cell lysis protocol is modified from Goffredi et al. (34) to maintain the integrity of the extracted RNA and allow for compatibility with SPR detection in SPIRIT.

1. Concentrate a known quantity of cells on a 5 μM Millipore filter in Swinnex syringe filter holder.
2. Remove filter from holder and insert into a 1.5 mL tube, adding 0.5 mL of GuSCN lysis buffer.
3. Heat at 85 °C for a total of 5 min, flicking tube exactly halfway through incubation.
4. After cooling, filter lysate through a 0.22 μM sterile Millipore syringe filter into a clean 2 mL microtube.
5. Add 1 mL of PBST to the sample.

3.5. Preparation of Sensor Surfaces

There are a variety of protocols that can be used for binding of probe sequences to sensor surfaces, including a streptavidin–biotin coupling method, and for this method of detection, a more favorable thiol immobilization was also utilized. SPR sensors are coated with probes outside of the instrument using a benchtop protocol.

3.5.1. Binding of PNA Probes to Streptavidin-Coated Sensors

1. Taking care not to touch the gold surface, rinse new BioCap SPR sensor with diH_2O.
2. Dry sensor surface using N_2.
3. Dilute biotin-labeled PNA probe to 1 μM concentration in 30 μL of PBS buffer.
4. Gently pipette the 1 μM probe solution onto the gold surface of the sensor chip, and taking care not to touch the pipette tip to the surface, spread the droplet across the surface.

5. Incubate sensor chip at room temperature in a dark humid chamber for 3.5 h.
6. Rinse sensor surface with diH_2O.
7. Dry sensor surface using N_2.
8. Chip is ready for use or can be stored with desiccant at 4 °C up to several months for future use.

3.5.2. Binding of PNA Probes via Thiol Immobilization to Sensors

1. Taking care not to touch the gold surface, rinse new SPR sensor with diH_2O.
2. Dry sensor surface using N_2.
3. Dilute cysteine-labeled PNA probe to 1 μM concentration in 30 μL of molecular biology grade water.
4. Gently pipette the 1 μM probe solution onto the gold surface of the sensor chip, and taking care not to touch the pipette tip to the surface, spread the droplet across the surface.
5. Incubate sensor chip at room temperature in a dark humid chamber for 3.5 h.
6. Rinse sensor surface with diH_2O.
7. Dry sensor surface using N_2.
8. Chip is ready for use or can be stored with desiccant at 4 °C up to several months.

3.6. Surface Plasmon Resonance

3.6.1. Preparation of Sensors for Assay

Prior to experimentation, sensors should be inserted into SPIRIT and washed with 5 mM NaOH for 1 min at 57 μL/min followed by PBST for several minutes to remove any loosely bound probe or organics on the surface.

3.6.2. Calibration of Instrument

Before starting each set of assays, SPIRIT must undergo a series of calibration steps to set the SPR detection parameters. The calibration must be repeated each time the instrument is powered on, but is not necessary between each individual sample that is tested.

3.6.3. Experimental Procedure

Assays reported here were carried out in PBS-Tween (0.1 %) running buffer at a flow rate of 57 μL/min at 25 °C (see Note 2). A 1.5 mL sample was flowed across the sensor surface for a total of 15 min for all of the oligonucleotide assays. At 15 min, running buffer was flushed through the sample loop and the SPR response was determined by subtracting reference sensors from experimental sensors after a stable signal was obtained (see Note 3).

3.6.4. Regeneration of Sensor Surfaces

Each assay is followed by a 2 min regeneration step using 7.5 mM NaOH to remove any bound target oligonucleotides while leaving the PNA probe layer intact (see Note 4). PBS–Tween (0.1 %) is then again flowed over the sensor surface for 1 min before starting a new assay cycle.

3.6.5. Sensor Storage

After experimentation, sensors should be removed from SPIRIT, rinsed thoroughly with diH_2O and dried with N_2. Store with desiccant at 4 °C protected from light.

3.6.6. SPIRIT Storage

After all experiments for the day have been completed, SPIRIT must be purged of all remaining buffers in order to ensure the integrity of the fluidics system and flushed out using diH_2O. Further care should be taken to purge and clean the fluidics for long-term storage using an additional EtOH flush (see Notes 5 and 6).

4. Typical Protocol Results

This method of combining the use of PNA probes with SPR can reliably detect and discriminate between the species of interest: *A. ostenfeldii* and *A. tamarense*. Figure 4 shows the SPR response of a sensor coated with the PNA probe specific for *A. tamarense* (PNA 008) to a 43-mer nucleotide sequence in which the first 15 nucleotides of the sequence are complementary to this probe. There was no detectable binding to a sensor coated with the negative control probe PNA 010. Assays have also been conducted to determine if increasing the concentration of target sequence present

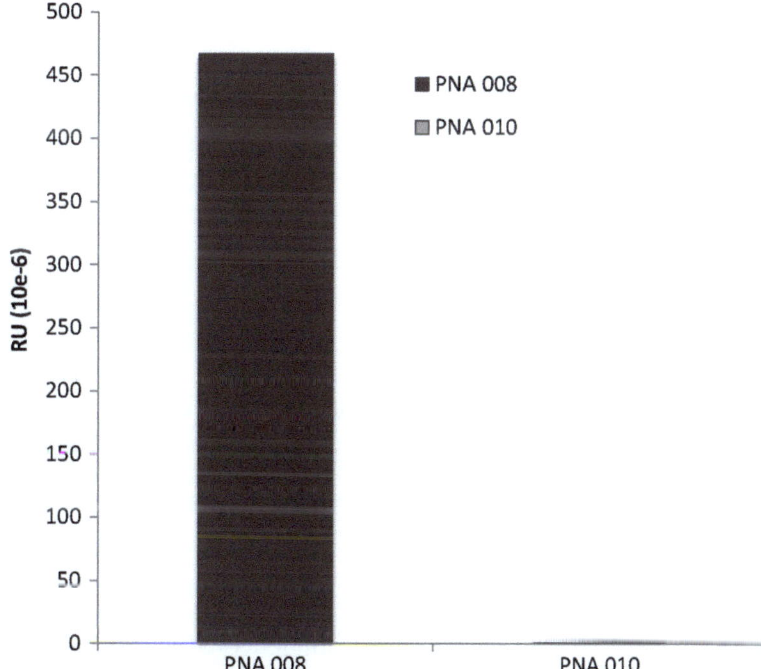

Fig. 4. A sensorgram of SPR detection of *Alexandrium tamarense* oligo by PNA 008 probe (designed specific to this species). There was no binding detected by the negative control probe PNA 010.

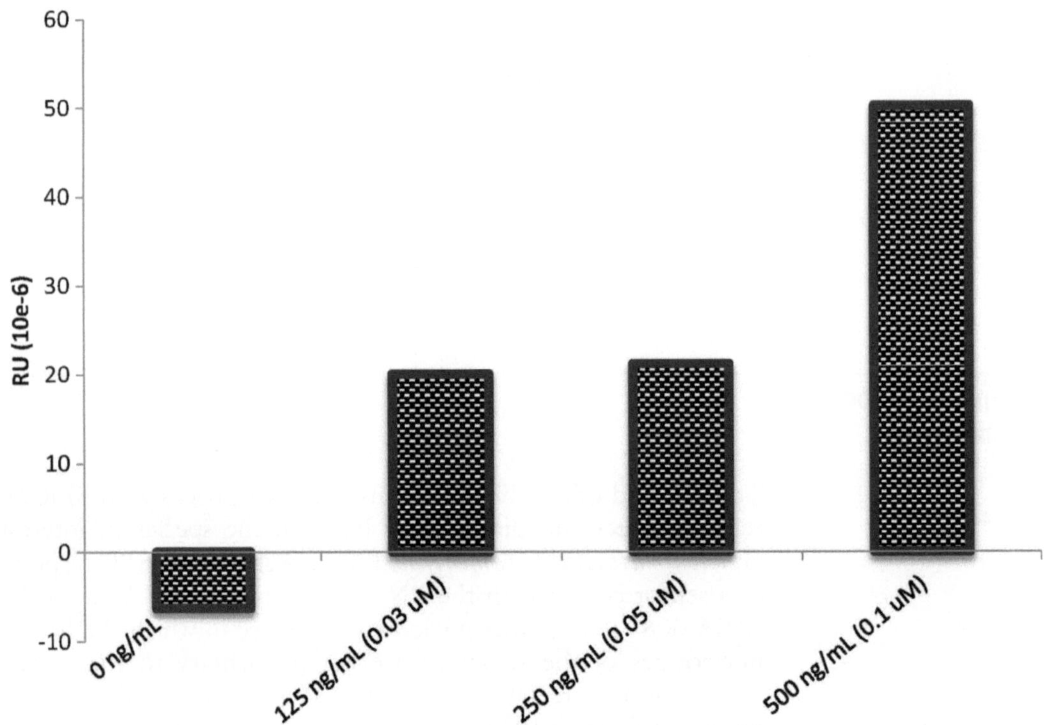

Fig. 5. SPR detection of *Alexandrium ostenfeldii* oligo at multiple concentrations (reference subtracted). Higher concentrations yield higher SPR responses.

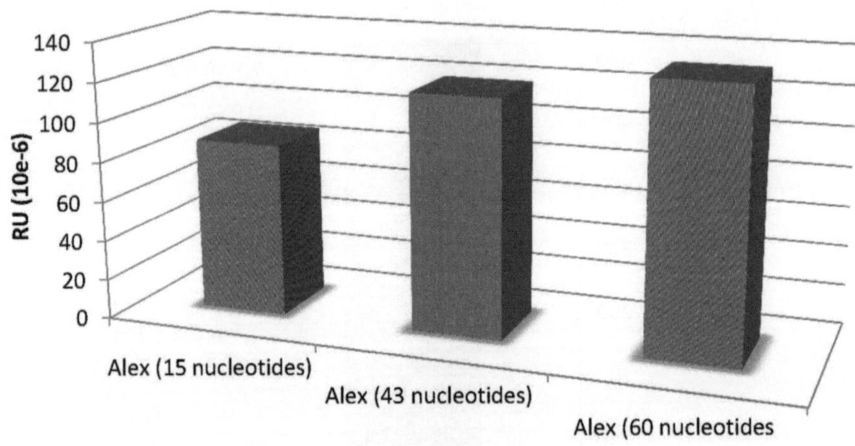

Fig. 6. SPR detection of *Alexandrium tamarense* oligonucleotides sequences of varying length (reference subtracted). Longer-length sequences yield higher SPR responses.

in the sample influences SPR response, and it was found that higher nucleotide concentrations yield higher SPR responses (Fig. 5). In order to investigate the SPR response to sequences of varying lengths, oligonucleotides of varying lengths that were designed to mimic *A. tamarense* were hybridized to probe PNA 011 (Fig. 6).

These results demonstrate that as target oligonucleotide sequences increase in length, SPR signal response also increases. These results are very promising for the use of this method in rapid detection of organisms based on genetic identification.

5. Notes

5.1. Sensor Maintenance

1. N_2 is optional, though extremely beneficial in preparation and maintenance of SPR sensor chips and its use is included in the protocols for chip preparation and storage.

5.2. Flow Rate and Time

2. Assays were conducted using decreased flow rates (6 μL/min) and shorter time periods for flowing samples across sensor surfaces (6 min), and SPR responses similar to the faster, longer flow rates and times were obtained. Though these times and rates will likely vary with changes in target sequences and hybridization parameters, for these methods using decreased flow rates dramatically reduces the time required for each assay cycle.

5.3. Reference Sensors

3. Though the Spreeta sensors contain three channels, the method of the probe immobilization causes all three channels to be coated with probe sequences and thus the SPR response is reported as a average of three channels to minimize any possible across-channel variance. Therefore, it is important to include a reference sensor during each assay that is coated with a probe that is known not to hybridize with the targeted sequence in order to eliminate any residual SPR responses due to the influence of buffers on the chip surface. The SPR signal from this reference sensor is then subtracted from the experimental sensor(s) to obtain the actual binding response during an assay.

5.4. Regeneration of Spreeta Sensor Surfaces

4. In order to keep costs per assay to a minimum, sensor surfaces are regenerated following each SPR-binding assay by removing bound target sequences while leaving the PNA probe layer intact. The effectiveness of regeneration is determined by the SPR response obtained after all regeneration solution has been flushed from the sensor surfaces, replaced with running buffer, and a stable response is achieved. Successful regeneration is achieved if the response returns to the same baseline level obtained prior to running a binding assay. Several different methods of achieving successful regeneration have been explored, including exposing the surface to 95 °C diH_2O as well as chemical treatments. Treating surfaces with 95 °C diH_2O was able to effectively remove target sequences from probes; however, this method was impractical for use inside SPIRIT due to the introduction of extreme heat to the instrument components.

Varying concentrations of HCl did not prove sufficient for removing all bound target sequences, while even extremely low concentrations of SDS removed all target sequences but also seemed to damage the PNA probe layer and decreased the binding efficiency in subsequent assays. It was found that for these assays, 7.5 mM NaOH is the most effective at removing bound target sequences without damaging the probe layer; lower concentrations of NaOH do not remove all of the target and higher concentrations adversely affect binding in subsequent assays.

5.5. Maintenance of SPIRIT

5. After use, it is important to ensure that all tubing within SPIRIT is sufficiently flushed to avoid crystallization of buffers or buildup of organics. This can be accomplished by replacing the buffer tube with a tube of diH_2O and allowing the diH_2O to flow for a period of 2 min to thoroughly flush out the instrument. Before resuming assays, ensure that any diH_2O has been flushed out and replaced with running buffer.
6. If SPIRIT is not used for several days, in addition to flushing the system out with diH_2O it is also recommended to follow with a 50 % EtOH flush. SPIRIT should be stored dry without any liquid in the buffer tubes.

6. Future Directions

The ability to rapidly detect harmful organisms in the ocean by genetic identification using portable SPR technology is an important advancement. This will allow users to run detection assays at potential HAB sites to determine the level of risk for exposure to algal toxins in that immediate area. Further, this technology is also currently being adapted for use on ocean gliders and buoys, allowing for autonomous detection at offshore sites. Additionally, the assay will be expanded for the detection of both organisms and toxins using the same platform, eliminating the need to run a separate assay for the detection of toxins if a harmful species is identified. These continuing advancements ensure that this method can be of benefit to a wide range of users, while maintaining the rapid and reliable detection attributes of an SPR assay.

Acknowledgements

Partial funding for this research was provided by NOAA MERHAB award NA05NOS4781232. This is Monitoring and Event Response for Harmful Algal Blooms (MERHAB) Publication Number 140 (L.C.). Partial funding support was also provided by the University of Maine's Graduate Student Government Grants-in-Aid of Research, NSF-IGERT, and the Phycological Society of America Grants-in-Aid of Research (A.R.B.). The authors would like to thank Clem Furlong and Scott Soelberg of Seattle Sensor Systems, P. Millard, J. Duy, R. Smith, and Chris Scholin of Monterey Bay Aquarium Research Institute.

References

1. Shumway S, Sherman-Caswell S, Hurst J (1988) Paralytic shellfish poisoning in Maine: monitoring monster. J Shellfish Res 7:643–652
2. Anderson D (1997) Bloom dynamics of toxic *Alexandrium* species in the northeastern U.S. Limnol Oceanogr 42(5/2):1009–1022
3. Hurst JW (1975) Proceedings of the first international conference on toxic dinoflagellate blooms. In: LoCicero VR (ed) Massachusetts Science and Technology Foundation, Wakefield, MA, pp 525–528
4. Armstrong IH, Coulson JC, Hawkey P, Hudson MJ (1978) Further mass seabird deaths from paralytic shellfish poisoning. Br Birds 71:58–68
5. Campbell RC, Teegarden GJ, Cembella AD, Durbin EG (2005) Zooplankton grazing impacts on *Alexandrium* spp. in the nearshore environment of the Gulf of Maine. Deep-Sea Res Pt II 52(19–21):2817–2833
6. Geraci J, Anderson D, Timperi R, St Aubin D, Early G, Prescott J, Mayo C (1989) Humpback whales (*Megaptera novaeangliae)* fatally poisoned by dinoflagellate toxin. Can J Fish Aquatic Sci 46:1895–1898
7. Guzman HM, Cortes J, Glynn PW, Richmond RH (1990) Coral mortality associated with dinoflagellate blooms in the eastern Pacific (Costa Rica and Panama). Mar Ecol 60:299–303
8. McGillicuddy D, Anderson D, Lynch D, Townsend D (2005) Mechanisms regulating large-scale seasonal fluctuations in *Alexandrium fundyonso* populations in the Gulf of Maine: results from a physical-biological model. Deep-Sea Res Pt II 52(19–21).2698–2714
9. Hallegraeff GM (2003) Harmful algal blooms: a global overview. In: Hallegraeff GM, Anderson DM, Cembella AD (eds) Manual on harmful marine microalgae. UNESCO, Paris, pp 25–49
10. Nielsen P, Egholm M, Buchardt O (1994) Peptide nucleic acid (PNA). A DNA Mimic with a Peptide Backbone. Bioconjugate Chem 5:3–7
11. Kretschmann E (1971) Determination of optical constants of metals by excitation of surface plasmons. Z Phys 241(4):313–324
12. Hu C, Gana N, Chenc Y, Bic L, Zhangc X, Songa L (2009) Detection of microcystins in environmental samples using surface plasmon resonance biosensor. Talanta 80:407–410
13. Soelberg SD, Stevens RC, Limaye AP, Furlong CE (2009) Surface plasmon resonance detection using antibody-linked magnetic nanoparticles for analyte capture, purification, concentration, and signal amplification. Anal Chem 81:2357–2363
14. Christensen LLH (1997) Theoretical analysis of protein concentration determination using biosensor technology under conditions of partial mass transport limitation. Anal Biochem 249:153–164
15. Richalet-Secordel PM, Rauffer-Bruyere N, Christensen LL, Ofenloch-Haehnle B, Seidel C, van Regenmortel MH (1997) Concentration measurement of unpurified proteins using biosensor technology under conditions of partial mass transport limitations. Anal Biochem 249: 165–173
16. Markgren P-O, Hämäläinen M, Danielson U (1999) Screening of compounds interacting with HIV-1 proteinase using optical biosensor technology. Anal Biochem 265:340–350
17. Strandh M, Persson B, Roos H, Ohlson S (1998) Studies of interactions with weak affinities and low-molecular-weight compounds using surface plasmon resonance technology. J Mol Recognit 11:188–190
18. Asai R, Ootani K, Nomura Y, Nakamura C, Ikebukuro K, Arikawa Y, Miyake J, Karube I (2003) PCR-based ribosomal DNA detection technique for Microalga (Heterosigma carterae) causing red tide and its application to a biosensor using labeled probe. Mar Biotechnol 5:417–423
19. Joung H, Lee N, Lee SK, Ahn J, Shin YB, Choi H, Lee C, Kim S, Kim M (2008) High sensitivity detection of 16 s rRNA using peptide nucleic acid probes and a surface plasmon resonance biosensor. Anal Chim Acta 630: 168–173
20. Persson B, Stenhag K, Nilsson P, Larsson A, Uhlen M, Nygren P (1997) Analysis of oligonucleotide probe affinities using surface plasmon

resonance: a means for mutational scanning. Anal Biochem 246:34–44

21. Anderson D, Kulis D, Keafer B, Gribble K, Marin R, Scholin C (2005) Identification and enumeration of *Alexandrium* spp. from the Gulf of Maine using molecular probes. Deep-Sea Res Pt II 52(19–21):2467–2490
22. Hosoi-Tanabe S, Sako Y (2005) Rapid detection of natural cells of *Alexandrium tamarense* and *A. catenella* (Dinophyceae) by fluorescence in situ hybridization. Harmful Algae 4(2):319–328
23. Dyhrman ST, Erdner D, Du JL, Galac M, Anderson DM (2006) Molecular quantification of toxic *Alexandrium fundyense* in the Gulf of Maine using real-time PCR. Harmful Algae 5(3):242–250
24. Handy SM, Demir E, Hutchins DA, Portune KJ, Whereat EB, Hare CE, Rose JM, Warner M, Farestad M, Cary SC, Coyne KJ (2008) Using quantitative real-time PCR to study competition and community dynamics among Delaware Inland Bays harmful algae in field and laboratory studies. Harmful Algae 7(5):599–613
25. Bowers HA, Tomas C, Tengs T, Kempton JW, Lewitus AJ, Oldach DW (2006) Raphidophyceae (Chadefaud Ex Silva) systematics and rapid identification: sequence analyses and real-time PCR assays. J Phycol 42(6): 1333–1348
26. Chinowsky TM, Soelberg SD, Baker P, Swanson NR, Kauffman P, Mactutis A, Grow MS, Atmar R, Yee SS, Furlong CE (2007) Portable 24-analyte surface plasmon resonance instruments for rapid, versatile biodetection. Biosens Bioelectron 22(9–10):2268–2275
27. Naimushin AN, Spinelli CB, Soelberg SD, Mann T, Chinowsky T, Kauffman P, Yee S, Furlong CE (2005) Airborne analyte detection with an aircraft-adapted surface plasmon resonance sensor system. Sens Actuators B 104:237–248
28. Naimushin AN, Soelberg SD, Bartholomew DU, Elkind JL, Furlong CE (2003) A portable surface plasmon resonance (SPR) sensor system with temperature regulation. Sens Actuators B 96:253–260
29. Stevens RC, Soelberg SD, Eberhart BTL, Spencer S, Wekell JC, Chinowsky TM, Trainer VL, Furlong CE (2007) Detection of the toxin domoic acid from clam extracts using a portable surface plasmon resonance biosensor. Harmful Algae 6:166–174
30. Jensen K, Orum H, Nielson P, Norden B (1997) Kinetics for hybridization of peptide nucleic acids (PNA) with DNA and RNA studied with the BIAcore technique. Biochemistry 36:5072–5077
31. Nielsen PE, Egholm M, Berg R, Buchardt O (1991) Sequence selective recognition of DNA by strand displacement with a thymine substituted polyamide. Science 254:1497–1500
32. Scholin C, Herzog M, Sogin M, Anderson D (1994) Identification of group-and strain specific genetic markers for globally distributed *Alexandrium* (Dinophyceae). II. Sequence analysis of a fragment of the LSU rRNA Gene. J Phycol 30(6):999–1011
33. John U, Cembella A, Hummert C, Elbrachter M, Groben R, Medlin L (2003) Discrimination of the toxigenic dinoflagellates *Alexandrium tamarense* and *A. ostenfeldii* in co-occurring natural populations from Scottish coastal waters. Eur J Phycol 38(1):25–40
34. Goffredi SK, Jones WJ, Scholin CA, Marin R III, Vrijenhoek RC (2006) Molecular detection of marine invertebrate larvae. Marine Biotechnol 8:149–160
35. Love JC, Estroff LA, Kriebel JK, Nuzzo RG, Whitesides GM (2005) Self-assembled monolayers of thiolates on metals as a form of nanotechnology. Chem Rev 105(4): 1103–1169

Chapter 8

Immobilization of Fluorescent Aptamer Biosensors on Magnetic Microparticles and Its Potential Application for Ocean Sensing

Po-Jung Jimmy Huang, Imran Khimji, and Juewen Liu

Abstract

Many important analytes are present in the ocean water and primary examples include various marine toxins. The unique marine environment possesses an extremely high ionic strength, posing a significant analytical challenge for biosensor design. Protein-based enzymes and antibodies are likely to denature under such non-physiological conditions. Aptamers are nucleic acid-based binding molecules that can be obtained using a combinatorial in vitro selection technique. Since such selections are carried out in the absence of living cells, it is possible to obtain aptamers that work optimally under high salt conditions. Similarly selections in low pH and high temperatures have already been carried out. The high salt concentration in marine samples may also cause significant fluorescence quenching, reducing the sensitivity of fluorescent aptamer sensors. We propose that this problem may be solved by immobilization of aptamer-based biosensors on magnetic microparticles, allowing spatial separation of the target binding and the fluorescence detection steps. In this chapter, we describe a protocol for the detection of adenosine and ATP in high salt buffers and in human blood serum. Compared to the non-immobilized sensor, more consistent results with reduced interference were achieved after immobilization. Future research directions of using such immobilized sensors for marine detection are also discussed.

Key words: Magnetic microparticles, DNA, Aptamers, Biosensors, Fluorescence, Real-time detection, Biomolecular recognition, Ocean sensing, Immobilized sensor, Marine detection

1. Introduction

Traditional analytical methods for small molecule detection employ mass spectrometry, NMR, and various chromatographic techniques. With their outstanding performance, these methods have established the current analytical standard. However, these instruments are available only in centralized laboratories, making on-site

Sonia M. Tiquia-Arashiro (ed.), *Molecular Biological Technologies for Ocean Sensing*, Springer Protocols Handbooks,
DOI 10.1007/978-1-61779-915-0_8,

and real-time detection difficult to achieve. Biosensors are devices based on biomolecular recognition. Protein enzymes, antibodies, nucleic acids, and even whole cells can be used for target recognition. The current biosensor market is dominated by protein-based assays. Primary examples include enzyme-based glucose sensors and antibody-based pregnancy tests (1). Ultrasensitive and selective detection of contaminants in water using biosensors has attracted a lot of research interests because it is highly desirable to obtain instant results for such applications (2, 3). The use of biosensors in various environmental water samples, such as rivers, lakes, and tap water, has already been demonstrated. Considering the extremely high ionic strength of the marine environment, biomolecular interactions that can occur under physiological conditions may be inactivated due to probe denaturation, posing a significant challenge for developing biosensors for related analytical applications. Table 1 lists the comparison of three metal ion concentrations of tap water, Lake Ontario water, and ocean water, where the ocean water ionic strength is three orders of magnitude higher. Many important targets need to be detected in ocean water including toxins, pollutants as well as nutrients. For example, saxitoxins produced by different species of algae can cause paralytic shellfish poisoning (PSP) which is life threatening. Spirolades were confirmed to be highly toxic through mouse bioassays (4). Other important targets include steroids, nitrate, pesticides, and organotin compounds such as tributyltin (TBT) (5).

One potential molecular biology solution to overcome the high ionic strength problem is to evolve functional biomolecules in the ocean environment and a viable approach is through aptamer selection. Aptamers are nucleic acid-based ligands that can bind to essentially any molecule of choice (9–11). Since aptamer selection does not involve living cells, rather harsh conditions can be used. Although no marine applications have been demonstrated yet, we believe that this molecular biology tool is feasible for such applications. In this chapter, we introduce aptamer selection, related fluorescent sensor design and in particular we emphasize the

Table 1
Concentrations of several common cations in different waters

Metal ions	Lake Ontario (6)	Tap water (7)	Ocean/seawater (8)
Na(I)	12.6	35	10,594
Mg(II)	8.6	10	12,171.9
Ca(II)	33.8	34	396.9

All values are in mg/L

importance of sensor immobilization. As an example to simulate the complex marine environment, the protocol of immobilizing a signaling aptamer on magnetic microparticles for detecting adenosine and ATP in human blood serum is described.

1.1. DNA Aptamers

DNA aptamers are DNA-based binding molecules. For environmental applications, DNA aptamers are preferred over their RNA counterpart for cost and stability reasons. Fluorophores and various attachment chemistries can be applied to DNA aptamers, popularizing aptamers as a choice for biosensor development. In the past 20 years, DNA aptamers have been isolated to bind to metal ions, small molecules, proteins, and even whole cells (12, 13). Many aptamer selection protocols have been reported and an example of a selection cycle is shown in Fig. 1. The selection starts from a single-stranded (ss) DNA library, typically containing 10^{14} to 10^{15} random DNA molecules. Each DNA has a randomized region (e.g., 20–80 nucleotides) flanked by constant regions for polymerase chain reaction (PCR) primer binding. Due to the large number of random strands produced, there is a high likelihood that a fraction of sequences exhibit binding affinity to the target. Small molecule targets are typically immobilized on a column to allow the retention of bound DNA. As binding affinity is expected

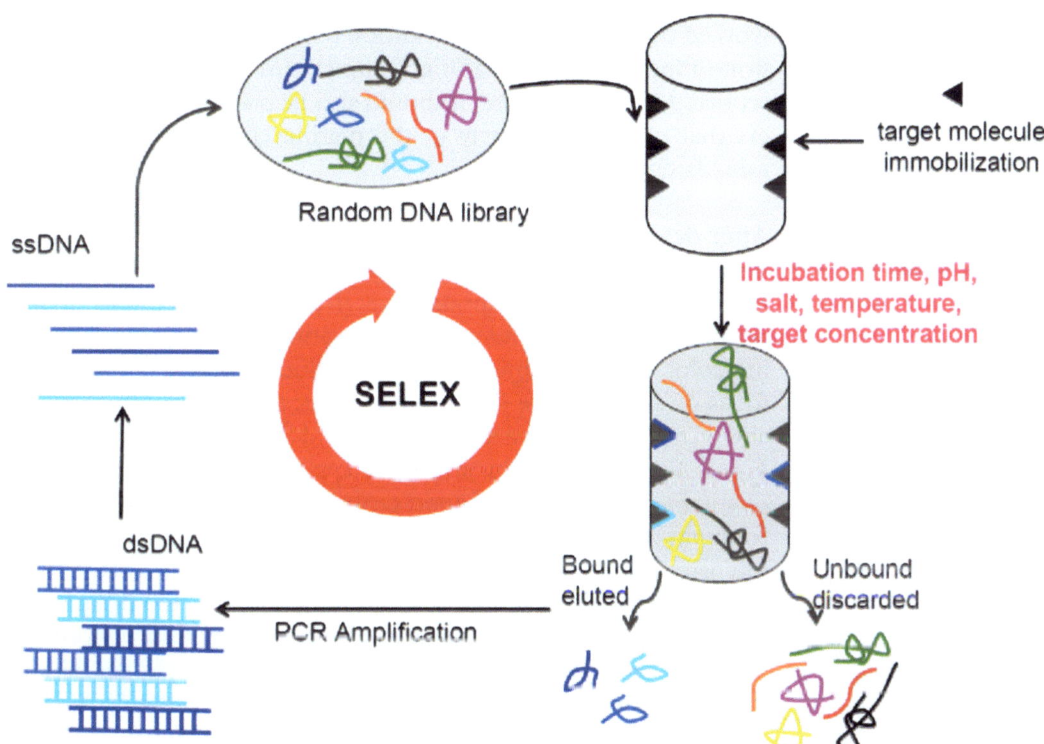

Fig. 1. Schematic presentation of a typical aptamer selection process for small molecule targets.

to depend on the external conditions, incubation should be carried out to mimic the expected test conditions. For example, in the case of marine detection, a buffer with very high salt concentration might be used. After a sufficient incubation period, unbound DNAs are washed away. The bound strands are then eluted and amplified via PCR, creating an enriched DNA pool for the next round of selection. Many methods exist for eluting bound DNA. The use of non-immobilized target for elution allows for good specificity. Subsequent rounds of selections may be carried out using more stringent conditions such as with a shorter incubation time to obtain aptamers with higher binding affinity. Once the library-binding activity stopped increasing, the DNAs are cloned and sequenced. Detailed protocols for aptamer selection have been reported a number of times in the literature and are not repeated here (14–16). This chapter focuses on the sensor design and detection part.

1.2. Aptamer Selection Under Non-Physiological Conditions

While most aptamer selections have been carried out under physiological conditions, a number of examples in the literature have demonstrated the feasibility of DNA selection under harsher conditions. For example, Li and coworkers performed a DNA selection experiment under various pH conditions ranging from 3.0 to 7.0 to obtain RNA-cleaving DNA (17). Active DNA sequences were obtained under all tested conditions and each selected DNA worked the best under the condition that it was selected. The selection of functional DNA at high temperature (e.g., 90 °C) was also reported (18). Therefore, it should also be possible to select aptamers that can work directly in the marine environment with high ionic strength.

1.3. Fluorescent Aptamer Sensors

After determination of the aptamer sequence, a signaling mechanism must be engineered to complete the biosensor design. A diverse range of methods have already been demonstrated to couple the aptamer binding event to various physical signals including color, electrochemistry, mass, and fluorescence (3, 19–22). We focus on fluorescence detection here because of its high sensitivity and versatility. Several commonly used designs are illustrated in Fig. 2. For the monochromophore approach (Fig. 2a), the change of aptamer conformation upon target binding alters the local environment of the attached fluorophore, leading to fluorescence intensity change (23). The amount of fluorescence change, however, was typically less than onefold for most reported designs. For bischromophore approaches shown in Fig. 2b, c, fluorophores and quenchers are arranged in a manner such that aptamer binding induces their distance change. Increasing the distance leads to fluorescence enhancement (Fig. 2b) and vice versa (Fig. 2c) (24–27). Unlike most of the assays that require covalent fluorophore labeling, another method to generate aptamer fluorescence signal is to

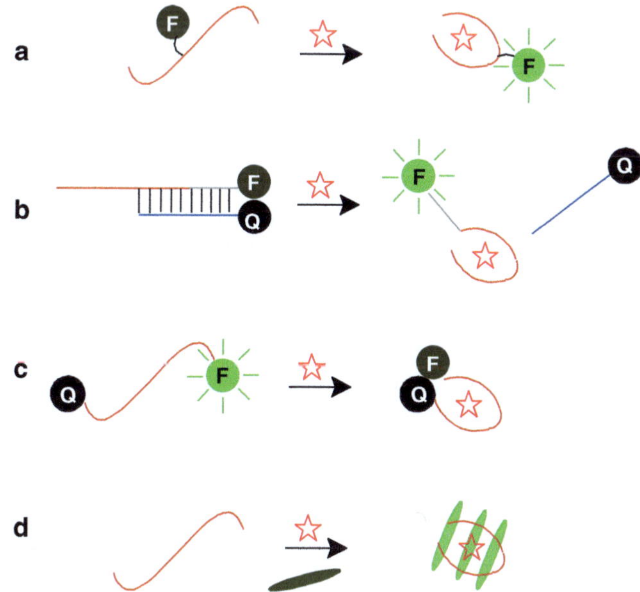

Fig 2. Rational design strategies for signaling aptamer (**a**) Monochromophore approach. (**b**, **c**) Bischromophore approach with covalently labeled fluorophore-quencher pairs. The design in (**b**) is typically called structure-switching aptamer. (**d**) A label-free design of fluorescent sensors using DNA binding dyes. In all the figures, *F* and *Q* denote for fluorophore and quencher, respectively.

use DNA-binding dyes (Fig. 2d) (28, 29). For instance, SYBR Green I dye is one of the commonly used stains available for detecting double-stranded DNA (dsDNA). The dye is commonly used in real-time polymerase chain reaction (RT-PCR) for monitoring DNA amplification. SYBR Green I exhibits minimal fluorescence when free in solution but when bound to dsDNA, the fluorescence can increase up to 1,000-fold. If aptamer binding is accompanied by a change in the DNA structure, such label-free detection can be achieved, allowing cost-effective detection.

1.4. Immobilized Sensors

Compared to homogeneous assays, immobilized biosensors have a number of advantages. For example, immobilization allows for signal amplification, target enrichment, sensor regeneration, and long-term storage. More importantly, the complex sample matrix in blood serum and in ocean water may complicate detection. For example, serum proteins scatter light very strongly while the high salt concentration in ocean water is likely to quench fluorophores. Immobilization allows the separation of target recognition event from signal generation. Aptamer immobilization onto various substrates has been reported. Materials like gold (13, 30, 31), silica (32), carbon (33, 34), hydrogels (35–38), and magnetic microparticles (39–43) are the common choices. Immobilization strategies can be generally classified into four different approaches: adsorption, entrapment, covalent linkage, and affinity binding. Adsorption is the simplest immobilization method since it does not require any nucleic acid modification. Adsorption process is based on ionic, hydrophobic, and Van der Waal's forces. Entrapment of aptamers can be achieved in a gel matrix. Covalent linkages usually require functionalization of either DNA or substrate or both. Primary examples include the thiol-Au chemistry or the coupling between amino-modified DNA and carboxyl substrate to form the amide bond. A primary example of affinity binding is the biotin–avidin interaction, which has an extremely high binding affinity and specificity; stable conjugation can be achieved after a simple mixing.

1.5. Magnetic Microparticles for Immobilization

Magnetic microparticles are ideal for immobilization because they can be conveniently manipulated using a magnet for separation and washing. Tagged samples can be accumulated in a smaller area to improve signal or detection. Signal amplification and mechanisms can also be used. Commercial microparticles are available where a large number of DNA molecules can be immobilized to achieve a high surface density. Commonly used signaling methods for homogeneous assays shown in Fig. 2 can all be applied for immobilized sensors. In addition, signal amplification mechanisms can be introduced. Several designs reported in the literature are shown in Fig. 3. We immobilized a signaling aptamer shown in Fig. 3a for detecting adenosine and ATP in blood serum (42). In the initial state, the particle was nonfluorescent because of the hybridized quencher. Addition of adenosine induced aptamer binding and releasing of the quencher to make the particle fluoresce green. Fig. 3b shows a label-free approach for signal generation and we used this method to detect mercury ions in water (43). Fig. 3c shows a method involving signal amplification (44). The target protein is designed to be sandwiched between two aptamers,

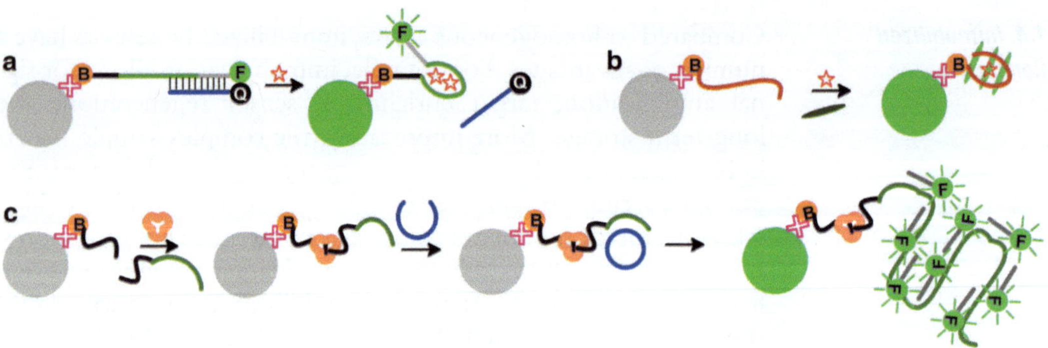

Fig. 3. Methods for signal generation and amplification on magnetic microparticles. All employed the biotin–avidin interaction for immobilization. (**a**) Bischromophore approach with quenching mechanism, (**b**) dye-staining approach, (**c**) a sandwiched assay system involving (RCA).

one of which is immobilized on a magnetic particle and the other contained a primer sequence shown in green. The primer can bind to an external template (in blue) to make it into a circular DNA in the presence of DNA ligase. The primer can then be extended using the rolling circle amplification (RCA) reaction in the presence of a DNA polymerase, allowing signal amplification upon addition of fluorophore-labeled DNA. In this chapter, we focus on the system shown in Fig. 3a.

2. Reagents and Equipment

2.1. Magnetic Microparticles

1. Dynabeads® M-270 Streptavidin 2.8 μm diameter (Life Technologies, 653-05)

2.2. DNA Samples

1. Aptamer DNA (5′-FAM-CACTGACCTGGGGGAGTATTGC GGAGGAAGGTTTTTT-biotin, purified by standard desalt) (Integrated DNA Technologies, Coralville, IA) (See Note 1)
2. Quencher-labeled DNA (5′-CCCAGGTCAGTG-Iowa Black FQ, purified by HPLC) (Integrated DNA Technologies, Coralville, IA)

2.3. Nucleosides and Nucleotides

1. Adenosine-free base (≥99 %, Mandel Scientific, Canada, AMR-0325-50G)
2. Cytidine (Ultra pure grade, Mandel Scientific, Canada, AMR-0182-25G)
3. Guanosine (Ultra pure grade, Mandel Scientific, Canada, AMR-0193-50G)
4. Uridine (Ultra pure grade, Mandel Scientific, Canada, AMR-0975-25G)
5. Adenosine 5'-monophosphate monohydrate from yeast (≥97 %, Sigma-Aldrich, A2252-5G)
6. Adenosine 5'-triphosphate disodium salt, Grade II (≥99 %, Sigma-Aldrich, A3377-1G)
7. Adenosine 3′,5′-cyclic monophosphate sodium salt monohydrate (Sigma-Aldrich, A6885-100MG)

2.4. Buffers and Salts

1. Sodium chloride (≥99.5 %, Mandel Scientific, Canada, AMR-X190-1KG)
2. Tris(hydroxymethyl)aminomethane (Tris) (Mandel Scientific, Canada, AMR-0497-500G)
3. 4-(2-Hydroxyethyl)-1-piperazineethanesulfonic acid (Mandel Scientific, Canada, AMR-0511-1KG)

4. 4-(2-Hydroxyethyl)piperazine-1-ethanesulfonic acid sodium salt (Mandel Scientific, Canada, AMR-0485-500G)
5. Ethylenediaminetetraacetic acid disodium salt dehydrate (≥99 %, Mandel Scientific, Canada, AMR-0105-1KG)

2.5. Blood Serum

1. Human blood serum (100 mL, Fisher Scientific, ICN2930249)

2.6. Water

1. MilliQ water (18.2 MΩ Millipore Milli-Q system, Billerica, MA)

2.7. Plastic Tubes

1. Microcentrifuge tubes (VWR, 10011-722)
2. 15 mL conical tubes (VWR, 89039-664)
3. 50 mL conical tubes (VWR, 89039-656)
4. 5 mL 12 × 75 mm polystyrene round-bottom BD Falcon tubes (VWR, CA60819-310)

2.8. Fluorometer

1. Perkin-Elmer LS-55 fluorometer

2.9. Fluorometer Cuvette

1. Micro-cuvette for fluorometer (Starna, 3-4.45/SOG/4)
2. Adaptor for micro-cuvette (Starna, FCA4)

2.10. pH Meter

1. Denver Instrument UltraBasic benchtop pH meters (Fisher Scientific, 02-226-213)

2.11. Analytical Balance

1. Analytical balance (Mettler Toledo, AL204)

2.12. Weighing Boat

1. Antistatic polystyrene weighing boat (VWR, 89106-764)

2.13. Syringes and Syringe Filters

1. Nalgene PTFE syringe filters (0.2 μm, VWR, 28195-868)
2. BD 10 mL syringe (10 mL, VWR, WLBBD309604)

2.14. Mini-Centrifuge

1. Mini-centrifuge (Fisher Scientific, 05-090-100)

2.15. Flow Cytometer

1. BD FACS Vantage SE flow cytometer

2.16. Zeta Potential Analyzer

1. Zetasizer nano ZS (Malvern)

2.17. Magnet

1. DynaMag—Spin magnet (Life Technologies, 123-20D)

2.18. Software for FACS Data Analysis

1. FCS Express 3 RUO Lite (De Novo software)

3. Protocol

3.1. Preparing Stock Solutions

3.1.1. Preparing 200 mL of Binding and Washing (B&W) Buffer (2×), pH 7.5. The buffer contains 10 mM Tris–HCl (pH 7.5), 1 mM EDTA, and 2 M NaCl

1. Weigh 23.36 g of NaCl, 0.074 g of EDTA, and 0.242 g of Tris and transfer it to a 250 mL PYREX bottle. Add 170 mL water to dissolve it.
2. Add 12 M HCl drop wise using a glass Pasteur pipette and monitoring the pH using a pH meter.
3. Stop the addition when pH reaches 7.5.
4. Fill with Milli-Q water to 200 mL.
5. Filter the buffer through a 0.22 μm syringe filter and store it at 4 °C.

3.1.2. Preparing 500 mL of 500 mM HEPES Buffers, pH 7.5

1. Weigh 29.79 g of HEPES-free acid and 32.54 g of HEPES salt in weighing boats.
2. Transfer it to a 500 mL PYREX bottle and dissolve it in 500 mL Milli-Q water.
3. Filter the solution through a 0.22 μm syringe filter and store it at 4 °C.

3.1.3. Preparing 50 mL of 1 M NaCl Solution

1. Weigh 2.92 g of NaCl and transfer it to a 50 mL conical tube.
2. Dissolve the salt in Milli-Q water.
3. Filter the solution through 0.22 μm syringe filter and store at 4 °C.

3.1.4. Preparing 200 mL of Buffer A (150 mM NaCl, 25 mM HEPES)

1. Transfer 30 mL of 1 M NaCl solution and 10 mL of 500 mM HEPES buffer, pH 7.5 to a 250 mL PYREX bottle.
2. Dilute the mixture to 200 mL with Milli-Q water.
3. Filter the solution through 0.22 μm syringe filter and store at 4 °C.

3.1.5. Preparing Adenosine and Other Nucleosides/Nucleotides Solution

1. Weigh 0.2 g of adenosine-free base and dissolve in 15 mL of Milli-Q water in conical tube. To fully dissolve adenosine, heat the solution to the boiling temperature.
2. Prepare other nucleosides/nucleotides solutions in a similar way. Due to solubility reasons, prepare cAMP and guanosine stock solutions at 25 mM (see Note 2).

3.2. DNA Hybridization

1. Incubate 15 μL of 1 μM of the aptamer DNA with 30 μL of 1 μM quencher-labeled DNA in 100 μL 1× B&W buffer at 40 °C for 10 min and gradually cool it to room temperature.
2. Store the conjugated DNA in ice bath before the immobilization step.

3.3. Preparing Magnetic Microparticles

3.3.1. Washing Stock Magnetic Microparticles to Remove Preservatives

1. Vortex/sonicate the stock bottle containing the magnetic microparticles for 1–2 min.
2. Pipette 40 μL of magnetic microparticles into a 0.7 mL microcentrifuge tube.
3. Place the centrifuge tube on a magnetic tube holder for 2 min and remove the supernatant.
4. Add 40 μL 1× B&W washing buffer to resuspend the microparticles.
5. Repeat steps 3–4 two more times and store them in 1× B&W buffer.

3.3.2. Sensor Immobilization

1. Incubate 10 μL of the washed magnetic microparticles with 20 μL of 1 μM hybridized DNA at room temperature for 30 min (see Note 3).
2. Wash away uncoupled DNA first with 30 μL 1× B&W twice followed by using 30 μL buffer A twice.
3. Resuspend the immobilized aptamer sensor in 10 μL buffer A.

3.4. Adenosine Detection Using the Non-Immobilized Sensor

1. Dilute the hybridized DNA from Sect. 3.2 in buffer A to a final aptamer DNA concentration of 10 nM and quencher-labeled DNA of 20 nM to a final volume of 400 μL.
2. Transfer the solution to a quartz micro-cuvette.
3. Measure the fluorescence spectrum with a fluorometer using 490 nm excitation and collect emission from 500 to 600 nm.
4. Add 10 mM adenosine and repeat the fluorescence measurement again after 5 min.
5. To detect in serum, include 30 % human blood serum in the sample and follow the same procedures.

3.5. Adenosine Detection Using the Immobilized Sensor

3.5.1. Immobilized Aptamer Sensor for Adenosine Detection

1. Prepare 49 μL of various concentrations of adenosine in buffer A or serum.
2. Add 1 μL of immobilized aptamer sensor into the adenosine solutions and incubate it for 20 min.
3. Dilute each sample to 300 μL with buffer A for flow cytometry analysis.

3.6. Flow Cytometry Analysis

1. Dilute 1 μL of unmodified magnetic microparticles into 300 μL buffer and transfer it to BD Falcon tube. This reference sample is used to define the gate for sample analysis.
2. During the live analysis, a boundary of the magnetic microparticles population in the forward (FSC)/side (SSC) scatter plot is drew. Only the events that fall within the boundary are counted. At least 20,000 events are counted for each sample.
3. The data collected can then be further analyzed with FCS Express software.

4. Typical Protocol Results

4.1. Non-Immobilized Sensor Response

The sensor is prepared by hybridizing the quencher-labeled DNA with the aptamer, resulting in an initially quenched fluorescence. Addition of adenosine induces a conformational change in the aptamer to allow target binding, resulting in the release of the quencher-labeled DNA and causing the fluorescence signal to increase. In the presence of 10 mM adenosine in buffer, the sensor can produce ~20-fold fluorescence enhancement (Fig. 4a). However, the fluorescence is largely obscured in 30 % serum with same concentration of the sensor and adenosine (Fig. 4b). In this case, the fluorescence enhancement is less than onefold due to the strong light scattering effect of serum proteins. With 80 % serum, fluorescence enhancement could barely be observed (data not shown).

4.2. Detecting Adenosine Using Immobilized Sensor

It appears that optical interference is unavoidable for direct detection of fluorescence in concentrated serum. We reason that if the target recognition and detection steps can be separated, high sensitivity can still be achieved. This goal can be achieved after immobilizing the sensor DNA on magnetic microparticles.

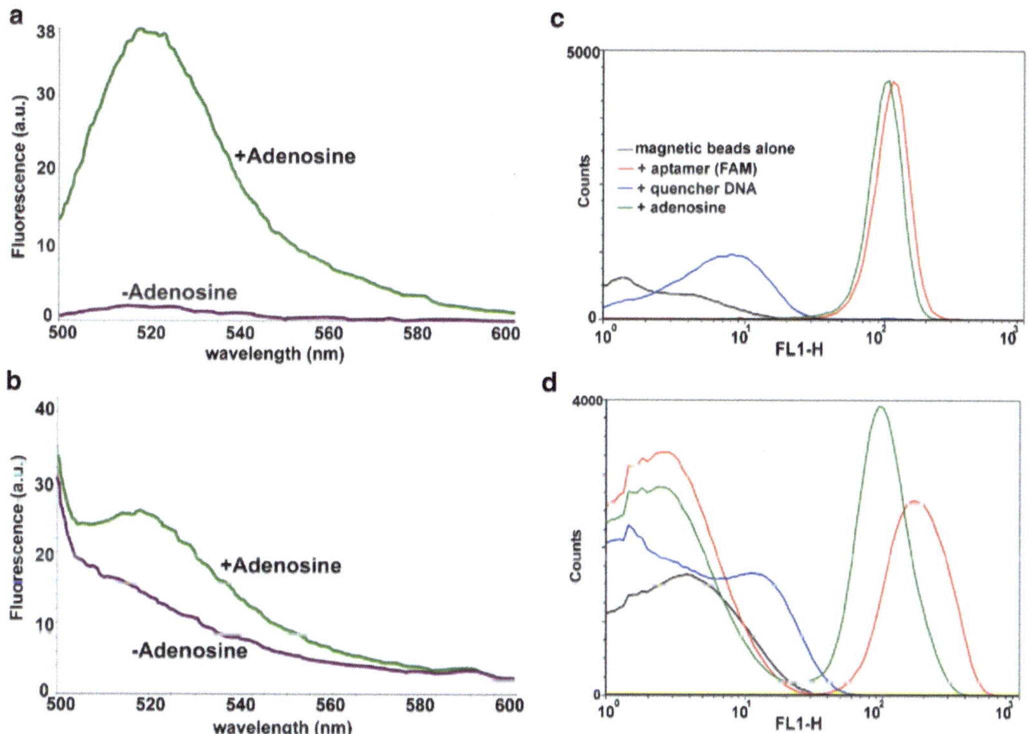

Fig. 4. Fluorescence spectra of the non-immobilized sensor in (**a**) buffer and (**b**) 30 % serum and flow cytometry histograms of the immobilized sensor in (**c**) 150 mM NaCl, 25 mM HEPES buffer and (**d**) 30 % serum. Copyright 2010 American Chemical Society. Reproduced from [Huang P. et al., 2010] with permission from American Chemical Society.

Flow cytometry analyzes particle fluorescence one-by-one in a capillary and is ideal for such applications. First, detection in buffer is carried out (Fig. 4c). The *x*-axis is the FAM fluorescence intensity and the *y*-axis is the number of particles. In buffer A, unmodified magnetic microparticles have an average fluorescence of ~2 (black curve). After adding the FAM-labeled aptamer DNA, the peak shifts to ~110 (red curve). With the quencher-labeled DNA, the fluorescence shifts back to ~8 (blue curve). The result corresponds to a 14-fold fluorescence difference. In the presence of 10 mM adenosine, the fluorescence signal is almost fully recovered (green curve). Similar degrees of fluorescence enhancement are achieved in the serum samples (Fig. 4d).

This improvement is attributed to two factors. First, adenosine binding to the DNA aptamer was carried out in a concentrated serum to release the quencher-labeled DNA. At this moment, the signal is already on the particle surface. Next the sample was diluted by buffer A and quickly analyzed using a flow cytometer. Although the target analyte concentration is also diluted, detection is finished before the quencher-labeled DNA can be rehybridized with the aptamer on the particle surface. Second, the detection happens in a small glass capillary with a diameter of 70 μm. This light path length is significantly reduced compared to that of 4 mm for the cuvette, further reducing the protein scattering effect.

4.3. Effect of Salt

For practical applications, the effect of ionic strength on sensor performance needs to be studied. For example, marine water samples contain more than 1 M salt. Since DNA is polyanion, the performance of such sensors is likely to be affected by salt. Here we compare the free and immobilized sensor as a function of NaCl concentration. The buffer contained 25 mM HEPES, pH 7.6, with which varying concentrations of NaCl were tested (25 mM to 1 M). For the non-immobilized sensor, the assays are performed in a cuvette with a flourometer (Fig. 5a). There is a clear trend that fluorescence signal was inhibited by the presence of high salt concentration. With 1 M NaCl, the signal decreased by ~70 % compared to the signal in 25 mM NaCl. On the other hand, the flow cytometry data shows a much less pronounced salt-dependent response for the immobilized sensor (Fig. 5b). In the range of 25–300 mM NaCl, the average fluorescence stayed at ~50 in the presence of 2 mM adenosine. The background signal in the absence of adenosine is more stable across the 1 M salt range tested. This study shows that such immobilized sensors can potentially be used even in the marine environment.

We attribute such a salt-independent performance of the immobilized sensor to the effect of charged surface. DNA immobilization results in a highly negatively charged surface that can attract counter ions (e.g., Na^+). This ability is reduced in high salt due to surface potential drop. Therefore, the overall effect is

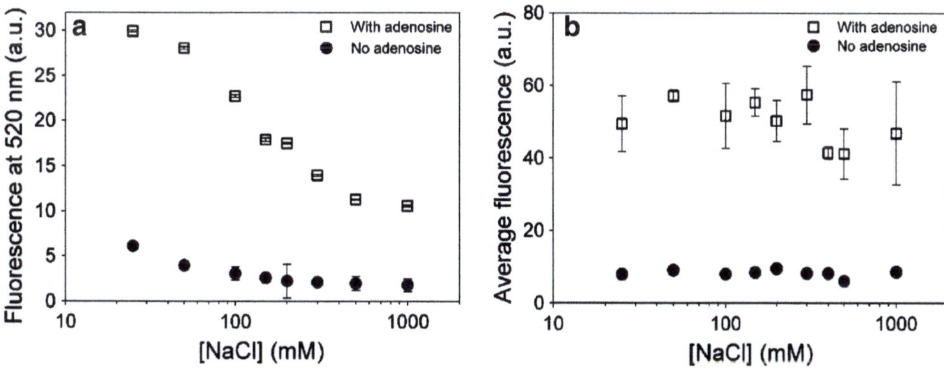

Fig. 5. Fluorescence signals of (**a**) non-immobilized and (**b**) immobilized sensor with increasing NaCl concentrations. In both plots, the sensor signals before and 20 min after addition of 2 mM adenosine were compared. Copyright 2010 American Chemical Society. Reproduced from [Huang P. et al., 2010] with permission from American Chemical Society.

relatively more cation accumulation at low salt and less accumulation at high salt.

4.4. Performance of the Immobilized Sensor in Buffer

With the optimized conditions, the sensitivity and selectivity of the sensor in buffer A is examined. The flow cytometry histograms of the sensor in the presence of varying concentrations of adenosine are presented in Fig. 6a. A gradual shift in the peak fluorescence can be clearly observed. If the average fluorescence in Fig. 6a is plotted against adenosine concentration, an overall sigmoid-shaped curve is obtained with a slow signal increase at low target concentrations followed by a sharp rise at high concentrations (Fig. 6b). Based on the 3σ/slope calculation, the low concentration region can be fit to a linear response (inset) with a detection limit of 178 μM. Various nucleosides (10 mM) are prepared to test the selectivity of the immobilized sensor. As shown in Fig. 6c, the fluorescence peak of the uridine, cytidine, and guanosine samples remain at the same position as the untreated sensor. The result indicated that the high sensitivity of the aptamer was not altered because of the immobilization step. Adenosine analogues like ATP, AMP, and cAMP were also tested (Fig. 6d). The sensor is also sensitive to these compounds but with slightly lower responses, which is consistent with literature reports (15).

4.5. Performance of the Immobilized Sensor in Serum

For quantitative analysis of adenosine in serum, samples with 30 % serum are spiked with varying concentrations of adenosine (see Note 4). Flow cytometry analysis shows the fluorescence signal increased with increasing adenosine (Fig. 7a). The detection limit of 167 μM is estimated, which is relatively close to the value obtained in buffer A. High selectivity of the immobilized aptamer in serum shown in Fig. 7b also indicated that serum did not compromise the target recognition property of the sensor. Therefore, combining immobilization and flow cytometry makes it possible to achieve sensitive detection even in very complex sample matrix.

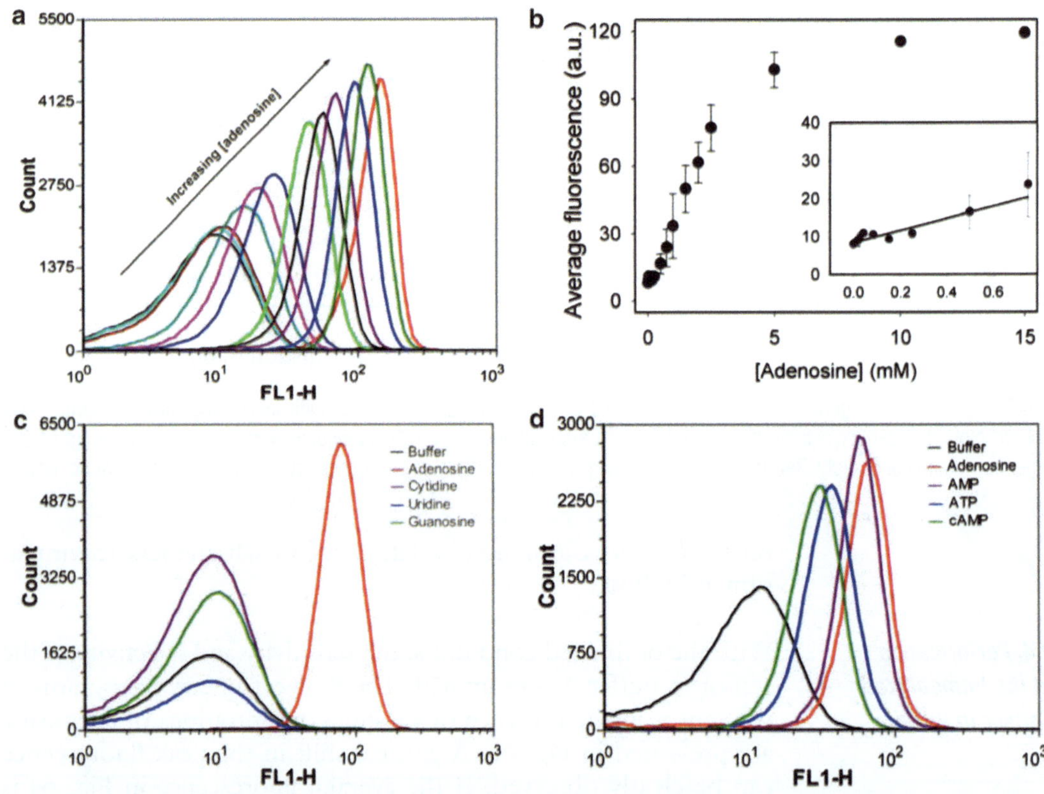

Fig. 6. (**a**) Flow cytometry histograms of the sensor in the presence of varying concentrations of adenosine. (**b**) Sensitivity and (**c**, **d**) selectivity of the immobilized adenosine sensor in buffer are also shown. In panels (**c**) and (**d**), 10 mM adenosine, adenosine analogues, or other nucleosides were used. Copyright 2010 American Chemical Society. Reproduced from [Huang P. et al., 2010] with permission from American Chemical Society.

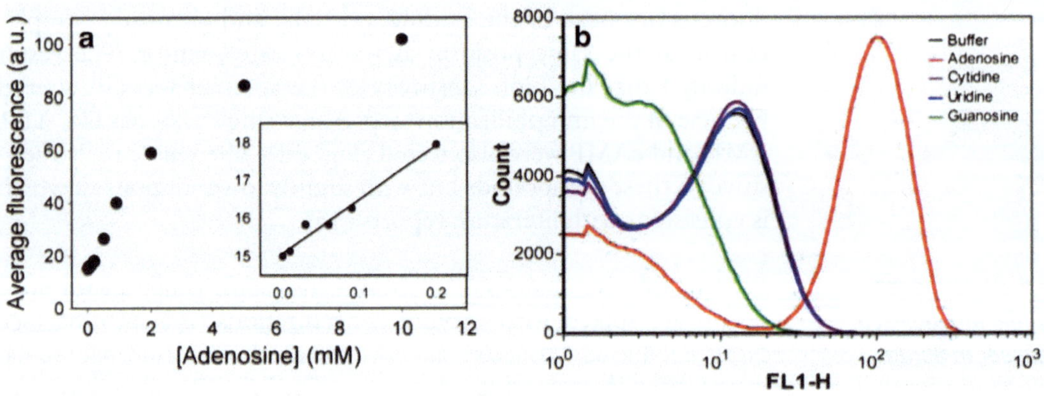

Fig. 7. (**a**) Sensitivity and (**b**) selectivity of the immobilized adenosine sensor in 30 % serum. Flow cytometry histograms of the sensor response to varying concentrations of adenosine in (**c**) 90 % serum and (**d**) 90 % serum after centrifugation to remove aggregated proteins in serum are also shown. Panel (**c**) shares the figure legend shown in panel (**d**), and the *inset* in panel (**c**) highlights the magnetic microparticles population. Copyright 2010 American Chemical Society. Reproduced from [Huang P. et al., 2010] with permission from American Chemical Society.

5. Notes

1. For sensor immobilization, five thymine nucleotides are introduced as a spacer between the biotin and the aptamer to minimize possible effects of surface immobilization on aptamer binding. Such spacers are typically used to separate the target recognition part from the surfaces.
2. Adenosine, guanosine, and cAMP are difficult to dissolve. The samples need to be heated to boiling temperature. Adenosine at 50 mM can remain dissolved for ~30 min while the other two recrystallize in a few minutes even with just 25 mM. After several times of heating, some white powders start to show up on the cap region. When this happens, it is recommended to prepare new samples to avoid errors in concentration determination.
3. It is important to estimate the DNA coverage on magnetic microparticles. For this purpose, mix a fixed amount of the FAM and biotinylated DNA (e.g., 16.5 pmol) with various amounts of particles. After 30 min incubation, supernatant fluorescence is measured. The fluorescence signal of the supernatant decrease with increasing amounts of particles. It is determined that 1 μL of magnetic microparticles ($\sim 6.5 \times 10^5$ particles) can take 1.31 ± 0.19 pmol DNA. The average number of DNA on each particle is thus 1.2×10^6. The experimental value is comparable to the value provided by the vendor ($\sim 1.8 \times 10^6$).
4. For flow cytometry experiments performed in the presence of human blood serum, we noticed that peak broadening sometimes occurred. This is attributed to the increased viscosity of the flow sample, which decreased particle flow rate in serum to produce a broader peak in the histogram.

6. Future Directions

In this chapter, we describe a protocol for immobilizing a fluorescent biosensor for adenosine and ATP detection on magnetic microparticles. High quality streptavidin-coated magnetic microparticles are commercially available, allowing for convenient attachment of biotinylated DNA. We find the following advantages for sensor immobilization. First, the target recognition and detection steps can be carried out independently, allowing detection even in concentrated human blood serum samples. Second, immobilization allows signal amplification. This has not been directly demonstrated in the current example, but in a related work, we have shown that by using a larger sample volume, sensitivity can be increased (43).

Other signal amplification mechanisms can also be found in the literature. Third, quite unexpectedly, we find that the immobilized sensor is less salt dependent. While both non-immobilized and immobilized sensors work even in 1 M NaCl, which is close to the salt concentration in the ocean water, the immobilized sensor performance is more consistent from low to high salt.

This work demonstrates that by combining aptamer sensor with magnetic microparticles, it is possible to achieve detection in complex sample matrixes. Future work will be focused on the following several aspects. First, aptamer selection work needs to be performed in buffers mimicking the marine environment for related targets such as the various marine toxins. Since aptamers are polyanions, their binding properties are likely to be affected by ionic strength. Selections carried out in ocean water conditions will likely generate aptamers that show optimal detection for practical marine applications. To demonstrate real applications of aptamers, sensors need to be able to detect several important targets at analytically relevant concentrations.

The current detection method relies on flow cytometry. While portable flow cytometers are being developed, alternative means of detection is desirable. For example, visual detection does not require any analytical instrument and is thus an important direction. This requires a relatively high DNA concentration of ~1 μM, which might be difficult to achieve for surface immobilized sensors. Therefore, three-dimensional immobilization is needed to achieve a higher sensor loading capacity. For example, we recently reported the use of hydrogel for visual detection of mercury and lead ions in water (38, 46). Incorporation of magnetic nanoparticles into hydrogels may achieve both magnetic separation and high loading capacity.

The current sensor has a relatively low sensitivity. For example, the adenosine aptamer has a dissociation constant of ~10 μM, but the immobilized sensor has a detection limit close to 200 μM. While the concentration of ATP is in the millimolar range inside a cell and this sensor is still useful for cellular analysis, detection in marine samples is likely to require much better sensitivity. Novel sensor design methods need to be developed to increase sensitivity.

References

1. Turner APF (2000) Techview: biochemistry: biosensors-sense and sensitivity. Science 290: 1315–1317
2. Rosi NL, Mirkin CA (2005) Nanostructures in biodiagnostics. Chem Rev 105:1547–1562
3. Liu J, Cao Z, Lu Y (2009) Functional nucleic acid sensors. Chem Rev 109:1948–1998
4. Metfies K, Huljic S, Lange M, Medlin LK (2005) Electrochemical detection of the toxic dinoflagellate Alexandrium ostenfeldii with a DNA-biosensor. Biosens Bioelectron 20: 1349–1357
5. Kroger S, Piletsky S, Turner APF (2002) Biosensors for marine pollution research,

monitoring and control. Mar Pollut Bull 45: 24–34

6. Dove A (2009) Long-term trends in major ions and nutrients in Lake Ontario. Aquat Ecosyst Health Manage 12:281–295
7. Azoulay A, Garzon P, Eisenberg MJ (2001) Comparison of the mineral content of tap water and bottled waters. J Gen Intern Med 16:168–175
8. Webb DA (1939) The sodium and potassium content of sea water. J Exp Biol 16:178–183
9. Tuerk C, Gold L (1990) Systematic evolution of ligands by exponential enrichment: RNA ligands to bacteriophage T4 DNA polymerase. Science 249:505–510
10. Ellington AD, Szostak JW (1990) In vitro selection of RNA molecules that bind specific ligands. Nature 346:818–822
11. Robertson DL, Joyce GF (1990) Selection in vitro of an RNA enzyme that specifically cleaves single-stranded DNA. Nature 344:467–468
12. Zhang X-B, Kong R-M, Lu Y (2011) Metal ion sensors based on DNAzymes and related DNA molecules. Annu Rev Anal Chem 4:105–128
13. Fang XH, Tan WH (2010) Aptamers generated from cell-SELEX for molecular medicine: a chemical biology approach. Acc Chem Res 43:48–57
14. Mayer G, Ahmed MSL, Dolf A, Endl E, Knolle PA, Famulok M (2010) Fluorescence-activated cell sorting for aptamer SELEX with cell mixtures. Nat Protoc 5:1993–2004
15. Sefah K, Shangguan D, Xiong XL, O'Donoghue MB, Tan WH (2010) Development of DNA aptamers using Cell-SELEX. Nat Protoc 5:1169–1185
16. Marshall KA, Ellington AD (2000) In vitro selection of RNA aptamers. RNA-Ligand Interact B 318:193–214
17. Liu Z, Mei SHJ, Brennan JD, Li Y (2003) Assemblage of signaling DNA enzymes with intriguing metal-ion specificities and pH dependences. J Am Chem Soc 125:7539–7545
18. Nelson KE, Bruesehoff PJ, Lu Y (2005) In vitro selection of high temperature Zn2+ dependent DNAzymes. J Mol Evol 61:216–225
19. Cho EJ, Lee J-W, Ellington AD (2009) Applications of aptamers as sensors. Annu Rev Anal Chem 2:241–264
20. Navani NK, Li Y (2006) Nucleic acid aptamers and enzymes as sensors. Curr Opin Chem Biol 10:272–281
21. Li D, Song SP, Fan CH (2010) Target-responsive structural switching for nucleic acid-based sensors. Acc Chem Res 43:631–641
22. Famulok M, Hartig JS, Mayer G (2007) Functional aptamers and aptazymes in biotechnology, diagnostics, and therapy. Chem Rev 107:3715–3743
23. Jhaveri SD, Kirby R, Conrad R, Maglott EJ, Bowser M, Kennedy RT, Glick G, Ellington AD (2000) Designed signaling aptamers that transduce molecular recognition to changes in fluorescence intensity. J Am Chem Soc 122:2469–2473
24. Nutiu R, Li Y (2004) Structure-switching signaling aptamers: transducing molecular recognition into fluorescence signaling. Chem Eur J 10:1868–1876
25. Nutiu R, Li Y (2003) Structure-switching signaling aptamers. J Am Chem Soc 125: 4771–4778
26. Yang CJ, Jockusch S, Vicens M, Turro NJ, Tan W (2005) Light-switching excimer probes for rapid protein monitoring in complex biological fluids. Proc Natl Acad Sci U S A 102:17278–17283
27. Stojanovic MN, de Prada P, Landry DW (2001) Aptamer-based folding fluorescent sensor for cocaine. J Am Chem Soc 123:4928–4931
28. Jiang Y, Fang X, Bai C (2004) Signaling aptamer/protein binding by a molecular light switch complex. Anal Chem 76:5230
29. Wang J, Liu B (2008) Highly sensitive and selective detection of Hg2+ in aqueous solution with mercury-specific DNA and Sybr Green I. Chem Commun:4759–4761.
30. Storhoff JJ, Mirkin CA (1999) Programmed Materials Synthesis with DNA. Chem Rev 99:1849–1862
31. Swearingen CB, Wernette DP, Cropek DM, Lu Y, Sweedler JV, Bohn PW (2005) Immobilization of a catalytic DNA molecular beacon on Au for Pb(II) detection. Anal Chem 77:442–448
32. Carrasquilla C, Li Y, Brennan JD (2011) Surface immobilization of structure-switching DNA aptamers on macroporous sol–gel-derived films for solid-phase biosensing applications. Anal Chem 83:957–965
33. Yang RH, Jin JY, Chen Y, Shao N, Kang HZ, Xiao Z, Tang ZW, Wu YR, Zhu Z, Tan WH (2008) Carbon nanotube-quenched fluorescent oligonucleotides: probes that fluoresce upon hybridization. J Am Chem Soc 130:8351–8358
34. Lu CH, Yang HH, Zhu CL, Chen X, Chen GN (2009) A graphene platform for sensing biomolecules. Angew Chem Int Ed 48:4785–4787
35. Zhu Z, Wu CC, Liu HP, Zou Y, Zhang XL, Kang HZ, Yang CJ, Tan WH (2010) An aptamer cross-linked hydrogel as a colorimetric

platform for visual detection. Angew Chem Int Ed 49:1052–1056

36. Yang HH, Liu HP, Kang HZ, Tan WH (2008) Engineering target-responsive hydrogels based on aptamer - target interactions. J Am Chem Soc 130:6320–6321
37. Liu J (2011) Oligonucleotide-functionalized hydrogels as stimuli responsive materials and biosensors. Soft Matter 7:6757–6767
38. Dave N, Huang P-JJ, Chan MY, Smith BD, Liu J (2010) Regenerable DNA-functionalized hydrogels for ultrasensitive, instrument-free mercury(II) detection and removal in water. J Am Chem Soc 132:12668–12673
39. Bruno JG, Kiel JL (2002) Use of magnetic beads in selection and detection of biotoxin aptamers by electrochemiluminescence and enzymatic methods. Biotechniques 32(178–80):82–83
40. Ma CP, Wang WS, Yang Q, Shi C, Cao LJ (2011) Cocaine detection via rolling circle amplification of short DNA strand separated by magnetic beads. Biosens Bioelectron 26:3309–3312
41. Bi S, Li L, Zhang SS (2010) Triggered polycatenated DNA scaffolds for DNA sensors and aptasensors by a combination of rolling circle amplification and DNAzyme amplification. Anal Chem 82:9447–9454
42. Huang PJJ, Liu JW (2010) Flow cytometry-assisted detection of adenosine in serum with an immobilized aptamer sensor. Anal Chem 82:4020–4026
43. Huang P-JJ, Liu J (2011) Immobilization of DNA on magnetic microparticles for mercury enrichment and detection with flow cytometry. Chem A Eur J 17:5004–5010
44. Lee J, Icoz K, Roberts A, Ellington AD, Savran CA (2009) Diffractometric detection of proteins using microbead-based rolling circle amplification. Anal Chem 82: 197–202
45. Huizenga DE, Szostak JW (1995) A DNA aptamer that binds adenosine and ATP. Biochemistry 34:656–665
46. Joseph KA, Dave N, Liu J (2011) Electrostatically directed visual fluorescence response of DNA-functionalized monolithic hydrogels for highly sensitive Hg2+ detection. ACS Appl Mater Interfaces 3:733–739

Chapter 9

Functional Gene Arrays for Analysis of Microbial Communities on Ocean Platform

Katelyn M. McKindles and Sonia M. Tiquia-Arashiro

Abstract

Recently there has been an increased use of microarray technology as a tool to determine the presence of functional genes in a population of hard to culture communities (e.g., soil, extreme environments). A functional gene array (FGA), or GeoChip, uses probes to screen for specific functional genes vital in biological systems such as nitrogen and carbon cycling, and has even been expanded to include aquatic conditions. However, the time delay from when the sample is taken from the ocean to evaluating the test results back in the lab still posed a problem. The Environmental Sample Processor (ESP) minimizes this time difference by housing a robotic system placed in the ocean for a long period of time that can collect a small sample, concentrate the DNA, run a microarray, and take a picture of the array before sending the data ashore to be evaluated by a researcher. The included protocol and reagents list goes through both lab microarray procedures as well as the procedures list for the ESP, which briefly mentions deployment and data acquisition. The protocols described here should advance applications in microbial oceanography using robotic instrumentation.

Key words: Autonomous platform, Functional gene array, Environmental sample processor, Microarray, Hybridization, Genosensor, DNA, GeoChip, Oligonucleotide, Biogeochemical cycling genes

1. Introduction

Over the past decade, environmental scientists have been casting a wider net in their attempts to understand complex environmental processes on a molecular scale. Microarray technology is used in gene expression studies of individual microorganisms (1–4), and has more recently been used to understand how genes are important regulators of earth-scale processes as carbon and nitrogen cycling (5, 6). DNA microarrays show great promise as a revolutionary tool for large-scale parallel analysis of microbial community structure and activities (5, 7–9).

Sonia M. Tiquia-Arashiro (ed.), *Molecular Biological Technologies for Ocean Sensing*, Springer Protocols Handbooks, DOI 10.1007/978-1-61779-915-0_9,

Different types of microarrays have been developed to monitor microbial community dynamics in environmental studies (Table 1) (10), including functional gene arrays (FGAs) or GeoChip (5–7, 10–15), community genome arrays (CGA) (16, 17), and phylogenetic oligonucleotide arrays (POAs) (10). FGA is a gene microarray chip designed to identify "functional genes" involved in important nutrient cycles. It allows the identification of genes in an environmental sample that regulate carbon fixation, decomposition, and atmospheric nitrogen fixation, to name a few. Understanding what functional genes are available in a system allows scientists to both understand the potential of that system for cycling nutrients and better predict how that system will respond to environmental change. Imagine a glass floor divided into hundreds of identical squares. Each of these squares contains a different fragment of DNA, reconstructed from known DNA sequences. To probe an environmental sample for specific DNA sequences, the samples are hybridized over the floor. Fragments of DNA will stick to their complementary sequence on the floor, causing a square to light up. The array can be

Table 1
Major differences of various types of microarrays for environmental studies (10)

	CGAs	PCR-product-based FGAs	Oligonucleotides-based FGAs	POAs
Probe size	Entire genomic DNAs	Individual functional genes (200–1,000 bp)	Individual functional genes (50–70 bp)	Ribosomal rRNA (18–25 bp)
Types of information provided	Phylogenetic	Functional	Functional	Phylogenetic
Construction of comprehensive arrays	More difficult	More difficult	Easier	Intermediate
Reagent handling and tracking	Intermediate	More difficult	Easier	Easier
Targeted microorganisms	Culturable	Culturable and non-culturable	Culturable and non-culturable	Culturable and non-culturable
Specificity	Species	<80–85 % sequence homology	<86–90 % sequence homology	Single nucleotide difference
Sensitivity (ng of pure genomic DNA)	~0.2	~1	~8	Undetermined
Quantitation	Yes	Yes	Yes	Unknown
Taxonomic resolution	Genus–species	Genus–species	Species–strains	Species–strains

read by identifying fluorescently lit spots where environmental DNA has attached. This information is then used to develop a picture of the functional genes present in that system (9).

In Antarctica, FGA has already been used to answer important ecological questions (18). For example, scientists are finding that genes for nitrogen fixation, the crucial ecosystem process that produces plant-useable nitrogen in the soil, occur in lichen-rich areas. Lichens are believed to be among the earliest land colonizers, and the ability of lichen-dominated systems to add nitrogen to the soil may be an important finding in reconstructing the early colonization of terrestrial systems. Other findings include carbon-fixation genes in plots that lack vegetation, indicating microbial communities that are able to perform some sort of photosynthesis in the absence of plants (18).

In two more recent studies, FGA technology was used in aquatic conditions, specifically in acid mine drainage (AMD) (15) and the hydrothermal vent at the Juan de Fuca Ridge (14). While both of these microbial communities have been thoroughly studied in the past, very little was understood about their functional gene and physiological diversity. The use of GeoChip 2.0 when studying the microbial community of an AMD found that almost all major metabolic processes could be found in this ecosystem, including carbon and nitrogen fixation, carbon degradation, methane metabolism, ammonification, and more (15). Similarly, GeoChip was used to study the community housed in the chemical and thermal gradients of a vent chimney at the Juan de Fuca Ridge (14), through which it was discovered and reaffirmed that high-throughput microarray technology has a great potential in understanding ecosystem dynamics. In the hydrothermal vent study, GeoChip revealed the presence of functional communities involved in CO_2 fixation, methane cycling, nitrogen cycling, and metal resistance (14). Unfortunately, there is one limitation of this method of research, and that is that DNA-based GeoChip analysis can only detect the functional potential of a community, not the specific population (15).

The Environmental Sample Processor (ESP) (Figs. 1a, b and 2) takes FGA technology in the ocean one step further by making it an almost fully automatic system. The ESP is a relatively new technology which enables scientists to analyze the microorganisms of the ocean in an almost real-time situation by remotely collecting samples from the oceans subsurface, administering reagents as needed for a selected few processes such as probe arrays, and collecting data to be sent as images to be processed ashore via radio mooring (19) (Figs. 3 and 4). This instrument has a uniform methodology, which makes it ideal in detecting a variety of targets using one system, and consists of three major sections: (1) the core sample processor (Fig. 5a), (2) the sampling modules, and (3) the analytical modules (20). The system uses a rotating carousel to house the sample chambers or "pucks," (Fig. 5b) which contain

Fig. 1. (**a**) 2 G ESP without pressure housing, showing pucks at bottom of clear plastic carousel tubes (*lower center*) and several reagent bags (*lower right*). (**b**) Field deployment of 2 G ESP contained in pressure housing and mounted with two battery packs (*orange boxes*) (64).

Fig. 2. The second-generation Environmental Sample Processor (2 G ESP) being tested in a seawater tank ahead of deployment in Monterey Bay. The instrument is moored subsurface and an electromechanical cable provides for communications between a remote station and the ESP's surface buoy. An integral conductivity-temperature depth (CTD) package is visible at left. The ESP operates on 12-V rechargeable batteries (at bottom, above the anchor). *Photo credit: Todd Walsh, Monterey Bay Aquarium Research Institute* (19).

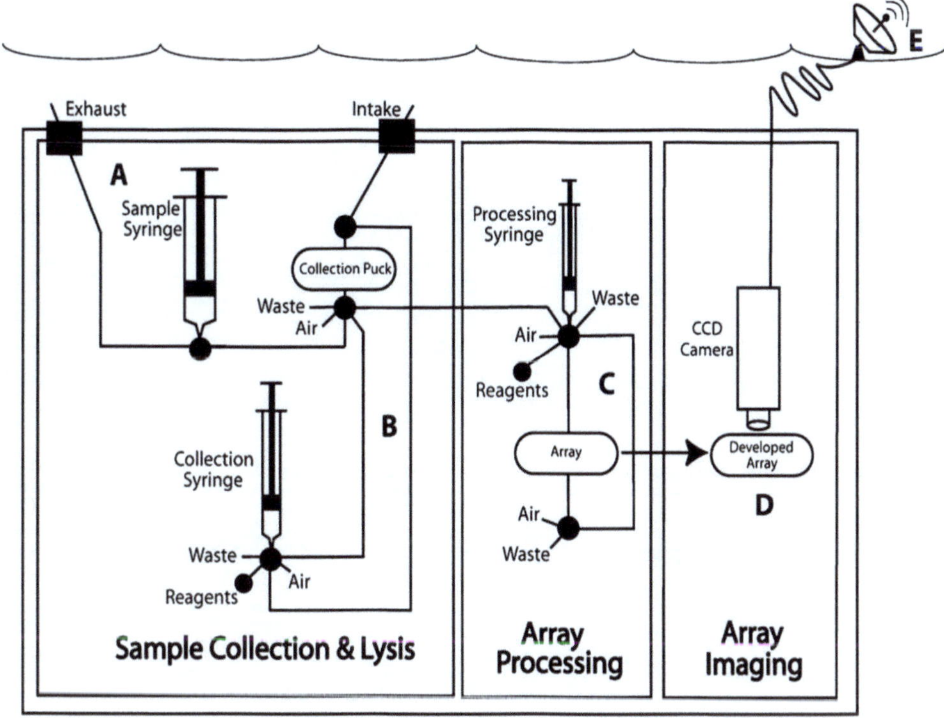

Fig. 3. Schematic diagram of the fluid path within the ESP instrument for sample collection and array processing. Seawater is brought into the instrument and filtered through a puck until the specified volume is reached or until the filter clogged. (**a**) The collection syringe presents various reagents to the particulates collected on the filter in the puck, resulting in cell lysis followed by dilution of the lysate. (**b**) The diluted lysate is passed to the processing syringe. The processing syringe delivers the lysate to a puck containing an array; then after incubation cleared to waste. The process is repeated for subsequent reagents. (**c**) The array is positioned under CCD camera and photographed. (**d**) The resulting image file is sent ashore via surface radio mooring. (**e**) Black circles represent valves that make connections between the syringes and puck, reagents, air or waste (22).

filter media and probe arrays, but utilize robotic mechanisms to load the pucks into their proper processing positions after the ESP has collected a sample (21). The samples are run through the sandwich hybridization assay (SHA) format, which detects 16S rRNAs indicative of phylogenetically distinct groups of marine bacterioplankton (22) (Fig. 6), a variety of invertebrates, and harmful algal species (23). In addition to collecting the samples, the ESP houses chemical and physical sensors which enable the samples to be evaluated in respect to the environmental conditions (23).

The technology originated as a means to study the emergence of harmful algal blooms (HABs) (24, 25), but has since included the detection of marine bacterioplankton (22). The HABs produce a toxin that severely disrupts both the ecosystem and the surrounding human population, a problem which demands further research. The use of the ESP in the ocean allowed for periodic sampling of the water during near real-time intervals which enabled researchers to find trends in phytoplankton abundances (24). Later deployments

Fig. 4. Vision of the components of an ocean-observing system, including cabled observatories, autonomous underwater vehicles, gliders, buoys, moorings, satellites, and a traditional observing platform (research vessel) (19).

of the ESP confirmed the hypothesis that environmental conditions, such as the strength of upwelling (a wind driven motion which greatly enhances nutrient supply to the surface of the ocean) and associated patterns in stratification effected the HAB species composition. *Alexandrium catenella*, a motile dinoflagellate, is favored by relatively strong stratification while *Pseudonitzschia* is favored by a strong upwelling pulse (25). The main goal of both of these studies was to be able to determine either the cause or any trends in HABs in order to effectively and efficiently monitor and predict their occurrence (25). The study done by Preston et al. (22) had a slightly different goal in mind to report the first in situ DNA probe-based detection of marine bacterioplankton. This study discusses the ability of the ESP to become a diverse detection tool, as well as the limitations of the system, such as its inability to detect low copy number targets. No matter what the target microbe the ESP has

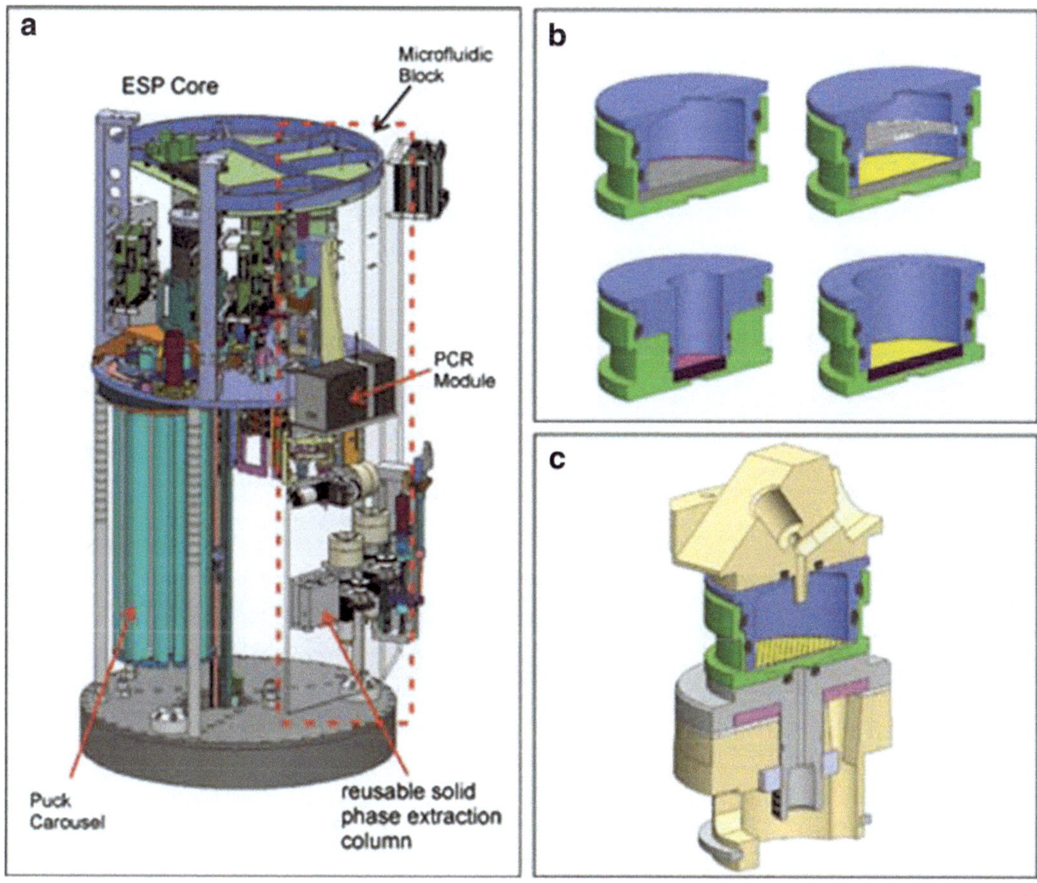

Fig. 5. SolidWorks® models showing: (**a**) the core ESP with a microfluidic block (MFB) and polymerase chain reaction module attached on the right-hand side, (**b**) different types of pucks used for processing samples, and (**c**) the sample-collection station clamp in the closed position holding a sample puck. Puck assemblies are specially designed for different operations. In (**b**), the pucks shown are designed for collecting and homogenizing large-volume samples (*top left*), archiving material for microscopy (*top right*), and developing probe arrays printed on 12-mm or 25-mm membranes (*bottom left* and *right*, respectively). All pucks conform to the same overall size and shape so that robotic systems used to move and utilize the pucks can be standardized against a constant form factor. ESP with MFB is ~0.5-m diameter and ~1-m tall. A puck is ~30-mm diameter × 17-mm tall (20).

been programmed to detect, this tool has such potential in real-world and autonomous applications.

The following sets of protocols are intended to serve as a basic introduction to microarray construction and the steps required in microarray experimental design. There are four fundamental steps required in oligonucleotide-based FGA construction and experimentation: (1) FGA microarray construction, (2) Labeling and quantitation of labeled DNA, (3) Hybridization, and (4) Image processing and data analysis. The schematic diagram for these steps is illustrated in Fig. 7. It is our hope that the methods presented here will serve as an initial and useful tool to study the functional gene profiles of microbes the ocean.

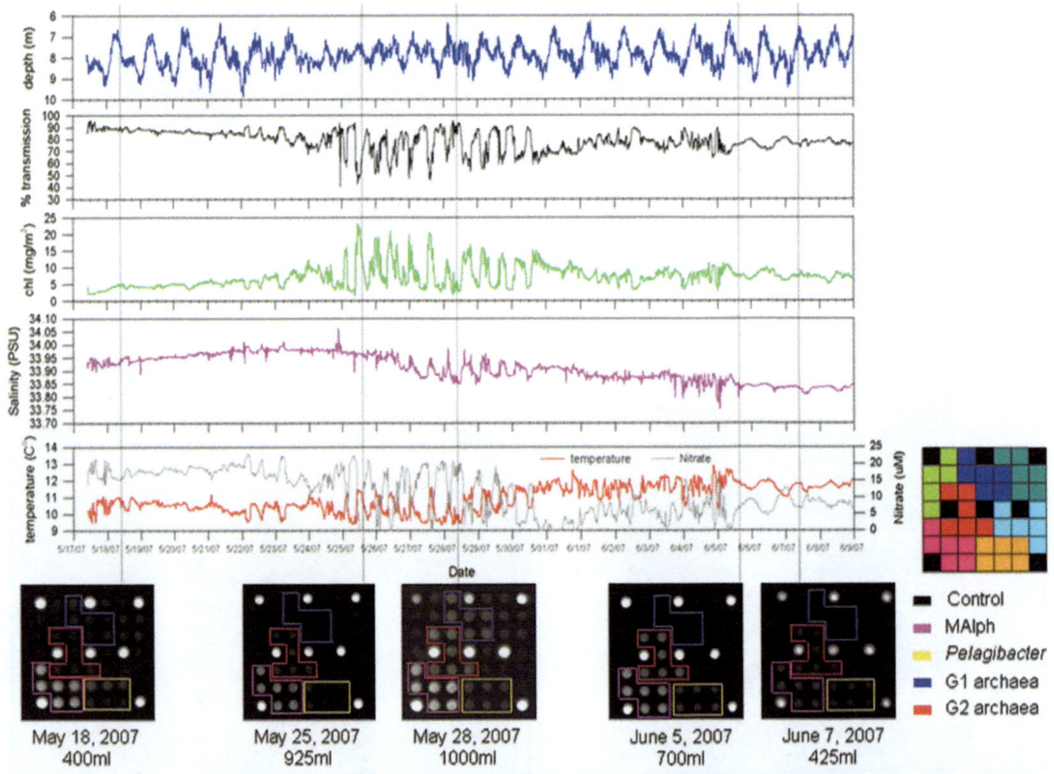

Fig. 6. In situ detection of marine bacterioplankton using ESP DNA probe arrays during a spring field deployment. The ESP was moored in Monterey Bay, California, 17 May to 11 June 2007. Top four graphs show physical and chemical data collected by contextual sensors on the ESP mooring during the deployment. The bottom images shows DNA probe arrays targeting rRNA indicative of various groups of marine bacterioplankton. Sample volume is shown underneath the array. The arrays shown are 15 mm × 15 mm (20).

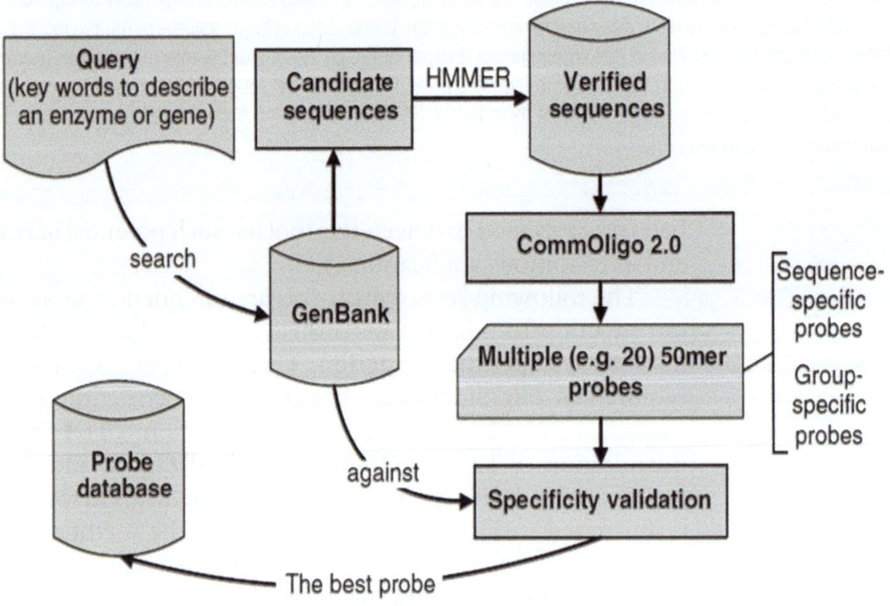

Fig. 7. The design pipeline for FGA construction (11).

2. Reagents and Equipment

2.1. Oligo Microarray Fabrication

2.1.1. Reagents

50 % D MSO printing buffer

0.1 % SDS buffer

2.1.2. Equipment and Materials

Aminosilane-coated glass slides (SuperAmine) (Telechem # SMA)

384-Well Printing Plate and lid

Orbital Shaker

Centrifuge with rotor for microtitre plates

Array Printer (PixSys 5500) (Cartesian Technologies, Irvine, CA)

200–500 μm spaced split pins

Slide box

GS Gene Linker® UV ChamberOligonucleotide probes

2.1.3. Software and Web-Based Resources

Genbank, EMBL, or Swiss Prot

UniGene (UniGene Laboratories, Boonton, NJ)

PRIMEGENS (Digital Biology Laboratory, University of Missouri-Columbia) Primer 3 (Whitehead Institute for Biomedical Research, Cambridge, MA), or Web Primer (Stanford University, Stanford, CA)

OligoArray (University of Michigan, Ann Arbor), Array Designer (PREMIER Biosoft, Palo Alto, CA), or Sarani (Strand Genomics, Burlingame, CA)

2.2. Target Preparation

Temliphi amplification kit

PicoGreen (Quant-iT™ PicoGreen® dsDNA kit; Invitrogen, Carlsbad, CA)

2.3. Labeling and Quantitation of Target DNA

2.3.1. Reagents

1 mM Cy3 or Cy5 dCTP (Amersham Pharmacia Biotech #PA55021)

RNase- and DNase-free water (Ambion, Inc. # 9931)

750 ng μL^{-1} random octamer primers (Invitrogen # Y01393)

5 mM dATP, dTTP, dGTP, and 2.5 mM dCTP (Biopioneer Inc., San Diego, CA)

40 U μL^{-1} Klenow fragment (Invitrogen # Y01396)

20 pmol PCR primers

25 mM dATP, dCTP, dGTP, and 15 mM dTTP (New England Biolabs)

10 mM aminoallyl-dUTP (Sigma, St. Louis, MO)

Taq DNA polymerase

0.1 M carbonate buffer (pH 9.0)

N-Hydroxy Succinimide esters Cy3 or Cy5 (NHS-Cy3 or Cy5; Amersham Pharmacia Biotech, Piscataway, NJ)

100 mM NaOAC

2.3.2. Equipment and Materials

QIAquick columns (Qiagen, Valencia, CA)

Speed-vac (e.g., SPD 1010 SpeedVac system; Thermo Savant Waltham, MA)

Spectrophotomer (NanoDrop^{TM} ND-1000 spectrophotomer; Nanodrop Technologies, Wilmington, DE)

2.4. Hybridization

2.4.1. Reagents

Hybridization solution: RNase-free water, formamide, 20× saline sodium citrate (SSC) solution, 5 % SDS buffer, Herring sperm DNA

SSC solution

0.2 % SDS buffer

2.4.2. Equipment and Materials

Hybridization chamber (product number 2551; Corning, Lowell, MA)

Hybridization oven (e.g., PersonalHyb Hybridization oven; Strategene, La Jolla, CA)

Array coverslip (Structure Probe, West Chester, PA)

Centrifuge with rotor for microtitre plates

2.5. Image Processing and Data Analysis

2.5.1. Equipment

Fluorescence reader (e.g., FLUOstar OPTIMA; BMG Labtech, Durham, BC)

2.5.2. Software

1. ScanArray 5000 System (GSI Lumonics, Watertown, MA).
2. ArrayStat^{TM} (Imaging Research, Inc., Ontario, Canada).
3. Imagene (BioDiscovery, El Segundo, CA), GenPix Pro (Axon Instruments, Union City, CA), Array Pro (Media Cybernetic, Carlsbad, CA), Quant Array (Packard Biosciences, Boston, MA), or TIGR Spot Finder (The Institute of Genomic Research TIGR, Rockville, MD).

2.6. Environmental Sample Processor

2.6.1. Reagents

Signal probe cocktail in 2 M guanidinium thiocyanate (GuSCN) signal buffer

Anti-Dig HRP (Pierce; Rockford, IL)

Stabilized diluent blocker (1 mL) (Pierce)

Substrate (Pierce SuperSignal West Femto Maximum Sensitivity Substrate: Stable Peroxidase Buffer and Luminol/Enhancer Solution mixed 1:1 before delivery to the array)

2.6.2. Equipment and Materials

First- or Second-Generation ESP.

5-μm pore size hydrophilic Durapore backing filter (Millipore).

Charge-coupled device (CCD) camera (1G ESP: Santa Barbara Instruments model ST-8EI [Santa Barbara, CA, USA] with a Fujinon [Japan] HF35A-2M1 lens; 2G ESP: Starlight Xpress model SXV-H9 [England] with a Fujinon model HF16HA-1B lens).

V++ Precision Digital Imaging System, v. 4.0 (Digital Optics, Auckland, NZ).

2.7. General Laboratory Supplies and Equipment Required

Pipettes (with appropriate tips)

0.1 ml PCR tubes

1.8 mL microcentrifuge tubes

Laminar flow hood or PCR workstation hood (e.g., AirClean 600 PCR Workstation, AirClean Systems, Raleigh, NC)

Thermocycler

Water bath

Heating block

Gel electrophoresis unit (with appropriate agarose, buffers, and staining supplies)

Ice bucket

Nuclease-free or PCR quality water

MilliQ water

Coplin jars or similar container

Diamond pen

Slide rack

3. Protocol

3.1. Oligo Microarray Fabrication

3.1.1. Oligo Design

The 50-mer FGAs can be constructed with the probes from sequences recovered from a variety of environments to represent the known microbial population diversity involved in the biogeochemical processes of interest. Sequences can be retrieved from public databases such as GenBank, EMBL, and SwissProt. However, these databases contain redundant sequences and it can be difficult to retrieve all sequences of interest. It might be better to retrieve sequences from the UniGene database (http://www.ncbi.nlm.nih.gov/UniGene/query_tips.html). This database is a collection of unique GenBank sequences grouped by organism and gene, and gives all sequences in one entry with links to the GenBank entries.

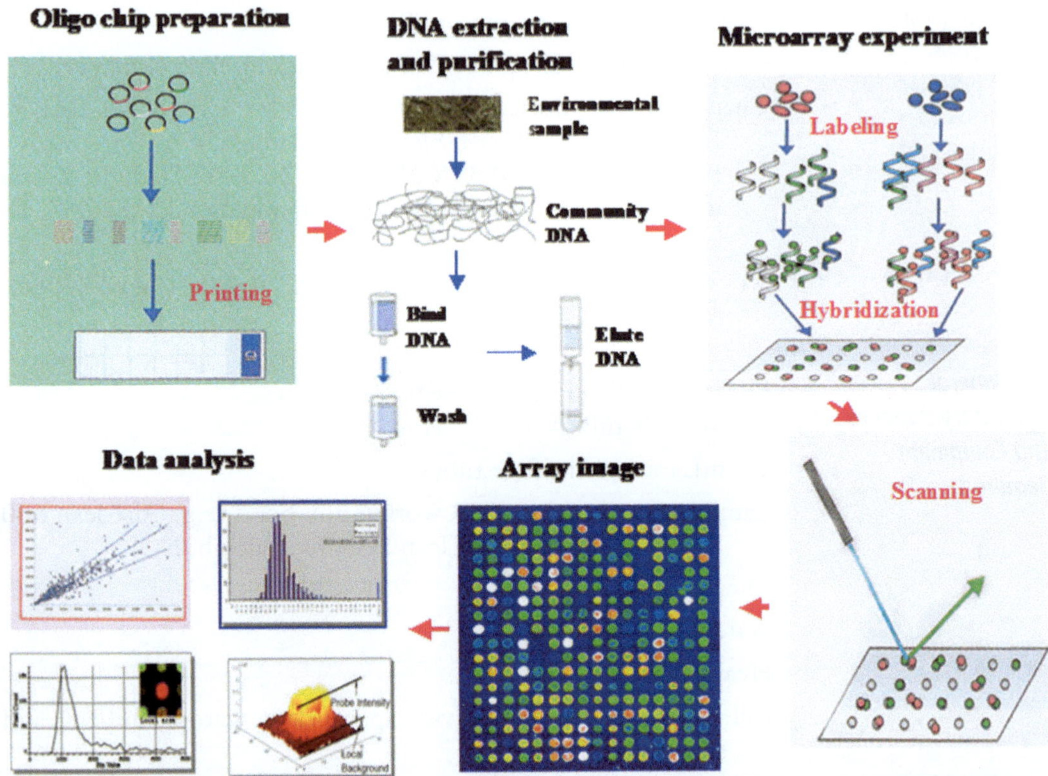

Fig. 8. Schematic diagram of microarray construction and experiments.

Alternatively, sequence retrieval can be carried out using the GeoChip design pipeline (11). The whole pipeline runs on a Web-based Common Gateway Interface (CGI) server and the scripts are written in Perl (Fig. 8). For each functional gene, a query of words is first submitted to GenBank Protein Database to fetch all candidate amino acid sequences. All candidate sequences for each functional gene are retrieved by key words and confirmed by HMMER 2.3.2. (Ashburn, VA, USA) (26), with seed sequences. In addition, all confirmed protein sequences are used to obtain the nucleic acid sequences from GenBank for probe design. A new version of CommOligo 2.0 with group-specific probe design features is used to design 50-mer oligonucleotide probes (both gene-specific and group-specific) using the same criteria as described for GeoChip 2.0 (13). After the specificity of all designed probes is computationally checked with currently available databases (GenBank), the best probe for each sequence or each group of sequences is selected to synthesize for GeoChip 3.0 construction. Because all seed sequences and key words are stored in databases, automatic updates can be performed in the future (see Note 1).

To design 50-mer oligonucleotide probes, oligo design software such as PRIMEGENS (http://compbio.ornl.gov/structure/primegens/), Primer 3 (http://www.bioinformatics.nl/cgi-bin/

primer3plus/primer3plus.cgi), and Web Primer (http://genome-www2.stanford.edu/cgi-bin/SGD/web-primer) can be used. These online programs have been used primarily for primer design but they can also be used to design oligo probes. The PRIMEGENS program has been used to design gene-specific primers for whole genome cDNA microarrays (27), and oligo probes for FGAs. The software initially compares each gene sequence against the entire sequence database using BLAST, and produces an alignment with the other sequences that have more than the desired threshold sequence similarity (e.g., 85 %) using dynamic programming. Based on the global optimal alignments, segments of 50 bp oligonucleotides with less than the threshold identity to the corresponding aligned regions of any of BLAST hit sequences are selected as potential probes. Among these identified potential probes, a final probe is selected by considering the GC content, melting temperature, and self-complementarity. Outputs of the designed probes are imported into Excel and a pivot table is constructed containing the sequence information of each probe. There are several free and commercial software packages for designing oligonucleotides. OligoArray (28) is a free software that designs gene-specific oligonucleotides for genome-scale microarray construction. Array Designer (Biosoft International, Palo Alto, CA) and Sarani (Strand Genomics, Burlingame, CA), are commercial softwares for automatic large-scale design of optimal oligonucleotide probes for microarray experiments. Thousands of gene sequences can be analyzed together, and the best available oligonucleotide probes with uniform thermodynamic properties and minimal similarity to nonspecific genes can be selected using these software.

Oligos are synthesized at the desired scale at the final concentration of 100 pmol ul^{-1} without any modification, and diluted to 30–40 pmol μl^{-1} with 50 % DMSO. Thereafter, oligonucleotides are printed onto aminosilane-coated glass slides such as SuperAmine (Telechem # SMA). SuperAmine slides contain covalent amine groups that allow stable attachment of nucleic acids.

3.1.2. Oligo Array Printing

1. Prepare printing oligo solution to a final concentration of 50 pmol μl^{-1} using 50 % DMSO in a 384-well printing plate (5 μl probe and 5 μl DMSO).
2. Cover the plate with plastic lid and mix in an orbital shaker at 700 rpm for 3 min.
3. Spin the printing plate using a centrifuge equipped with a rotor for microtitre plates at 500 rpm for 5 min.
4. Setup the array printer (PixSys 5500 printer; Cartesian technologies, Inc., Irvine, CA) and print slides according to the manufacturer's protocol. The ideal relative humidity should be between 40 and 60 % at room temperature (20–25 °C). The spot size should be approximately 100–150 μm, with 200–500 μm spacing distance using split pins from Telechem.
5. Allow the slides to dry for 2 h prior to UV cross-linking.

Reproducibility is one of the most critical requirements for microarray fabrication. For reliable and reproducible data, the uniformity of individual spots across the entire array is crucial for simplifying image analysis and enhancing the accuracy of signal detection. Several factors will affect the uniformity of spots, including array substrate, pins, printing buffer, and environmental controls. For instance, significant variations could be caused by pin characteristics due to the mechanical difference in pin geometry, pin age, and sample solutions. Movement of the pin across the surface in the *XY* direction may cause the tip to bend. Tapping the pins on the surface may result in deformation of the pin tips. Also, dragging the pin tip across the surface may cause clogging of the pin sample channel. Therefore, great care is needed in handling pins. Pins should be cleaned with an ultrasonic bath after each printing.

Environmental conditions have significant effects on spot uniformity and size (29). Humidity control is crucial to prevent sample evaporation from source plates and the pin channel during the printing process. Sample evaporation can cause changes in DNA concentration and viscosity. As a result, the quality of the deposited DNA will be changed. Also, reducing evaporation can help the spotted volume of DNA to have more time to bind at equal rates across the entire spot. As a result, DNA spots of increased homogeneity will be obtained (30). The printing buffer is also critical for obtaining homogeneous spots. With the widely used SSC buffer, the spot homogeneity as well as binding efficiency is often poor. Using the printing buffer containing 1.5 M betaine improves spot homogeneity as well as binding efficiencies (30). This is because betaine increases the viscosity of a solution and reduces the evaporation rate. More uniform spots can also be obtained with the printing buffer containing 50 % DMSO (dimethyl sulfoxide) (9, 29).

3.1.3. UV Cross-Linking and Slide Processing

At the end of the print, remove slides from the printer, label each slide with an identifier and the slide number by writing on the edge of the slide with a diamond pen and place slides in a dust-free slide box. It is useful to etch a line, which outlines the printed area of the slide, onto the first slide. This serves as a guide to locate the area after the slides have been processed.

1. Expose the slides, printed face up, to a 80 mJ dose of ultraviolet irradiation in a GS Gene Linker® UV Chamber for 30 s.
2. Wash slides at room temperature first with 0.1 % SDS and then with water:

		Washing time
Wash 1	0.1 % SDS	4 min
Wash 2	Water	2 min

3. Transfer slides to a ten-slide glass rack and place the rack into a glass tank.
4. Remove the slides and spin using a centrifuge equipped with a rotor for microtitre plates at 500 rpm for 5 min to dry.
5. Transfer the slides to a clean, dust-free slide box and let it stand overnight before hybridization.

3.2. Target Preparation

Once the oligo microarrays are printed, targets are prepared for hybridization. For community analysis of environmental samples, genomic DNAs from pure cultures or environmental clones are normally used as target, and human genes as controls (9). Successful application of microarray for microbial community analysis relies on the effective recovery of nucleic acids from the environment. Hurt et al. (31) and Zhou et al. (32) pointed out some criteria for ideal recovery of DNA or RNA from environmental samples: (1) the nucleic acid recovery efficiency should be high and not biased so that the final nucleic acids are representative of the total nucleic acids within the naturally occurring microbial community; (2) the DNA should be of sufficient purity for reliable hybridization; (3) the extraction and purification protocol should be robust and reliable. The DNA extraction and purification protocol described by Hurt et al. (31) fulfills the above criteria (see Note 2). Of course it should be possible to substitute other protocols that meet these criteria.

The FGA requires 2–5 μg of genomic DNA for hybridization. Depending on the amount of DNA available, whole genome amplification (WGA) may be required. WGA can be performed using either phage Φ29 (16, 33) or *Bacillus stearothermophilus* DNA polymerases (34, 35). Amplification using Φ29 (Templiphi, GE Healthcare; Piscataway, NJ) for WGA of microbial community DNA has been systematically evaluated and shown to provide sensitive (10 fg detection limit) and representative amplification (<0.5 % of amplified genes showed more than twofold different from unamplified) (16). The following protocol uses the Templiphi amplification kit (GE Healthcare) and is based on a previously published protocol (16). All steps should be carried out in a laminar flow hood or PCR workstation hood.

1. Add 10 μl of sample buffer (supplied with kit) to a PCR tube or micro-well plate.
2. Transfer 10–100 ng DNA to the sample buffer. The total volume of DNA added should be no more than 5 μl and the sample volume should be the same for all samples. Use nuclease-free water to bring the volume up, if necessary. See Note 3.
3. Mix the DNA and buffer thoroughly and incubate 10 min at room temperature.
4. While DNA and buffer are incubating, prepare the Templiphi premix [for each reaction: 10 μl reaction buffer, 0.6 μl enzyme

mixture (both supplied in the kit), and single-stranded binding protein (USB; Cleveland, OH) and spermidine to a final concentration of 260 ng μl^{-1} and 0.1 μM, respectively].

5. Transfer 12.85 μl of the Templiphi premix to the DNA/buffer mixture (or the equivalent volume for one sample).
6. Incubate the reaction at 30 °C for 3 h and then heat-inactivate the enzyme at 65 °C for 10 min.
7. To evaluate the amplification quality, run approximately 2 μl of amplified product on a gel. The product should produce a smear rather than a single band.
8. Quantify the amplified DNA using a dye-binding assay, such as PicoGreen (Quant-iT™ PicoGreen® dsDNA kit; Invitrogen, Carlsbad, CA). The amplified product cannot be measured using 260/280 ratios due to primers and dNTPs remaining in the sample. There should be at least 2 μg of amplified DNA. If there is less than this, the amplification should be repeated.

3.3. Labeling and Quantitation of Target DNA

DNA for hybridization is generally labeled using fluorescent dyes, primarily Cy3 or Cy 5. The DNA can be labeled directly (dyes are directly integrated into the target DNA) or indirectly (targets are labeled after hybridization) (see Note 4).

3.3.1. Labeling

Random primer and PCR amplification labeling with Cy3 or Cy5 fluorescent dyes are the most common means used for target detection in environmental samples (9). Random primer labeling with Klenow fragment of DNA polymerase I is particularly useful for labeling genomic DNA fragments. Targets can also be labeled by PCR using gene-specific primers. PCR labeling targets using gene-specific primers is particularly important for increasing detection sensitivity.

Random Priming Labeling Method

1. In a 0.2 ml PCR tube combine:
   (a) 1–2 μg purified community DNA (in 10 μl RNase-free water)
   (b) 20 μl (750 ng μl^{-1}) random octamer primers (Invitrogen # Y01393)
   (c) 5 μl DNase- and RNase-free water
2. Mix them well and denature at 100 °C for 5 min.
3. Place immediately on ice for at least 30 s.
4. In a 1.5-ml microcentrifuge tube, combine:
   (a) 0.2 μl dNTP's (5 mM dATP, dTTP, dGTP, and 2.5 mM dCTP)
   (b) 0.4 μl (1 mM) Cy3 or Cy5 dCTP
   (c) 1 μl (40 U μl^{-1}) Klenow fragment (Invitrogen # Y01396)
   (d) 13.4 μl DNase- and RNase-free water

5. Add this mixture to the 0.2 ml PCR tube that contains DNA (volume = 35 μl).
6. Mix well and incubate at 37 °C for 3 h or overnight.
7. After incubation, boil the mixture at 100 °C for 5 min and chill on ice.
8. Purify labeled target DNA using QIAquick columns according to the manufacturer's instructions (Qiagen, Valencia, CA).

PCR Amplification Specific Labeling Method

1. In a PCR tube, combine the following and make up to 30 μl volume using RNAse-free water:

   10 pg of plasmid containing the desired target gene

   20 pmol PCR primers (specific primers for gene of interest)

   25 mM of dATP, dCTP, dGTP, 15 mM dTTP (New England Biolabs),

   10 mM aminoallyl-dUTP (Sigma, St. Louis, MO)

   0.5 U *Taq* DNA polymerase
2. Place PCR mixture in a thermocyler using the following amplification conditions: 1 cycle at 80 °C for 30 s, 94 °C for 2 min followed by 25 cycles of 94 °C for 30 s, 57 °C for 1 min, and 72 °C for 1 min, with a final extension step at 72 °C for 7 min. Note that the annealing temperature may vary depending on primers used.
3. Purify PCR product using QIAquick columns (Qiagen, Valencia, CA).
4. Dry PCR product in speed-vac for 30 min and resuspend in 4.5 μl 0.1 M carbonate buffer (pH 9.0).
5. Mix the solution with (4.5 μl) *N*-hydroxy succinimide esters Cy3 or Cy5 (NHS-Cy3 or Cy5; Amersham Pharmacia Biotech, Piscataway, NJ) and incubate in the dark for 1 h.
6. After incubation, add 35 μl of 100 mM NaOAC (pH 5.2).
7. Purify labeled target PCR products using QIAquick columns (Qiagen, Valencia, CA).

3.3.2. Quantifying the Amount and Specific Activity of Cy-Labeled DNA Targets

Labeling is a critical step for obtaining high-quality microarray data. The experimental problem most often encountered is that microarray hybridization signal varies greatly from time to time. In many cases, poor hybridization signal results from poor dye incorporation. Decreased dye incorporation (<1 dye per 100 nucleotides) gives unacceptably low hybridization signals. However, studies have shown that very high-dye incorporation (e.g., >1 dye molecules per 20 nucleotides) is also not desirable, because high-dye incorporation significantly destabilizes the hybridization duplex (36). Thus, it is important to measure dye incorporation efficiency prior to hybridization. The specific activity of dye incorporation can be determined by measuring the absorbance at wavelengths of 260 and 550 nm for Cy3 or 650 for Cy5. A suitable labeling reaction should have 8–15 A_{260}/A_{550} ratio for Cy3 and 10–20 A_{260}/A_{650} for Cy5.

1. Use a spectrophotometer to quantify the OD at 550 for Cye 3 and OD 650 for Cy5. Also, measure OD at 230, 260 and 280 to assess purity.
2. Take 1 μl of the labeled DNA OD using NanoDrop™ ND-1000 spectrophotometer (NanoDrop Technologies, Inc., Montchanin, DE).
3. Calculate the amount of DNA and as well as the specific activity of the labeled DNA (see Note 5). The specific activity is calculated as follows:

$$\text{Specific activity} = \frac{\text{amount of target DNA } \times 1{,}000}{\text{pmol of dye incorporated } \times 324.5}.$$

4. Dry in speed-vac (no heat) for 1–2 h. Do not use high heat or heat lamps to accelerate evaporation. The fluorescent dyes could be degraded.

3.4. Hybridization

Hybridizations using glass arrays can be carried out manually or using automated or semi-automated hybridization stations. Manual hybridizations are performed using a water bath or hybridization oven and specially designed hybridization chambers that maintain humidity levels within the chamber. Several hybridization stations provide incubation at controlled temperatures and mixing (e.g., Mail Tai from SciGene, SlideBooster from Advalytix, Maui from BioMicro Systems). Washing after hybridization can be accomplished manually or using an automated wash station (e.g., Maui Wash Station, BioMicro Systems). Other systems are completely automated from pre-hybridization through post-hybridization wastes (e.g., Tecan HS4800Pro, TECAN, USA).

An array covered by a 22 × 22 mm coverslip will require ~15 μl of hybridization solution. For a 15 μl hybridization solution, combine the following components (see Note 6):

Hybridization buffer	Volume (μl)	Final concentration
a. RNase-free water	2.5	
b. Formamide	7.5	50 %
c. 20× SSC	2.5	3.25×
d. 5 % SDS	2.0	0.31 %
e. Herring sperm DNA (Promega)	1.2	0.775 μg

3.4.1. Hybridization Protocol

1. Heat the hybridization solution at 95 °C for 2 min in a thermocycler, cool quickly to 25 °C, and spin down at 14,000×*g* for 5 min (see Note 7).
2. Deposit the hybridization (15 μl) solution directly onto the immobilized DNA prior to placing a cover slip (6.25 mm×8 mm) over the array, avoiding bubble formation. It is helpful to practice this operation with buffer and plain slides before attempting actual samples (see Notes 8 and 9).
3. Put the slide in the hybridization chamber.
4. Dispense 20 μl of 3× SSC solution into the hydration wells on both sides.
5. Close the hybridization chamber. Make sure the seal is formed along the O-ring.
6. Incubate the chamber in a 50 °C water bath for 12–15 h (see Notes 10 and 11).

3.4.2. Post-hybridization Wash

1. Place slides, with the coverslips still affixed, in a jar filled with 1× SSC and 0.2 % SDS buffer and wash for 5 min. Allow the coverslips to fall from the slide and then remove the coverslips from the jar with forceps.
2. Transfer the slides to a fresh jar filled with 0.1× SSC and 0.2 % SDS wash buffer. Wash the slides for 5 min.
3. Wash slides with 0.1× SSC for 30 s.
4. Transfer the slides to a slide rack and immediately spin the slides dry at 600 rpm for 5 min in a centrifuge with a horizontal rotor for microtitre plates. As the rate of drying can be quite rapid, it is suggested that the slide be placed in the centrifuge immediately upon removal from the jar.
5. Slides are ready for scanning (see Note 12).

3.5. Image Processing and Data Analysis

The objective of microarray image processing is to measure and quantify the relative abundance of the signal intensity of the arrayed spots. It is therefore important that the spots on the array image be correctly identified. Microarray images are comprised of arrays of spots arranged in grids. An ideal microarray image for easy spot detection should have the following properties: (1) the location of spots should be centered on the intersections between the row and column lines, (2) the spot size and shape should be circular and homogeneous, (3) the location of the grids on the images should be fixed, (4) the slides should have no dust or other contaminants, and (5) the background intensity should be low and uniform across the entire image.

3.5.1. Image Acquisition and Processing

1. Scan the slide initially at a low resolution of 50 μm to obtain a quick display image and then at 5 μm using for instance the

ScanArray 5000 System (GSI Lumonics, Watertown, MA). The emitted fluorescent signal is detected by a photomultiplier tube (PMT) at 570 nm (Cy3) or 670 nm (Cy5) (see Note 13).

2. Save the scanned display as a 16-bit TIFF image file and quantify the intensity of each spot. Many methods are available for resolving the spot location errors, spot size, shape irregularities, and contamination problems (37) to accurately estimate spot intensities. Typically, a user-defined gridding pattern is overlaid on the image and the areas defined by patterns of circles are used for spot intensity quantification.
3. Assess spot quality and reliability, and perform background subtraction of the microarray data. Because of the inherently high variation associated with array fabrication, hybridization, and image processing, the intensity data for some spots may not be reliable. Thus, the first step in data processing is to assess the quality of spots and to remove unreliable, poor spots prior to data analysis. Also, in many cases, because of slide quality, background and contamination, the quality of data can vary significantly among different slides (38). Be sure to subtract local background for each spot and then flag and remove poor quality spots from the data set for further analysis.
4. Compute signal-to-noise ratio (SNR) for each spot to discriminate true signals ($SNR \geq .3.0$) from noise ($SNR < 3.0$) (39). The SNR ratio is calculated as follows:

$$SNR = \frac{\text{Signal mean} - \text{Background mean}}{\text{Background standard deviation}}.$$

Remove outlying spots (outliers) prior to data analysis using ArrayStat™ (Imaging Research, Inc., Ontario, Canada). Outliers are extreme values in a distribution of replicates. Outlying spots could be caused by uncorrected image artifacts such as dust or by the factors undetectable by image analysis such as cross-hybridization. Thus, removal of outlying spots is an important step for pre-data analysis. However, distinguishing outliers is very challenging, because there is no general definition for outliers.

Besides Imagene software, there are other software packages available for image processing, spot identification, quantitation, and normalization. These imaging softwares include GenPix Pro (Axon Instruments, Union City, CA), Array Pro (Media Cybernetic, Carlsbad, CA), Quant Array (Packard Biosciences, Boston, MA), and TIGR Spot Finder (The Institute of Genomic Research TIGR, Rockville, MD).

3.5.2. Data Analysis

Data analysis is the most challenging aspect of FGA because of the large amount of data generated. Several methods have been frequently used in FGA studies (Table 2). These include various

Table 2
Microarray data analysis methods

Analysis methods	Information provided
Descriptive statistics	
Richness, evenness, diversity	Commonly used descriptive methods. For microarray data, functional genes (or probes) would be considered "species" and signal intensity would be used for abundance
Relative abundance	Percent of all genes detected that belong to a certain functional group of gene. Signal intensity of gene number can be used for this calculation
Shared/unique genes	Percent of all genes detected that were found in two or more samples. Unique genes are those that are only detected in one sample
Response ratios	Determine changes in gene abundance based on different treatments or conditions by comparing the signal intensities between two samples, generally control vs. treatment (46)
Ordination techniques	
Principal component analysis (PCA)	Ordination method that reduces the number of variables needed to explain the data and highlight the variability between samples. In the ordination plot, the distance between sample points indicates how similar or dissimilar samples are
Detrended correspondence analysis (DCA)	Similar to the PCA but uses detrending to remove artefacts (i.e., the arch effect) typically found in correspondence analysis (65)
Parallel coordinate analysis	Plots microarray data such that data points that showed similar signal intensities are clustered together
Methods for environmental data	
Canonical correspondence analysis (CCA)	Provides information on how abiotic and biotic factors impact and drive the community structure. Ordination plots show similarity between samples based on distance and how much influence environmental variables (shown as arrows) have on a given sample
Variation partitioning analysis (VPA)	Uses data obtained in the CCA to determine the relative influence of environmental variables on the microbial community. Data is shown as a percentage of variation
Similarity comparisons of microarray data	
Euclidean distance	Uses the square root of the summation of the squares of the differences between all pair-wise comparisons
Pearson correlation coefficient	Correlates between two variables X and Y to identify profiles with similar shapes
Other multivariate statistical analyses	
Cluster analysis	Group samples based on overall similarity of gene patterns or profiles
Neural network analysis	Based on the random matrix theory (66) and creates microbial ecological networks to visualize relationships between genes or gene groups
Self-organizing maps (SOMs)	Data points are mapped onto a grid and clustered in such a way that those points closest to each other are the most closely related
Mantel test	A multivariate correlation analysis used to compare environmental factors and functional genes

diversity indices (e.g., richness, evenness, diversity) based on the number of functional genes detected. The relative abundance of specific gene groups can be determined based on the total signal intensity of the relevant genes, or the number of genes detected.

The percent of genes shared by different samples can also be calculated to compare communities. The response ratio can be used to determine changes in gene abundance based on different treatments or conditions. For statistical analysis of FGA data, several methods are commonly used. These include ordination techniques such as principal component analysis (PCA) or detrended correspondence analysis (DCA), cluster analysis (CA), neutral network analysis (NNA), and parallel coordinate analysis (5, 40). PCA and DCA are multivariate statistical methods that reduce the number of variables needed to explain the data and highlight the variability between samples. They provide an easy way of identifying outliers in the data such as genes that behave differently than most of the genes across a set of experiments (41, 42). It also can be used to visualize clusters of genes that behave similarly across different experiments. CA groups samples based on the overall similarity of gene patterns. It has been used to identify groups of genes, or clusters that have similar expression profiles (43). Subsequently, the clusters and genes within them can be examined for commonalities in functions as well as sequences for better understanding of how and why they behave similarly. Cluster analysis can also help establish functionally related groups of genes and can predict the biochemical and physiological roles of functionally unknown genes (44). NNA is used to visualize relationships between genes or gene groups. Parallel coordinates technique is a multivariate visualization technique that allow for efficient analysis and understanding of complex data (45). In this study, parallel coordinates technique was used to cluster microarray data and determine the most dominant genes at different depths. This technique uses hierarchical clustering algorithm that aims at grouping items so that items in a cluster are similar as possible and as different from data items in the other clusters as possible. In order to determine the most dominant genes in a sample, microarray data with similar signal intensities were clustered.

Response ratios compare the signal intensities between two samples, generally control versus treatment (46). If environmental data is available, several statistical methods are available to correlate environmental variables with functional community structure. These include canonical correspondence analysis (CCA) (47), variable partitioning analysis (VPA) (48, 49), self-organizing maps (SOMs) or other correlation analyses (e.g., Mantel test). CCA has been used many times in FGA studies to better understand how environmental factors are affecting the community structure (6, 50, 51). Based on the results of the CCA, the relative environmental variables on the microbial community can be determined using VPA. SOMs are a more robust and accurate method for grouping large data sets (52). In this analysis, the data points are mapped onto a grid and the positions of the representative points are iteratively relocated in a way that each center has one representative point. Clusters close to each other in the grid are more similar to each other than those further apart. Further correlations can be made with Mantel test (6, 13, 50, 51). For similarity comparisons of microarray data, two approaches are generally used for quantifying the relationships among different genes. One approach is to use Euclidean distance, which is defined as the square root of the summation of the squares of the differences between all pair-wise comparisons (53). The other approach is the Pearson correlation coefficient, which is ideal for identifying profiles with similar shape (54, 55).

There are software packages available to facilitate statistical analyses of array data. For instance, ArrayStat (Imaging Research, Inc., Ontario, Canada) allows analysis of statistical significance, *p*-values, and standard deviation of microarray data. GeneSpring (Silicon Genetics, CA) permits the analysis of array data for scatter plot, cluster analysis, PCA, and SOMs. Free-computer programs (i.e., CLUSTER and TREEVIEW) to ascertain hierarchical relationships of different spots are also available (43).

3.6. Application of Microarray Technology in Autonomous Platform

The following protocol was taken from Greenfield et al. (56), except where noted, which briefly outlines the use of an ESP, including the deployment (Fig. 9), array processing, and sample archiving.

1. When the ESP is deployed in the past, it has been fielded different platforms, including moorings, piers, remotely operated vehicles (ROVs) and benthic "elevators" for a period of around 30 days (57). So far, all of the ESP deployments have taken place in Monterey Bay, California, and at a depth range of the surface to 1,000 m for the ESP and continuing down to 4,000 m for the D-ESP (23). For further references on ESP deployment, see Preston et al. (22), Scholin et al. (23), and Jones et al. (58).
2. Printed arrays prepared above are loaded into a clean array puck on top of a 5-μm pore size hydrophilic Durapore backing filter (Millipore), leaving a 0.009-in. gap above the array, and the puck is then placed into the instrument's rotating carousel.
3. An aliquot of sample is drawn into the ESP's processing syringe. The ESP then loads the puck to the SHA (Sandwich Hybridization Assay) processing position and heats the puck to 25–30 °C. The 1G and 2G ESPs add sample and reagents in the same sequence, but the 1G uses 2 ml of each whereas the 2G uses 1 ml each. Lysate and processing solutions (see Goffredi et al. (59), except where noted) are applied automatically as follows: lysate, 20 min; 1× wash, 2 min; signal probe cocktail in 2 M GuSCN signal buffer, 10 min; 2× wash, 2 min each; anti-Dig HRP (Pierce; Rockford, IL) diluted 1:1,500 in stabilized diluent blocker (1 mL) (Pierce), 5 min; 4× wash,

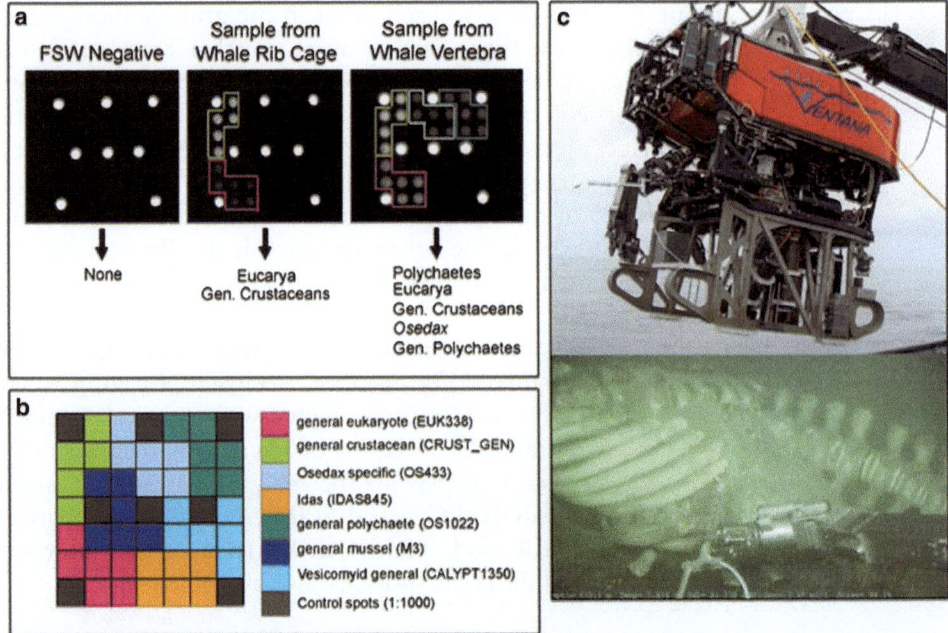

Fig. 9. Application of ESP for detection of invertebrates associated with a whale fall at 633-m depth in Monterey Bay, California, in August 2007. (**a**) From *left* to *right* are arrays from pre-deployment negative control (filtered seawater [FSW]) and two arrays from material collected from different portions of the carcass. Different sets of probes reacted positively depending on sample source. Actual size of the arrays is ~15 mm × 15 mm. (**b**) Array key showing locations of probes for different invertebrate rRNA sequences, including universal probe for Eucarya. *Colored boxes* surrounding probe spots on arrays (**a**) correspond to invertebrate species detected. (**c**) The *top picture* shows deployment of ROV Ventana with the D-ESP mounted below; the sampling wand is held in a robotic arm. The *bottom picture* is a video frame grab showing the sampling wand extended during sampling of the rib cage (23).

2 min each; HRP substrate (Pierce SuperSignal West Femto Maximum Sensitivity Substrate: Stable Peroxidase Buffer and Luminol/Enhancer Solution mixed 1:1 before delivery to the array), 10 s.

4. Afterward, the puck is immediately positioned beneath the camera and the image is captured using a CCD camera (1G ESP: Santa Barbara Instruments model ST-8EI [Santa Barbara, CA, USA] with a Fujinon [Japan] HF35A-2M1 lens; 2G ESP: Starlight Xpress model SXV-H9 [England] with a Fujinon model HF16HA-1B lens).

5. The final phase of ESP array processing, image analysis, is done using V++ Precision Digital Imaging System, v. 4.0 (Digital Optics, Auckland, NZ). The grand mean (±standard error [SE]) spot intensity per DNA probe is determined for each image by measuring a 10-by-10 pixel area per spot then recording average (±standard deviation [SD]) pixel intensity. An array spot with a grand mean intensity significantly higher than background (array region where no probe spotting occurred)

indicates a positive reaction for that probe. Background intensity is determined as above by taking the grand mean of three randomly selected unspotted regions of the imaged array.

6. The resulting image taken by the CCD camera is sent ashore via surface radio mooring (22). An electromechanical cable provides for communications between a remote station and the ESP's surface buoy (19) (Fig. 4).

4. Typical Protocol Results

The hybridization image indicated that the 50-mer oligonucleotide arrays hybridized well with the DNAs from marine sediment (Fig. 10). The DNA content of the sediments ranged from 1.69 to 18.10 μg, with highest yield from sediment samples collected at 0–0.5 cm. Within the mixed zone (0–25.5 cm depth), the DNA content ranged between 5.23 and 18.10 μg, whereas it was between 1.69 and 3.56 μg in the unmixed zone (50–84.5 cm depth). The hybridization image indicated that the microarrays hybridized reasonably with DNA's from Puget Sound marine sediments (Fig. 10). Strong signals were obtained with some nitrogenases (*nif*H) dissimilatory sulfate reductase (*dsr*AB), ammonia monooxygenase (*amo*A), methane monooxygenase (*pmo*A), and nitrite reductase (*nirK/*S). Microarray analysis indicated difference in the number of positive hybridization signals and signal intensity between sediments from shallow (bioturbation zone) and deeper

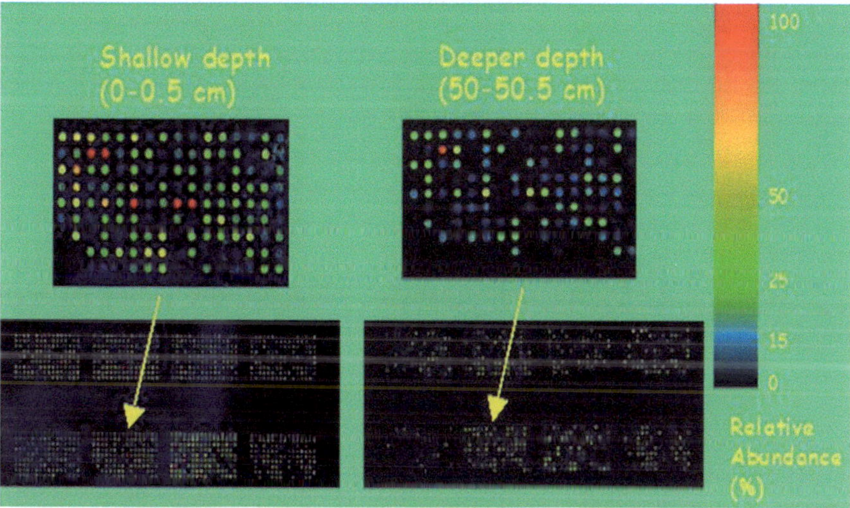

Fig. 10. Hybridization images hybridization images showing the profiles of different *dsr*AB, *nir*S, *nir*K, *nif*H, *amo*A, and *pmo*A genes. Community DNA (2 μg) from marine sediment was labeled with Cy5 using random primer labeling method, and hybridized at 50 °C for 15 h to the oligonucleotide arrays printed in replicate.

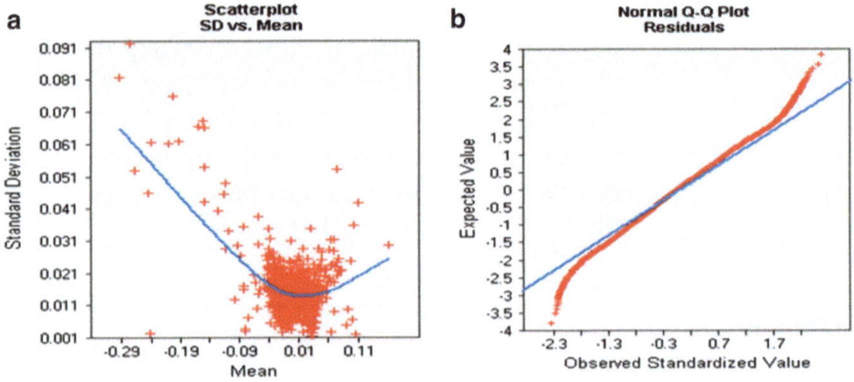

Fig. 11. Quality of microarray data. (**a**) Proportional model showing the relationship between standard deviation and the mean. (**b**) *Q–Q* plot displays.

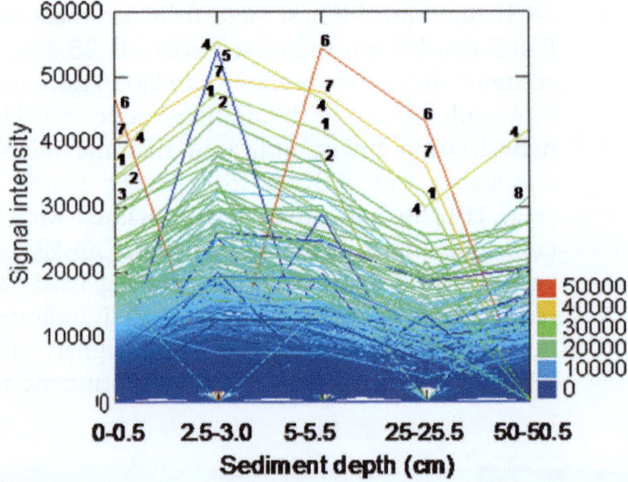

Fig. 12. Ordinate plots from principal component analysis based on the amount of *amoA*/pmoa, *dsrAB*, *nirS*, *nirK*, and *nifH* genes at different sediment depths. Values in *parentheses* indicate percent of total variances of PCA derived from the amount of individual functional gene group data and the combine data.

(below bioturbation zone) depths. The standard variation of the mean is relatively small for most of the gene probes (Fig. 11a). The *Q–Q* (quantile–quantile) plots of the observed standardized residual versus the expected values showed that the majority of the expected values fell closely along the observed values in a wide dynamic range, and only 6.4 % of the spots were outliers (Fig. 11b). Therefore the designed 50-mer FGAs should be useful in monitoring the composition, structure, activities, and dynamics of microbial populations involved in these functional processes across different natural environments. Using the hierarchical rule induction method, the test generated five layers of hierarchy corresponding to five signal intensity groups (1×10^4, 2×10^4, 3×10^4, 4×10^4, and 5×10^4) (Fig. 12). Eight most abundant functional genes (signal intensities between 3×10^4 and 5×10^4), similar to those found in groundwater, terminate gut, sediments, and known cultivable bacteria. Some of these genes were dominant in all samples collected at different depths, while others are abundant only within the bioturbation zone or below the bioturbation. For example, the nitrite reductase gene similar to *nir*K_NKTH17 was found to be dominant across all samples; the nitrogenase gene similar to that isolated from termite gut (*nif*H_3157499), the nitrogenase-containing bacterium *Pseudanabaena* sp. (*nif*H_1698867), and the ammonia monooxidase gene similar to amoA_E03A16280 were only detected within the bioturbation zone; and nitrite reductase gene similar to *nir*K_WA20 was dominant only below the bioturbation zone.

5. Notes

5.1. Notes for DNA Microarray Fabrication

1. Regardless of how comprehensive the FGA is, sequences are constantly being added to public databases, leading to an exponential increase in the number of functional genes as well as the number of sequences for each particular functional gene. As such, continual updates of the FGA are necessary. Even with advances in probe design software, this process is still time consuming due to the large number of sequences and probes that must be designed and tested.

5.2. Notes for Target Preparation

2. The extracted DNA should be purified as soon as possible after extraction to prevent degradation. We have observed degradation of raw DNA extracts after as little as 1 month, even at −20 °C. Ideally, DNA should have a 260:280 ratio ≥1.8 and 260:230 ratio ≥1.7. Gel purification of soil and sediment DNA has worked very well for our lab. A column purification kit can also be used, especially for samples with low DNA yields. While these steps can be tedious, it is critical that DNA used for hybridizations be as pure as possible; therefore, it is beneficial (both in terms of time management and cost) to spend the time necessary on DNA preparation before proceeding with subsequent steps. Any impurities remaining in the DNA can interfere with amplification, labeling, and hybridization.
3. The best amplification results will be obtained using freshly extracted, high molecular weight DNA of the highest quality obtainable. However, DNA from samples with very low biomass, limited sample size, or that would be impossible to replace may not be of optimum quality yet are important to analyze. These samples can still be amplified, but may require some additional steps. If the DNA sample is very dilute, the DNA can be concentrated so that more DNA can be added to the reaction. If no or poor amplification occurs, try decreasing the amount of sample volume used to dilute out any inhibitors that may be

present. Serial dilution (2–3 dilution steps) of the sample can be used to "wash" the DNA. This approach has been used successfully in our lab for a variety of samples. Serial dilution can also be used if the DNA quality is above the recommended thresholds, but the amplification results are poor. Other options that can be tried to include re-precipitating the DNA (using an ethanol or isopropanol protocol) to try and remove any inhibitors, increasing amplification time or performing multiple amplifications of low product samples and combining the products to increase the total amount of amplified DNA.

5.3. Notes on Labeling and Quantitation of Target DNA

4. High quality, fresh reagents should be used to insure the best possible results (9). The higher the labeling efficiency, the better the hybridization results will be. Cy dyes are light sensitive and should be protected from the light as much as possible. We have stored dried, labeled products for months with no apparent loss of efficacy.
5. Corning (http://www.corning.com/cmt) recommends not using labeled target if the specific activity is more than 75. Check specific activity of labeled target before use. Specific activity should be one dye molecule per 25–50 nucleotides for good hybridization.

5.4. Notes on Hybridization

6. Since microarray hybridization is generally performed in the absence of mixing, the hybridization solution should be mixed well so that the labeled targets are evenly distributed on the array surface to obtain optimal target–probe interactions across the entire microarray. Otherwise, the availability of the labeled target molecules to the arrayed spots could be significantly different across the microarray surface.
7. One common problem in microarray hybridization is the quality of fluorescent dyes. The labeling efficiency and hybridization vary significantly sometimes from batch to batch, especially Cy5. It is very important to use fresh reagents to achieve highly sensitive detection (9).
8. Labeled target molecules may be depleted in some areas, yet abundant in others. As a result, significant differences in signal intensity could be observed. Nonuniform hybridization is a common problem associated with microarray experiments. Thus, it is essential to have replicate spots well separated on a slide. It is also imperative to determine the volume of hybridization solution required.
9. The volume of the hybridization solution is critical. When too little solution is used, it is difficult to place the coverslip without introducing air bubbles over some portion of the arrayed oligos. If the coverslip is bowed toward the slide in the center, there will be less labeled DNA in that area and the hybridization will

be nonuniform. When too much volume is applied, the coverslip will move easily during handling, which may lead to misplacement relative to the arrayed oligos, and non-hybridization in some areas of the array may occur.

10. Labeled DNA or RNA is suspended in hybridization buffer. FGA can be hybridized at 42–50 °C and 50 % formamide (5–7, 13, 51, 60, 61). The hybridization temperature and formamide concentration can be adjusted to increase or decrease stringency in order to detect more or less diverse sequences. The effective hybridization temperature can be increased be the use of formamide (0.6 for every 1 %).
11. To reduce handling time of the slides at room temperature after hybridization, take out only one hybridization chamber at a time from the water bath.
12. Completed arrays should be protected from the light until imaged. A black or foil wrapped slide box works well. Imaging should be done within a few hours of removing the arrays from the hybridization chamber to minimize loss of signal.

5.5. Notes on Image Processing and Data Analysis

13. While increasing the PMT increases signal intensity, this also increases background. The hybridization signal can be photobleached, so the number of scans should be limited. The hybridization quality should be evaluated both during scanning (e.g., presence of positive control spots, even hybridization signals across the array, minimal background intensity) and after image analysis (e.g., presence of weak or poor spots). Several sources are available that provide more information regarding image preprocessing and analysis (13, 62, 63).

6. Future Directions

This paper focused on applying the DNA microarray technology in detecting functional genes and attempted to describe the potential ESP technology to extend the methodology in autonomous platform. The availability and the reliability of commercially available instruments, methods, and supplies have made it possible to dig into the molecular underpinnings of just about everything that is "environmental." These achievements, coupled with advances in ocean observatory technology and plans to extend those networks, have fueled the idea of applying molecular sensors in remote settings. In that regard, there is considerable potential to explore the ESP technology for DNA microarray applications as it is a highly configurable platform. Sampling and processing protocols can be tailored to user specifications. For example, assay chemistry can be modified, new probes can be developed, and sampling/processing procedures (sample volume, archiving, filtration, lysis, etc.) can be amended (56). Since the first-generation ESP trials, the system has evolved to take into account deep-sea deployments on ROVs and benthic observatories. The deep-sea ESP (D-ESP) represents the next step in autonomous platform sampling, but still more work is being done to extend deployment duration, geographic coverage, depth rating, and analytical capacity (23). In the long term, it is our hope that this project will stimulate and inspire ocean scientists and engineers, who will in turn, contribute the major breakthroughs needed to make an impact in this field.

Acknowledgements

This research was part of S.M. Tiquia-Arashiro work at Oak Ridge National Laboratory. We gratefully acknowledge financial support from the Office of the Vice President (OVPR) Grants, The University of Michigan-Ann Arbor and the United States Department of Energy (US DOE).

References

1. Beliaev AS, Thompson DK, Fields MW, Wu L, Lies DP, Nealson KH, Zhou J (2002) Microarray transcription profiling of a *Shewanella oneidensis etr*A mutant. J Bacteriol 184:4612–4616
2. Liu Y, Zhou J, Omelchenko M, Beliaev A, Venkateswaran A, Stair J, Wu L, Thompson DK, Xu D, Rogozin IB, Gaidamakova EK, Zhai M, Makarova KS, Koonin EV, Daly MJ (2003) Transcriptome dynamics of *Deinococcus radiodurans* recovering from ionizing radiation. Proc Natl Acad Sci U S A 100:4191–4196
3. Haveman S, DiDonato R, Villanueva L, Shelobolina E, Postier B, Xu B, Liu A, Lovley D (2008) Genome-wide gene expression patterns and growth requirements suggest that Pelobacter carbinolicus reduces Fe(III) Indirectly via sulfide production. Appl Environ Microbiol 74:4277–4284
4. Borneman A, Bartowskey E, McCarthy J, Chambers P (2010) Genotypic diversity in Oenococcus oeni by high-density microarray comparative genome hybridization and whole genome sequencing. Appl Environ Microbiol 86:681–691
5. Tiquia SM, Gurczynski S, Zholi A, Devol A (2006) Diversity of biogeochemical cycling genes from Puget Sound sediments using DNA microarrays. Environ Technol 27:1377–1389
6. Van Nostrand JD, Wu WM, Wu L, Deng Y, Carley J, Carroll S et al (2009) GeoChip-based analysis of functional microbial communities during the reoxidation of a bioreduced uranium contaminated aquifer. Environ Microbiol 11:2611–2626
7. Tiquia SM, Wu L, Chong SC, Passovets S, Xu D, Xu Y et al (2004) Evaluation of 50-mer oligonucleotide arrays for detecting microbial populations in environmental samples. Biotechniques 36:664–675
8. Guschin DY, Mobarry BK, Proudnikov D, Stahl DA, Rittman BE, Mitzabekov AD (1997) Oligonucleotide microarrays as genosensors for determinative environmental studies in microbiology. Appl Environ Microbiol 63: 2397–2402
9. Wu LY, Thompson DK, Li G, Hurt RA, Tiedjie JM, Zhou J (2001) Development and evaluation of functional gene arrays for detection of selected genes in the environment. Appl Environ Microbiol 67:5780–5790
10. Zhou J (2003) Microarrays for bacterial detection and microbial community analysis. Curr Opin Microbiol 6:288–294
11. He Z, Deng Y, Van Nostrand JD, Tu Q, Xu M, Hemme CL, Li X, Wu L, Gentry TJ, Yin Y, Liebich J, Hazen TC, Zhou J (2010) GeoChip 3.0 as a high-throughput tool for analyzing

microbial community composition, structure and functional activity. ISME J 4:1167–1179

12. Van Nostrand JD, Khijniak TV, Gentry TJ, Novak MT, Sowder AG, Zhou JZ, Bertsch PM, Morris PJ (2007) Isolation and characterization of four Gram-positive nickel-tolerant microorganisms from contaminated sediments. Microb Ecol 53:670–682
13. He Z, Gentry TJ, Schadt CW, Wu L, Liebich J, Chong SC, Huang Z, Wu W, Gu B, Jardine P, Criddle C, Zhou J (2007) Geochip: a comprehensive microarray for investigating biogeochemical, ecological and environmental processes. ISME J 1:67–77
14. Wang FP, Zhou H, Meng J, Peng X, Jiang L, Sun P, Zhang C, Van Nostrand JD, Deng Y, He Z, Wu J, Zhou J, Xiao X (2009) GeoChip-based analysis of metabolic diversity of microbial communities at the Juan de Fuca Ridge hydrothermal vent. Proc Natl Acad Sci U S A 106:4840–4845
15. Xie J, He Z, Liu X, Liu X, Van Nostrand JD, Deng Y, Wu L, Zhou J, Qiu G (2011) GeoChip-baded analysis of the functional gene diversity and metabolic potential of microbial communities in Acid Mine Drainage. Appl Environ Microbiol 77:991–999
16. Wu L, Liu X, Schadt CW, Zhou J (2006) Microarray-based analysis of submicrogram quantities of microbial community DNAs by using whole-community genome amplification. Appl Environ Microbiol 72:4931–4941
17. Krause DO, Smith WJM, McSweeney CS (2004) Use of community genome arrays (CGAs) to assess the effects of Acacia angustinssima on rumen ecology. Microbiology 150:2899–2909
18. Yergeau E, Kang S, He Z, Zhou J, Kowalchuk GA (2007) Functional microarray analysis of nitrogen and carbon cycling genes across an Antarctic latitudinal transect. ISME J 1: 163–179
19. Paul JH, Scholin C, van Den Engh G, Perry MJ (2007) *In situ* instrumentation. Oceanography 20:58–66
20. Scholin C, Jensen S, Roman B, Massion E, Marin R III, Preston C, Greenfield D, Jones W, Wheeler K (2006) The Environmental Sample Processor (ESP): an autonomous robotic device for detecting microorganisms remotely using molecular probe technology. Paper presented at OCEANS 2006 MTS/IEEE conference, Boston, MA,18–21 September 2006. Marine Technology Society, Columbia, MD
21. Scholin C (2009) What are "ecogenomic sensors?" A review and thoughts for the future. Ocean Sci Discuss 6:191–213
22. Preston CM, Marin R III, Jensen S, Feldman J, Birch J, Massion E, DeLong E, Suzuki M, Wheeler K, Scholin C (2009) Near real-time, autonomous detection of marine bacterioplankton on a coastal morring in Monterey Bay, California, using rRNA-targeted DNA probes. Environ Microbiol 11:1168–1180
23. Scholin C, Doucette G, Jensen S, Roman B, Pargett D, Marin R III, Preston C, Jones W, Feldman J, Everlove C et al (2009) Remote detection of marine microbes, small invertebrates, harmful algae and biotoxins using the Environmental Sample Processor (ESP). Oceanography 22:158–167
24. Greenfield D, Marin R III, Doucette GJ, Mikulski G, Jensen S, Roman B, Alvarado N, Scholin CA (2008) Field applications of the second-generation Environmental Sample Processor (ESP) for remote detection of harmful algae: 2006–2007. Limnol Oceanogr Methods 6:667–679
25. Ryan J, Greenfield D, Marin R III, Preston C, Roman B, Jensen S, Pargett D, Birch J, Mikulski C, Doucette G, Scholin C (2011) Harmful phytoplankton ecology studies using an autonomous molecular analytical and ocean observing network. Limnol Oceanogr 56: 1255–1272
26. Eddy SR (1998) Profile hidden Markov models. Bioinformatics 14:755–763
27. Xu D, Li G, Wu L, Zhou J, Xu Y (2002) PRIMEGENS: a computer program for robust and efficient design of gene-specific probes for microarray analysis. Bioinformatics 18: 1432–1437
28. Rouillard JM, Herbert CJ, Zuker M (2002) OligoArray: genome-scale oligonucleotide design for microarrays. Bioinformatics 18:486–487
29. Hegde P, Qi R, Abernathy K, Gay C, Dharap S, Gaspard R, Hughes JE, Snesrud E, Lee N, Quackenbush J (2000) A concise guide to cDNA microarray analysis. Biotechniques 29:548–560
30. Diehl F, Grahlmann S, Beier M, Hoheisel JD (2001) Manufacturing DNA microarrays of high spot homogeneity and reduced background signal. Nucl Acid Res 29:E38
31. Hurt RA, Qui X, Wu L, Roh Y, Palumbo AV, Tiedje JM, Zhou J (2001) Simultaneous recovery of RNA and DNA from soils and sediments. Appl Environ Microbiol 67:4495–4503
32. Zhou J, Bruns MA, Tiedje JM (1996) DNA recovery from soils of diverse composition. Appl Environ Microbiol 62:461–468
33. Dean FB, Nelson JR, Giesler TL, Lasken RS (2001) Rapid amplification of plasmid and

phage DNA using Phi29 DNA polymerase and multiply-primed rolling circle amplification. Genome Res 11:1095–1099

34. Lage JM, Leamon JH, Pejovic T, Hamann S, Lacey M, Dillon D, Segraves R, Vossbrinck B, Gonzalez A, Pinkel D, Albertson DG, Costa J, Lizardi PM (2003) Whole genome analysis of genetic alterations in small DNA samples using hyperbranched strand displacement amplification and array-CGH. Genome Res 13:294–307
35. Aviel-Ronen S, Zhu CQ, Coe BP, Liu N, Watson SK, Lam WL, Tsao MS (2006) Large fragment Bst DNA polymerase for whole genome amplification of DNA from formalin fixed paraffin embedded tissues. BMC Genomics 7:312–321
36. Worley J, Bechtol K, Penn S, Roach D, Hanzel D, Trounstine M, Barker D (2000) A systems approach to fabricating and analyzing DNA microarrays. In: Schena M (ed) Microarray biochip technology. Eaton Publishing, Natick, MA, pp 65–85
37. Zhou YX, Kalocsai P, Chen JY, Shams S (2000) Information processing issues and solutions associated with microarray technology. In: Schena M (ed) Microarray biochip technology. Eaton Publishing, Natick, MA, pp 167–200
38. Tseng GC, Oh MK, Rohlin L, Liao JC, Wong WH (2001) Issues in cDNA microarray analysis: quality filtering, channel normalization, models of variations and assessment of gene effects. Nucl Acid Res 29:2549–2557
39. Verdnik D, Handran S, Pickett S (2002) Key considerations for accurate microarray scanning and image analysis. In: Shah S, Kamberova G (eds) DNA array image analysis - nuts & bolts. DNA Press, LLC, Eagleville, PA, pp 83–98
40. He Z, Van Nostrand JD, Wu L, Zhou J (2008) Development and application of functional gene array for microbial community analysis. Trans Nonferrous Met Soc Chin 18: 1319–1327
41. Hilsenbeck SG, Friedrichs WE, Schiff R, O'Connell P, Hansen RK, Osborne CK, Fuqua SAW (1999) Statistical analysis of array expression data as applied to the problem of tamoxifen resistance. J Natl Cancer Inst 91:453–459
42. Raychaudhuri S, Stuart JM, Altman RB (2000) Principal components analysis to summarize microarray experiments: application to sporulation time series. Pacific symposium on biocomputing, pp 455–466
43. Eisen MB, Spellman PT, Brown PO, Botstein D (1998) Cluster analysis and display of genome-wide expression patterns. Proc Natl Acad Sci U S A 95:14863–14868
44. Wen X, Fuhrman S, Michaels GS, Carr DB, Smith S, Barker JL, Somogyi R (1998) Large-scale temporal gene expression mapping of central nervous system. Proc Natl Acad Sci U S A 95:334–339
45. Inselburg A (1998) Visual data mining with parallel coordinates. Comput Stat 13:47–63
46. Luo Y, Hui D, Zhang D (2006) Elevated CO_2 stimulates net accumulations of carbon and nitrogen in land ecosystems: a meta-analysis. Ecology 87:53–63
47. ter Braak CJF (1986) Canonical correspondence analysis: a new eigenvector technique for multivariate direct gradient analysis. Ecology 67:1167–1179
48. Okland RH, Eilertsen O (1994) Canonical correspondence analysis with variation partitioning: some comments and applications. J Veg Sci 5:117–126
49. Ramette A, Tiedje JM (2007) Multiscale responses of microbial life to spatial distance and environmental heterogeneity in a patchy ecosystem. Proc Natl Acad Sci U S A 104:2761–2766
50. Wu L, Kellogg L, Devol AH, Tiedje JM, Zhou J (2008) Microarray-based characterization of microbial community functional structure and heterogeneity in marine sediments from the Gulf of Mexico. Appl Environ Microbiol 74:4516–4529
51. Waldron PJ, Van Nostrand JD, Watson DB, He Z, Wu L, Jardine PM, Hazen TC, Zhou JZ (2009) Functional gene array-based analysis of microbial community structure in groundwaters with a gradient of contaminant levels. Environ Sci Technol 43:3529–3534
52. Kohonen T (1997) Self-organizing maps, 2nd edn. Springer, Berlin
53. Knudsen S (2002) A biologist's guide to analysis of DNA microarray data. Wiley, New York
54. Chen Y, Bittner ML, Dougherty ER (1999) Issues associated with microarray data analysis and integration. Nat Genet 22:213–215
55. Jagota A (2001) Microarray data analysis and visualization. Bioinformatics by the Bay Press, Santa Cruz, CA
56. Greenfield DI, Marin R III, Jensen S, Massion E, Roman B, Feldman J, Scholin C (2006) Application of Environmental Sample Processor (ESP) methodology for quantifying Pseudo-nitzschia australis using ribosomal RNA-targeted probes in sandwich and fluorescent in situ hybridization formats. Limnol Oceanogr Methods 4:426–435
57. Birch J, Preston C, Pargett D, Jensen S, Roman B, Everloce C, Marin III R, Orphan V, Girguis P, Scholin C (2010) Searching for microbes in deep-sea seep and hydrothermal vents using the Environmental Sample Processor. Presented at the Astrobiology Science Conference 2010

58. Jones WJ, Preston C, Marin R III, Scholin C, Vrijenhoek R (2008) A robotic molecular method for in situ detection of marine invertebrate larvae. Mol Ecol Resour 8:540–550
59. Goffredi SK, Jones W, Scholin CA, Marin R III, Vrijenhoek RC (2005) Molecular detection of marine larvae. Mar Biotechnol 8:1–12
60. Mason OU, Di Meo-Savoie CA, Van Nostrand JD, Zhou J, Fisk MR, Giovannoni SJ (2009) Prokaryotic diversity, distribution, and insights into their role in biogeochemical cycling in marine basalts. ISME J 3:231–242
61. Liang T, Li G, Van Norstrand JD, He Z, Wu L, Deng Y, Zhang X, Zhou J (2009) Microarray-based analysis and microbial diversity along an oil contaminated gradient in oil field. FEMS Microbiol Ecol 70:324–333
62. Zhou J, Thompson DK (2002) Microarrays: application in environmental microbiology. In: Bitton G (ed) Encyclopedia of environmental microbiology, Vol 4. Wiley, New York, pp 1968–1979
63. Gentry TJ, Schadt CW, He Z, Zhou J (2007) Functional gene arrays for microbial community analysis. In: Hurst CJ, Crawford RL, Garland JL, Lipson DA, Mills AL, Ststzenbach LD (eds) Manual of environmental microbiology, 3rd edn. ASM, Washington, DC, pp 1052–1062
64. Doucette G, Mikulski C, Jones K, King K, Greenfield D, Marin R III, Jensen S, Roman B, Elliott C, Scholin C (2009) Remote, subsurface detection of the algal toxin domoic acid onboard the Environmental Sample Processor: assay development and field trials. Harmful Algae 8:880–888
65. Hill MO, Gauch HG Jr (1980) Detrended correspondence analysis, an improved ordination technique. Vegetatio 42:47–58
66. Sawa T, Ohno-Machado L (2003) A neutral network-based similarity index for clustering DNA microarray data. Comput Biol Med 33:1–15

Chapter 10

Bioluminescence Detection for ATP Quantification Using Microfluidic Device

Tatsuhiro Fukuba and Teruo Fujii

Abstract

An IISA-ATP (integrated in situ analyzer for ATP) was developed for the quantitative determination of ATP (adenosine-5′-triphosphate) in ocean environments, including the deep-sea. Total (intracellular and dissolved) ATP was quantitatively determined using a simple luciferin-luciferase (L-L) bioluminescence assay. A microfluidic device was incorporated as a core functional element in the IISA-ATP for performing cell lysis and carrying out the L-L reaction. The IISA-ATP has the capability to process environmental samples in the laboratory and in situ. Here, we provide an overview of the performance of the microfluidic device and the IISA-ATP on the basis of results of an in situ trial.

Key words: ATP, Bioluminescence, Luciferin-luciferase assay, Microfluidic device, Ocean sensing, In situ detection, Microfluidic chip, IISA-ATP, Microbial activity, Microbial intracellular ATP

1. Introduction

A large number of prokaryotic microbes such as bacteria and archaea inhabit ocean environments. These organisms play important roles in the local and global circulation of materials through ecological functions such as primary production or final decomposition. Furthermore, their large biomass in ocean environments is recognized as one of the major pools of biologically essential elements such as carbon, nitrogen, and phosphorus on Earth. Therefore, estimation of their biomass is essential for elucidating the contribution of microbes to material circulation and stocks.

Adenosin-5′-triphosphate (ATP), a biogenic molecule well known as "energy currency," is frequently used as a proxy of microbial activity in the field of environmental sciences. The quantity of

Sonia M. Tiquia-Arashiro (ed.), *Molecular Biological Technologies for Ocean Sensing*, Springer Protocols Handbooks, DOI 10.1007/978-1-61779-915-0_10,

microbial intracellular ATP (particulate ATP: P-ATP) can be converted into carbon weight to obtain a semiquantitative estimate of the microbial biomass (1). P-ATP in ocean environments is widely dispersed with concentrations of 10^{-11} to 10^{-8} M at shallow depths (0–100 m) and 10^{-13} to 10^{-11} M at large depths (>1,000 m) (1). Near geochemically active sites such as hydrothermal vents, the P-ATP concentration shows apparent anomalies (e.g., 10^{-9} M near the hydrothermal vent at Galápagos Rift (2)). Anomalies in P-ATP concentration have also been observed in the deep-sea halocline layer of the Bannock Basin, where the chemocline supports the formation of a microbial biomass-rich layer (3). Therefore, P-ATP concentrations in seawater can be utilized as markers or indicators of novel geochemically active sites and to elucidate the extent of their influences on the surrounding environments. Dissolved ATP (D-ATP) found in seawater is recognized as one of the important carbon and phosphorus sources for marine microbes, and its concentration is closely related to microbial activity through uptake and release processes. Azam and Hodson reported D-ATP concentrations of up to 10^{-10} M in relatively oligotrophic surface seawater and 10^{-11} M in deep water (~500 m) (4). Hodson et al. determined that the percentages of D-ATP in total ATP (T-ATP: sum of P-ATP and D-ATP) were up to 27.9 % at the surface of the western wall of the Gulf Stream (5). Furthermore, Björkman and Karl reported that D-ATP concentrations generally exceed P-ATP concentrations in the oligotrophic North Pacific Ocean (6). Because both P-ATP and D-ATP concentrations are closely linked to microbial activity, quantitative determination of the T-ATP concentration in seawater can be employed as an operational strategy to elucidate the spatiotemporal distributions of potential activities of marine microbial communities.

The most widely accepted method for quantitative determination of ATP is the bioluminescence-based luciferin-luciferase (L-L) assay. Owing to its simplicity and sensitivity, the L-L assay is used for monitoring food hygiene (7, 8) and cleanliness (9, 10), as well as in environmental microbiological studies. Today, a variety of L-L assay kits and desktop apparatuses are commercially available for measuring the bioluminescence intensity, and these are commonly used for scientific studies and practical tests. Even though some of these apparatuses are highly portable, to the best of our knowledge, in situ apparatus that can be remotely operated in ocean environments has not been developed yet. Automated and portable in situ apparatus for the quantitative determination of ATP concentrations will facilitate microbial activity assays in ocean environments without the risk of contamination or degradation of samples. Furthermore, real-time and spatiotemporally resolved measurements of ATP concentrations will enable the detailed determination of multidimensional distribution of marine microbial communities and biogenic phosphorus pools.

Microfluidic technology can be employed to realize highly portable, automated, and functionally integrated in situ apparatus for L-L bioluminescence assays. Micrometer-scale channels or reaction chambers enable rapid and appropriate mixing of reagents in a limited space and time. Further, the use of microfluidic devices eliminates the need for complicated tubing, thus improving the reliability of the system and reducing dead volumes. Previously, we reported and evaluated the performance of a microfluidic device for L-L bioluminescence assays (11). Here, we describe the performance of microfluidic devices for L-L bioluminescence assays and introduce the complete in situ apparatus named integrated in situ analyzer for ATP (IISA-ATP) for the quantitative determination of T-ATP in ocean environments.

2. IISA-ATP System

2.1. System Overview

The IISA-ATP is a functionally integrated in situ instrument used for the quantitative determination of T-ATP by a simple method employing a commercially available L-L bioluminescence assay kit (CheckLite HS Set, Kikkoman Co., Japan). In this system, a seawater sample is mixed with reaction reagents without pre-concentration or precipitation processes for condensation and desalting. For highly sensitive and quantitative determination of P-ATP concentrations in seawater, a pre-concentration process of microbial cells using a filter is commonly employed. Because L-L reactions are inhibited by elevated ionic strength in reaction mixtures (12), desalting of the seawater sample is generally recommended to achieve the best sensitivity. Besides the pre-concentration process, the filtration process also results in the desalting of microbial cells. D-ATP can be highly condensed and desalted using the modified magnesium-induced co-precipitation procedure reported by Björkman and Karl (6). These pretreatment procedures improve the sensitivity and reliability of the instrument when applied to laboratory-based analysis; however, they are too complicated for application to in situ analysis. Therefore, the simplest method for ATP quantification, direct mixing of seawater samples with the reaction reagents, was selected for the IISA-ATP. Even though the bioluminescence intensity is reduced by five to ten times when seawater-based samples are used instead of freshwater-based samples, the sensitivity of the assay is still acceptable (10^{-12} M, data not shown).

A flow diagram, schematic illustration, and photograph of the IISA-ATP are shown in Fig. 1, Fig. 2, and Fig. 3, respectively. The most important element of the IISA-ATP is the microfluidic device, which has the ability to conduct microbial cell lysis and an L-L

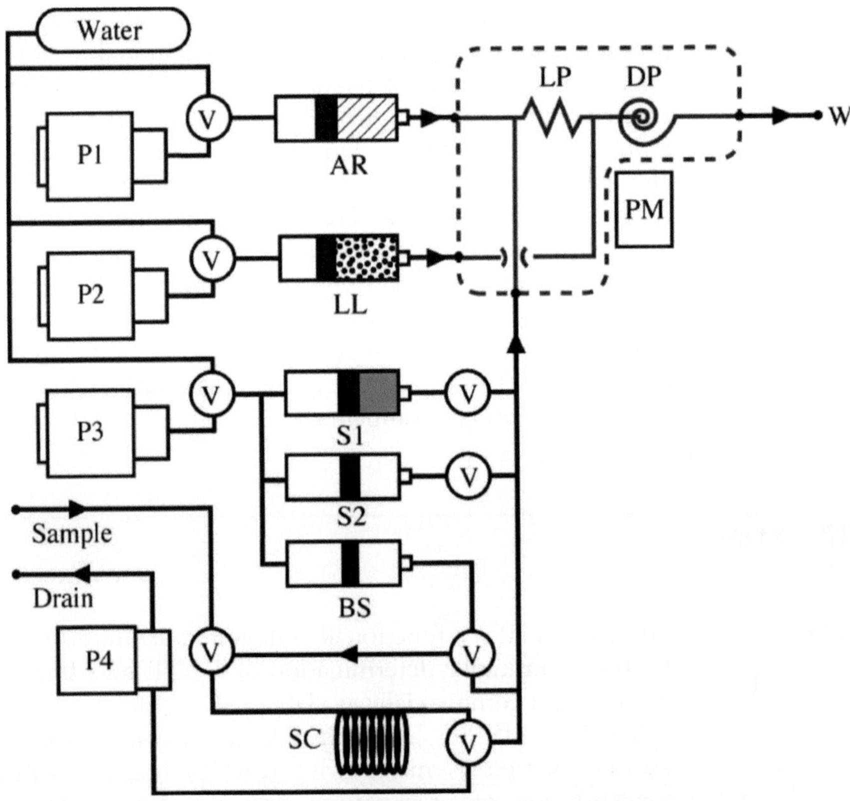

Fig. 1. Flow diagram of IISA-ATP. *P1–3* miniature syringe pumps, *P4* DC motor pump, *V* 2-way or 3-way valves, *AR* ATP-releasing reagent, *LL* luciferin-luciferase mix, *S1 and 2* ATP standards, *BS* blank artificial seawater, *SC* sample coil, *LP* lysis part, *DP* detection part, *PM* photomultiplier tube, *W* waste. Parts enclosed within the *dashed line* are included in the microfluidic device.

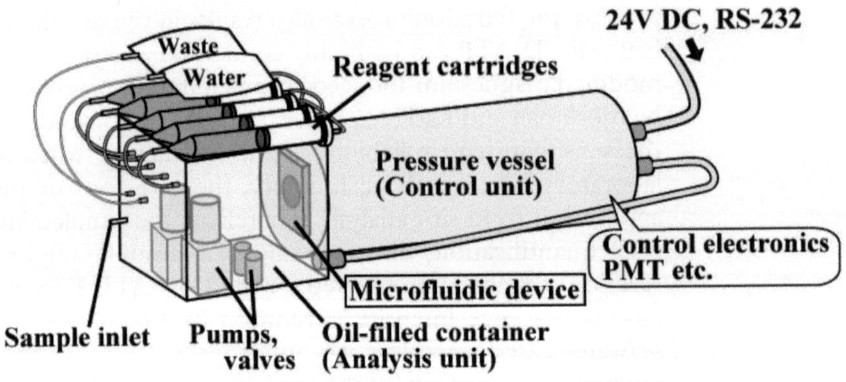

Fig. 2. Schematic illustration of IISA-ATP. The microfluidic device is placed in front of the optical window on the pressure vessel for measuring the bioluminescence intensity using the PMT placed inside the pressure vessel.

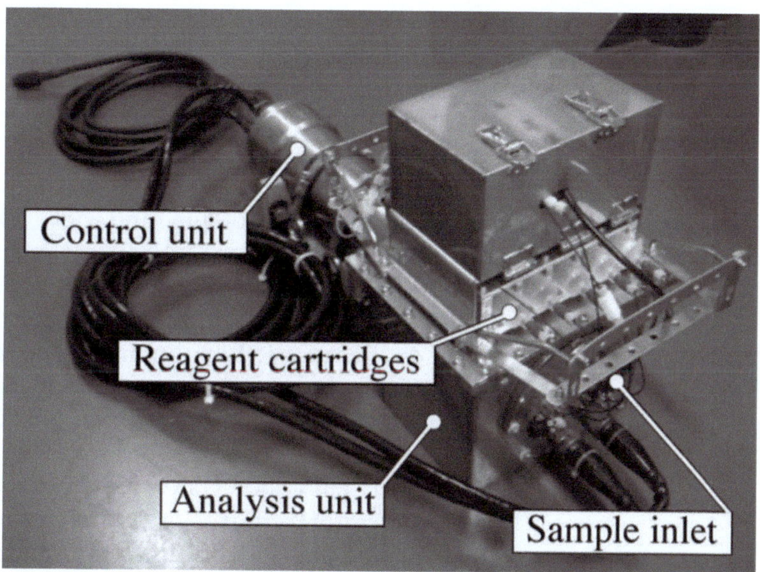

Fig. 3. Photograph of IISA-ATP. Plastic bags for DI water and waste storage are enclosed in a metal cage fixed on the analysis unit, for protection.

bioluminescence reaction. All reagents (ATP-releasing (cell lysis) reagent, L-L reagent, ATP standards, and blank seawater) are stocked in individual reagent cartridges and introduced into the microfluidic device by miniature syringe pumps in parallel. Additionally, 1 ml of sample seawater is stocked in a sample coil (a coiled Teflon FEP tube) and introduced into the microfluidic device. Reagents are pumped indirectly using a displacement fluid (DI water) that is separated from the reagent by a plastic piston to prevent contamination. Seawater samples are indirectly pumped by double-sterilized and filtered artificial seawater as a displacement fluid that is also used as a blank standard. The pumps, valves, and microfluidic device are immersed in fluorinated oil (Fluorinert FC-43, 3M) and enclosed in a metal box that provides electrical insulation and pressure compensation. The reagent cartridges and plastic bags for DI water and waste are stored outside of the oil-immersed metal box. L-L bioluminescence intensity is measured using a photomultiplier tube enclosed in a cylindrical pressure vessel with a pressure-resistant optical window. Control electronics are also placed in the pressure vessel. Electric power (24VDC) is supplied from a surface boat or an underwater platform such as an ROV. RS-232 communication with a PC is necessary to control the IISA-ATP and monitor the data in real time. The analysis of a single seawater sample or measurement for calibration is completed in approximately 10 min.

2.2. Microfluidic Device

The most important feature of the IISA-ATP is utilization of the microfluidic device as the core functional element. The microfluidic device is composed of a polydimethylsiloxane (PDMS) microfluidic chip and a glass substrate with integrated heaters and a temperature sensor (Fig. 4) (see Note 1). The transparent microfluidic chip has a multilayer format as shown in Fig. 5. The upper layer contains a narrow (100 μm) and wide (800 μm) serpentine microchannel (200-μm deep) for reagent mixing and cell lysis, respectively. The lower layer has a spiral microchannel (800-μm wide, 400-μm deep) as the detection part for the L-L bioluminescence (Fig. 6). Because L-L bioluminescence reactions are temperature dependent, the temperature of the microfluidic device should be maintained under optimum conditions (25 °C here, see Note 2). Accordingly, transparent heaters and a temperature sensor made of Indium Tin Oxide (ITO) are fabricated on the glass substrate for temperature control of the microfluidic device. The PDMS microfluidic chip is fabricated by a replica-molding method (13) using a negatively patterned mold master. ITO heaters and temperature sensors are fabricated by a wet etching method (14).

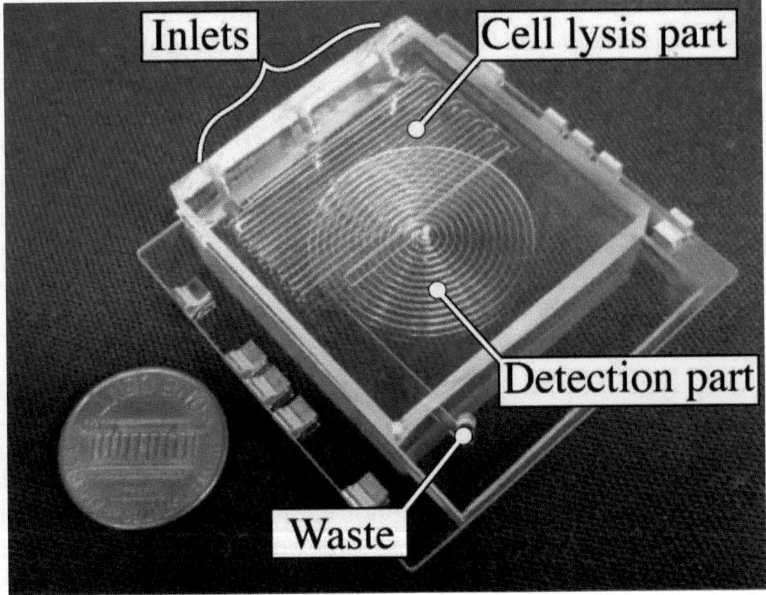

Fig. 4. Photograph of multilayer microfluidic device used in IISA-ATP. All the materials used in the microfluidic device are transparent to enable bioluminescence intensity measurements.

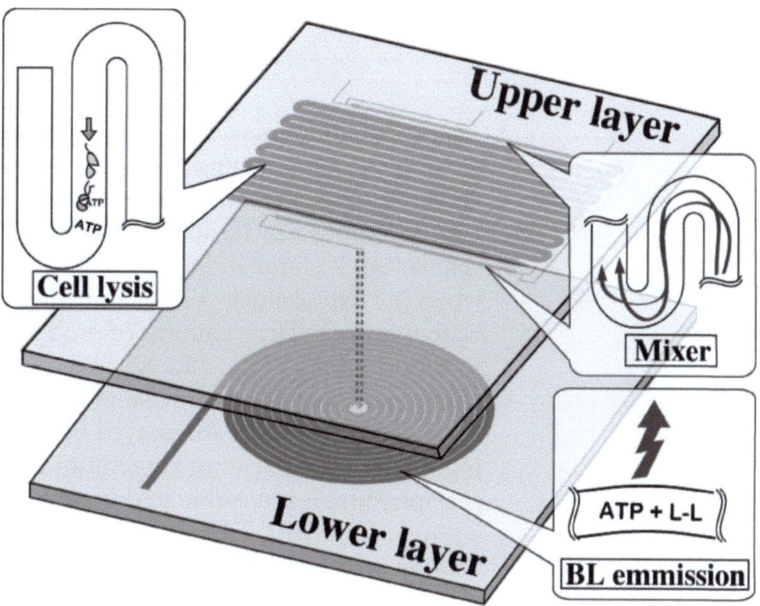

Fig. 5. Schematic illustration of PDMS multilayer microfluidic chip used for quantitative determination of ATP.

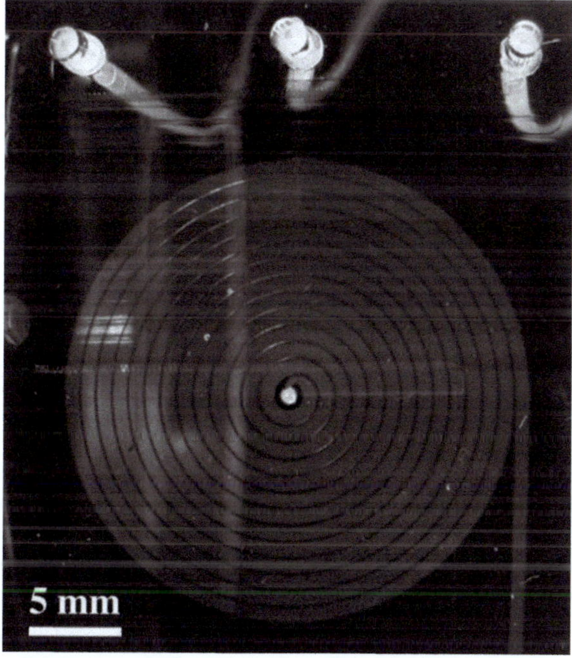

Fig. 6. Bioluminescence emission on spiral-shaped detection part (contrast- and brightness-adjusted image).

3. Reagents

The L-L bioluminescence assay kit (CheckLite HS Set) used for the IISA-ATP contains extracellular ATP-eliminating enzymes, ATP-releasing (cell lysis) reagent, and L-L reagent. The P-ATP concentrations can be specifically quantified by adding the extracellular ATP-eliminating enzyme to a liquid sample. However, when the extracellular ATP-eliminating enzyme is added to a seawater sample, a large amount of precipitation occurs, which causes serious clogging of the microchannel. Accordingly, use of the ATP-eliminating enzyme is not suitable for microfluidic devices and thus, T-ATP contents are assayed by the IISA-ATP. Although mixing of the ATP-releasing reagent and seawater sample also causes precipitation, it is possible to decrease the amount of precipitate by adding 10 mM EDTA to the ATP-releasing reagent, thereby preventing microchannel clogging. To avoid the adsorption of L-L reagent contents such as luciferin and luciferase onto a hydrophobic PDMS microchannel surface, 2 % (v/v) Tween 20 (see Note 3) is added to the L-L reagent. ATP standard solutions (2×10^{-12} to 2×10^{-8} M) for calibration are prepared by diluting the ATP standard solution (Kikkoman) with double-autoclaved and filtered artificial seawater (Daigo's artificial seawater SP for marine microalgae medium, Nihon Pharmaceutical Co. Ltd., Japan).

4. Protocol

4.1. ATP Quantification Using Desktop Apparatus

Fresh ATP-releasing reagent and L-L reagent were prepared according to the manufacturer's instructions and EDTA and Tween 20 were added to the ATP-releasing reagent and L-L reagent, respectively. Hundred microliters of sample or ATP standard was introduced into a plastic test tube (see Note 4). Then, 100 μl of the ATP-releasing reagent was added to the sample or ATP standard and mixed thoroughly using the vortex mixer. After a few seconds of the mixing for the ATP release, 100 μl of the L-L reagent was added and mixed thoroughly using the vortex mixer. Immediately after the mixing, the test tube was applied to the luminometer (Lumitester C-110, Kikkoman) to record the bioluminescence intensity data (see Note 5). For all samples and ATP standard solutions, triplicate measurements were conducted.

4.2. ATP Quantification Using Microfluidic Device

Fresh ATP-releasing reagent and L-L reagent were prepared and modified as described in Sect. 1. ATP-releasing reagent, L-L reagent, ATP standard solutions (2×10^{-12} to 2×10^{-8} M), and blank artificial seawater were filled into individual 1-ml plastic syringes aseptically. Temperature of the microfluidic device was controlled at 25 °C using the integrated heaters and temperature sensor. After the temperature had been stabilized, ATP-releasing reagent, L-L reagent, and ATP standard solution or blank seawater were derived to the microfluidic device at a flow rate of 160 μl/min each (480 μl/min in total) using a syringe pump (KD-230, KD Scientific Inc., USA). Bioluminescence intensity data was recorded during 100 s after the bioluminescence reached a plateau. For all samples and ATP standard solutions, triplicate measurements were conducted.

4.3. Environmental Sample Analysis Using IISA-ATP

Fresh ATP-releasing reagent and L-L reagent were prepared and modified as described in Sect. 4.1. ATP-releasing reagent, L-L reagent, ATP standard solutions (2×10^{-11} and 2×10^{-9} M), and blank artificial seawater were filled into individual reagent cartridges aseptically (see Note 6). The microfluidic device and the cartridges were set to the IISA-ATP. Before deployment of the IISA-ATP, temperature control was started using the integrated heaters and temperature sensor. Immediately after the IISA-ATP reached to the sea bottom, calibration was conducted using the ATP standards and blank artificial seawater. The IISA-ATP was placed on the desired location followed by introduction of an environmental sample into the system using the pump for 2–3 min. The sample processing was started immediately after the sample introduction. Typical flow-rate setting for the analysis is 160 μl/min each for all of the reagents (480 μl/min in total). The bioluminescence data were recorded during approximately 7 min of the processing time.

5. Typical Protocol Results

5.1. ATP Quantification Using Microfluidic Device

A typical calibration curve for ATP standards of 2×10^{-12} to 2×10^{-8} M was used (Fig. 7). The bioluminescence intensity was linearly correlated with the ATP concentration (R2 > 0.99). The detection limit calculated from these data was 1.1×10^{-11} M, which is sufficient for microbial biomass assays in most ocean environments, except for those with extremely low-ATP concentrations such as deep-sea and deep-subsurface environments. When compared to the results obtained from conventional manual procedures using the desktop apparatus and test tubes, the detection limit was approximately five times higher in microfluidic device. This is because of the weaker bioluminescence emissions caused by the difference in the total volume of the final reagent that can be used for bioluminescence intensity measurements (178.3 μl for the microfluidic device and 300 μl for the conventional method using the test tubes).

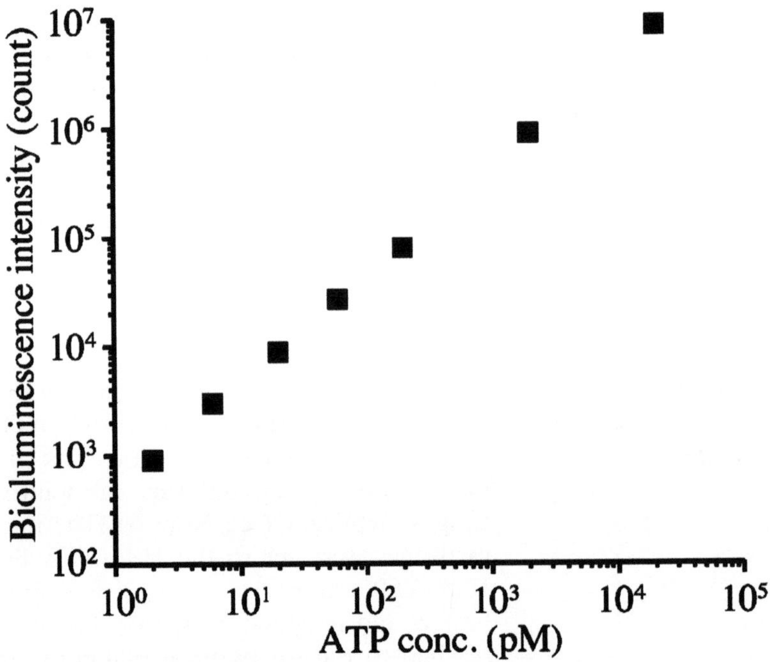

Fig. 7. Typical calibration curve for ATP quantification using microfluidic device (adapted from ref. (15) with permission).

5.2. Results of Environmental Sample Analysis

To demonstrate the performance of the IISA-ATP, a surface seawater sample collected from an artificial shore area in Tokyo Bay, Japan, was analyzed using the desktop apparatus and the IISA-ATP in the laboratory. The sample was collected using a plastic test tube and stored for several hours at room temperature until analysis in the laboratory. The time course bioluminescence intensity data is shown in Fig. 8. As shown here, the bioluminescence intensities that correspond to the ATP standard solutions, the blank seawater, and the T-ATP contents in the environmental sample can be successfully measured by the IISA-ATP. The time course bioluminescence data from the environmental samples appeared to fluctuate more than that of the ATP standard solutions and the blank seawater. This may have been caused by the nonuniformity of the actual seawater samples owing to the formation of microbial aggregates. The average T-ATP concentration calculated from the data obtained using the IISA-ATP was determined to be 2.7×10^{-10} M. On the other hand, $3.3 \pm 0.9 \times 10^{-10}$ M of T-ATP was determined from the data obtained by the desktop apparatus. As shown here, the IISA-ATP can process and analyze environmental samples appropriately, and reliable T-ATP concentration data can be acquired.

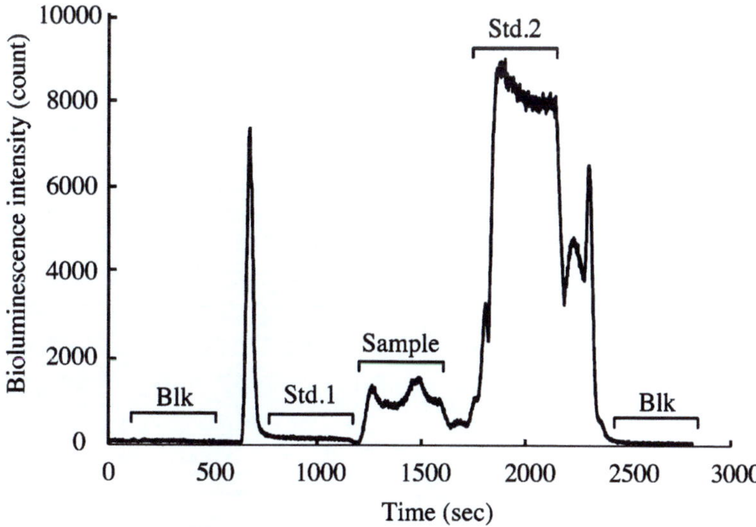

Fig. 8. Results of sample processing in laboratory using IISA-ATP. *Blk* blank seawater, *Std.1* 2×10^{-11} M ATP, *Std.2* 2×10^{-9} M ATP, *Sample* Environmental sample collected from Tokyo Bay (adapted from ref. (15) with permission). Spike noises observed occasionally during the analysis are caused by instability of the pumping system.

5.3. Field Deployment Result

A field trial of the IISA-ATP system was conducted in a shallow submarine hydrothermal area in the subtropical coral reef off Taketomi Island, Okinawa, Japan, in March 2009. A surface boat with a power supply and divers was employed for the field trial (Fig. 9). The IISA-ATP was connected to a rope with an electrical cable for power supply and communication. In situ calibration and analysis were conducted at the prominent gas bubbling site depth of approximately 13 m (16). The sample inlet of the IISA-ATP was directly subjected to a stream of hydrothermal fluid. The hydrothermal fluid at the same position was collected manually using disposable plastic syringes by the diver for control analysis with the IISA-ATP and the desktop apparatus in the laboratory. The sample was stored at room temperature for approximately 4 h prior to conducting measurements in the laboratory. Representative results are summarized in Table 1. The in situ T-ATP value was lower than that measured by the IISA-ATP and the desktop apparatus in the laboratory. These results suggest that the microbial activity changed during sample transportation and emphasize the importance of in situ microbial activity assays in ocean environments to elucidate their biomasses and contribution to the circulation of materials with higher reliability.

Fig. 9. IISA-ATP deployed by diver at submarine hydrothermal site at coral reef off Taketomi Island.

Table 1
In situ T-ATP concentration measured by IISA-ATP at bubbling site

	IISA-ATP (in situ)	IISA-ATP (lab.)	Conventional (lab.)
ATP (M)	3.4×10^{-10}	1.2×10^{-9}	1.4×10^{-9}

Data obtained by laboratory analysis of environmental samples collected by the IISA-ATP and the conventional method are also shown (adapted from ref. (15) with permission)

6. Notes

1. All of the materials for the microfluidic device are autoclavable. The microfluidic device should be autoclaved prior to integrating with IISA-ATP system. After the autoclave, the PDMS material becomes cloudy white color because of penetration of water vapor in the material. The PDMS returns to a transparent state by keeping it in a dry condition for a few hours.

2. At the low temperature conditions without the temperature control, the bioluminescence intensity becomes apparently lower than that measured under the optimum temperature condition. For instance, the bioluminescence emission under 4 °C becomes approximately 4 % of bioluminescence intensity compared to 25 °C.
3. Tween 20 should be sterilized using an autoclave prior to adding to the L-L reagent.
4. The plastic test tubes should be handled with care to avoid contaminations of microbes or ATP to the assay. Wearing disposable plastic gloves is recommended.
5. After adding the L-L reagent to the tube, bioluminescence emission is immediately started, and it decays gradually. Therefore, the time lag between the mixing and the measurement of the bioluminescence intensity should be reproducible as possible.
6. The plastic pistons and barrels that are used as the reagent cartridges are not compatible with autoclave. Sometimes, apparently high background signal is observed because of native contamination of the pistons and barrels. In such a case, replacement of the reagent cartridges is necessary.

7. Future Directions

For quantitative estimation of the microbial activity, biomass, and dissolved phosphorus pools, independent measurements of P-ATP and D-ATP are preferable. For example, further optimization of the L-L assay kit to prevent the occurrence of precipitation when a seawater sample is mixed with the extracellular ATP-eliminating enzyme will enable specific quantification of P-ATP. To compensate for the inhibitory effect of seawater contents such as salt on the L-L reaction system for reliable assays, internal standards should be utilized for in situ calibrations of the IISA-ATP (17). For further miniaturization of the IISA-ATP system, highly miniaturized pumping components such as micropumps, on chip microvalves, and micro flow-rate regulators (18) will be integrated in the future. The IISA-ATP can be operated at hydrostatic pressures as high as 60 MPa. Experimental results obtained using a high-pressure experimental tank in our laboratory showed that the L-L reaction could be successfully conducted at hydrostatic pressure of 60 MPa without inhibition (data not shown). An improved IISA-ATP will be deployed in a variety of ocean environments, including deep-sea areas, in the near future.

Acknowledgements

The authors are grateful to Dr. H. Yamamoto (Japan Agency for Marine-Earth and Technology: JAMSTEC) and Dr. Y. Furushima (JAMSTEC) for their assistance during the in situ trial of the IISA-ATP in Okinawa. This study was supported by a grant-in-aid by the Scientific Research (S) and Development of Fundamental Tools Program for underwater resource survey from the Japanese Ministry of Education, Culture, Sports, Science and Technology (MEXT).

References

1. Karl DM (1980) Cellular nucleotide measurements and applications in microbial ecology. Microbiol Rev 44(4):739–796
2. Karl DM, Wirsen CO, Jannasch HW (1980) Deep-sea primary production at the Galapagos hydrothermal vents. Science 207:1345–1347
3. Daffonchio D, Borin S, Brusa T, Brusetti L, van der Wielen PWJJ, Bolhuis H, Yakimov MM, D'Auria G, Giuliano L, Marty D, Tamburini C, McGenity TJ, Hallsworth JE, Sass AM, Timmis KN, Tselepides A, de Lange GJ, Hübner A, Thomson J, Varnavas SP, Gasparoni F, Gerber HW, Malinverno E, Corselli C (2006) Biodeep Scientific Party, Stratified prokaryote network in the oxic-anoxic transition of a deep-sea halocline. Nature 440(7081):203–207
4. Azam F, Hodson RE (1977) Dissolved ATP in the sea and its utilisation by marine bacteria. Nature 267(5613):696–698
5. Hodson RE, Maccubbin AE, Pomeroy LR (1981) Dissolved adenosine triphosphate utilization by free-living and attached bacterioplankton. Mar Biol 64(1):43–51
6. Björkman KM, Karl DM (2001) A novel method for the measurement of dissolved adenosine and guanosine triphosphate in aquatic habitats: applications to marine microbial ecology. J Microbiol Methods 47(2):159–167
7. Bautista DA, McIntyre L, Laleye L, Griffiths MW (1992) The application of ATP bioluminescence for the assesment of milk quality and factory hygiene. J Rapid Methods Autom Microbiol 1(3):179–193
8. Hawronskyj J-M, Holah J (1997) ATP: A universal hygiene monitor. Trends Food Sci Technol 8(3):79–84
9. Venkateswaran K, Hattori N, La Duc MT, Kern R (2003) ATP as a biomarker of viable microorganisms in clean-room facilities. J Microbiol Methods 52(3):367–377
10. Aycicek H, Oguz U, Karci K (2006) Comparison of results of ATP bioluminescence and traditional hygiene swabbing methods for the determination of surface cleanliness at a hospital kitchen. Int J Hyg Environ Health 209(2):203–206
11. Aoki Y, Fukuba T, Yamamoto T, Fujii T (2009) Design optimization and evaluation of a bioluminescence detection part on a microfluidic device for *in situ* ATP quantification. IEEJ Trans SM 129(3):73–76
12. Denburg JL, McElroy WD (1970) Anion inhibition of firefly luciferase. Arch Biochem Biophys 141(2):668–675
13. Fujii T (2002) PDMS-based microfluidic devices for biomedical applications. Microelec Eng 61–62:907–914
14. Fukuba T, Yamamoto T, Naganuma T, Fujii T (2004) Microfabricated flow-through device for DNA amplification—towards *in situ* gene analysis. Chem Eng J 101(1–3):151–156
15. Fukuba T, Aoki Y, Fukuzawa N, Yamamoto T, Fujii T (2011) Microfluidic *in situ* analyzer for ATP quantification in ocean environments. Lab Chip 11:3508–3515
16. Hirayama H, Sunamura M, Takai K, Nunoura T, Noguchi T, Oida H, Furushima Y, Yamamoto H, Oomori T, Horikoshi K (2007) Culture-dependent and -independent characterization of microbial communities associated with a shallow submarine hydrothermal system occurring within a coral reef off Taketomi Island. Jpn Appl Environ Microbiol 73(23):7642–7656
17. Fukuba T, Fukuzawa N, Glutz LS, Miyaji A, Fujii T (2007) Development of an integrated *in situ* analyzer for quantitative analysis of microbial ATP in aquatic environments.

Underwater technology and workshop on scientific use of submarine cables and related technologies, Tokyo, Japan. p 210

18. Kinoshita H, Atsumi T, Fukuba T, Fujii T (2010) Active micro flow-rate regulation technique based on soft membrane deformation using miniaturized electroosmotic pumps. 14th international conference on miniaturized systems for chemistry and life sciences, Groningen, The Netherlands. pp 390–392

Chapter 11

Use of Biosensors as Alternatives to Current Regulatory Methods for Marine Biotoxins

Luis M. Botana, Natalia Vilariño, Amparo Alfonso, M. Carmen Louzao, Mercedes R. Vieytes, Ana M. Botana, and Carmen Vale

Abstract

Marine toxins need a continuous monitoring system that is usually implemented in most producer and consumer countries. This is needed to protect consumers, as marine toxins are very toxic compounds. Up to now, the mouse bioassay has been the method of choice, but there is a legal requirement in Europe to replace the bioassay for mass spectrometry with chromatographic separation. This method is rather complex and expensive, and therefore biosensors are an option that provides an easy, reliable, and affordable fast system. A biosensor commonly used for marine toxin developments is the optical, label free, surface plasmon resonance (SPR). We provide in the chapter a discussion about SPR and a comparison among several technologies, and also a detailed example of an SPR-based method for saxitoxin and analogs. We also provide a method that uses another approach, the resonant mirror, in this case for yessotoxin. Both resonant mirror and SPR are good options for monitoring purposes for marine toxins.

Key words: Resonant mirror, SPR, Marine toxin, Alternative method, Food safety, Biosensor, PSP detection, Ocean sensing, Biorecognition system, Real-time detection

1. Introduction

High-throughput screening methods for lead compound identification is a resource that drug companies have been using for years to speed up the flow of information and determine in a fast pace what compounds are interesting and deserve further resources to be invested (1). A similar approach has been suggested in many occasions for marine toxins, but to date, there has been no success. This chapter explores the reasons to the current situation on marine toxin monitoring, and what the use of biosensors will deserve for the future.

Sonia M. Tiquia-Arashiro (ed.), *Molecular Biological Technologies for Ocean Sensing*, Springer Protocols Handbooks,
DOI 10.1007/978-1-61779-915-0_11,

Marine toxins, or phycotoxins, are a group of compounds, very complex in their biochemical nature, which are produced as secondary metabolites by about 100 different species of dinoflagellates or diatoms (2, 3). Their ecological role is not known yet, but since dinoflagellates are accumulated by filter-feeding molluscs in the digestive tract, shellfish are natural concentrators of marine toxins, and their concentration reaches deadly levels within a short period of time. Since some of these toxins are very potent, some of them being the most potent compounds in nature, as it is the case of palytoxin (4, 5) or maitotoxin (6), then it is evident that the need for a continuous control of their presence in seafood products intended for human consumption.

Ever since the very beginning of marine toxin control, the mouse bioassay has been the method of choice (7) given its combined easiness of use, low cost, and accessibility. But as animal protection issues and legal requirements (8) have increased the pressure on the use of animals, a recent European Food Safety Agency (EFSA) working group has provided opinions on the legal, toxicological, methodological, and risk status for most of the marine toxins (9–20). The conclusions from this working group were used as base for a new Regulation issued in Europe that requires after June 2011, and coexisting with the mouse bioassay for 3 years, the reference method for marine toxins will be liquid chromatography coupled to mass spectrometry (LC-MS) (21). This same regulation allows the use of screening methods, and here a window for biosensors is open, as long as they are proven to be equivalent to the reference method in terms of protection to consumers, and the method is validated. Although this approach is not necessarily the preferred option in other parts of the world, and it is not the approach being followed by the Codex Alimentarius, in terms of trading with Europe it is the one to use, and therefore, since Europe is a major consumer of bivalves, it seems logical to expect that LC-MS will be the standard for trading in the future. The need to convert LC-MS results into values of a reference compound for each group makes necessary the use of toxic equivalent factor (TEF) to translate concentration values of one compound into the equivalent of the reference toxin. This is a complex issue that is not clarified yet, as TEF are unknown for most of the toxin analogs (22). Therefore, those methods that use a biorecognition system that identifies the TEF are the best for monitoring purposes, of them, obviously, the receptor itself is the best choice, but this is not always possible. This chapter focuses on a method that is probably the most widely explored, the surface plasmon resonance (SPR) biosensor. The method uses antibodies for the detection of one of the most dangerous marine toxins, the saxitoxin analogs. The article also describes the use of a resonant mirror biosensor method for yessotoxin, that we chose because it uses a totally different approach than the SPR method, with receptors as biorecognition system. Both methods, SPR and resonant mirror, with antibodies and receptors, respectively, show that biosensors are a fast, reliable, and affordable way of monitoring marine toxins.

1.1. Marine Toxins Diversity

There are probably a thousand marine toxins being produced by dinoflagellates and diatoms. Only a few compounds are quantitatively relevant, and they can be grouped on few toxin groups. The classification of the marine toxins has been always a matter of debate, but it is clear that the best approach is to assign a group name to a set of compounds with the same target and mode of action (22–24). Based on this, marine toxins present in shellfish can be classified as:

1. Saxitoxin and analogs (gonyautoxins, decarbamoyl analogs, benzoate analogs, neosaxitoxin), also known as PSP group (paralytic shellfish poison). With about 30 analogs described (25), they target the site 1 of the sodium channel (26). Tetrodotoxin is an analog produced by bacteria (27) with a similar action, but it is not produced by dinoflagellates, and it is commonly present in fish (28). Nevertheless, its presence and a severe human intoxication have been identified recently in gasteropods (29).
2. Azaspiracids (30) is a group with three major analogs and up to 30 compounds detectable in traces (31–33). Their mechanism of action is not yet known (34, 35).
3. Palytoxin is a unique, very large molecule that has been identified recently as a problem in bivalves (4). There are many potential stereoisomers (5), and the target is the sodium-potassium ATPase (36, 37).
4. Brevetoxins are activators of the site 5 at the sodium channel (38), and there are about 15 analogs described (39).
5. Ciguatera toxins, with about 20 analogs, (40, 41) target the same receptor as brevetoxins (42–44).
6. Domoic acid (amnesic shellfish poison, ASP) activates the kainate receptor (45, 46), and there are 11 analogs reported (47).
7. Cyclic imines, this group is the most complex in terms of chemical diversity, with several subgroups (19, 48) (gymnodimine, spirolide, pinnatoxin, pteriatoxin, pinnamine, prorocentrolide, spiroprorocentrimine, symbioimine), each one with several analogs. They target the cholinergic receptor, either nicotinic (49) or muscarinic (50). The use of this receptor as a reagent for manipulation in method development is rather feasible, as it is very stable if kept in membrane vesicles (49, 51, 52).
8. Okadaic acid (OA) and analogs (dinophysistoxins), known as diarrheic shellfish poison (DSP), with about 12 analogs, it is the most common marine toxin worldwide, and inhibits protein phosphatases 1 and 2A (53, 54).

9. Pectenotoxins (20 analogs) are inhibitors of F-actin formation (55–57).
10. Yessotoxin, this group has about 50 analogs (58, 59), and its mechanism of action is not fully elucidated yet (60–62).

1.2. Biosensors

A biosensor can be defined as a system that converts a biochemical reaction into an electrical signal. This wide definition allows many possible setups, but essentially, the biological recognition of the compound is achieved by means of an antibody or a receptor. The term receptor is here again used in a broad sense, as both enzymes modifying a substrate or constitutive macromolecules that trigger transduction signals could be used for the purpose of biorecognition for a biosensor. The electrical device, the transducer, that converts the biological recognition to a signal is again open to many possibilities, as this is a field of research that is constantly expanding its scope, and there are many technical solutions in the market (63, 64), which range from ion-sensitive electrodes, to optical, piezoelectric or calorimetric systems (63). Marine toxins are a proven difficult sort of compounds to detect, since the bivalve matrix effect is rather important, and therefore, although there are many methods described, very few sustaine real chances of success.

Biosensors offer simplicity, low reagent use, and very high sensitivity in their usually simple experimental design (65). Figure 1 summarizes the strategies to use for the development of a biosensor.

1.2.1. Biorecognition

Any receptor has three properties which are essential to their role; they are selective, specific, and sensitive. Therefore, they are able to discriminate among many compounds the structures related to the target group and they always respond and bind to the compound in the same way. The definition of receptor used here covers enzymes and protein complexes that trigger a mechanism of transduction, and do not necessarily correspond to the concept of receptor being used in therapeutics. These set of characteristics are essential to understand why this would be the best approach for marine toxins. The use of a receptor to a marine toxin group would mean that all the analogs of that group sharing the same mechanism of action would be detected by the receptor. Therefore, they would be identified within a matrix mixture. Nevertheless, matrices, with high salt concentrations, are a common source of interferences; and in general any method newly developed must be accompanied by a specific extraction protocol (24). Both enzymes and receptors share similar capabilities to recognize the molecules of the target group, although in general it becomes easier to work with isolated enzymes than with isolated receptors. For this reason, receptors are usually semipurified and contained within membrane vesicles (52, 66, 67).

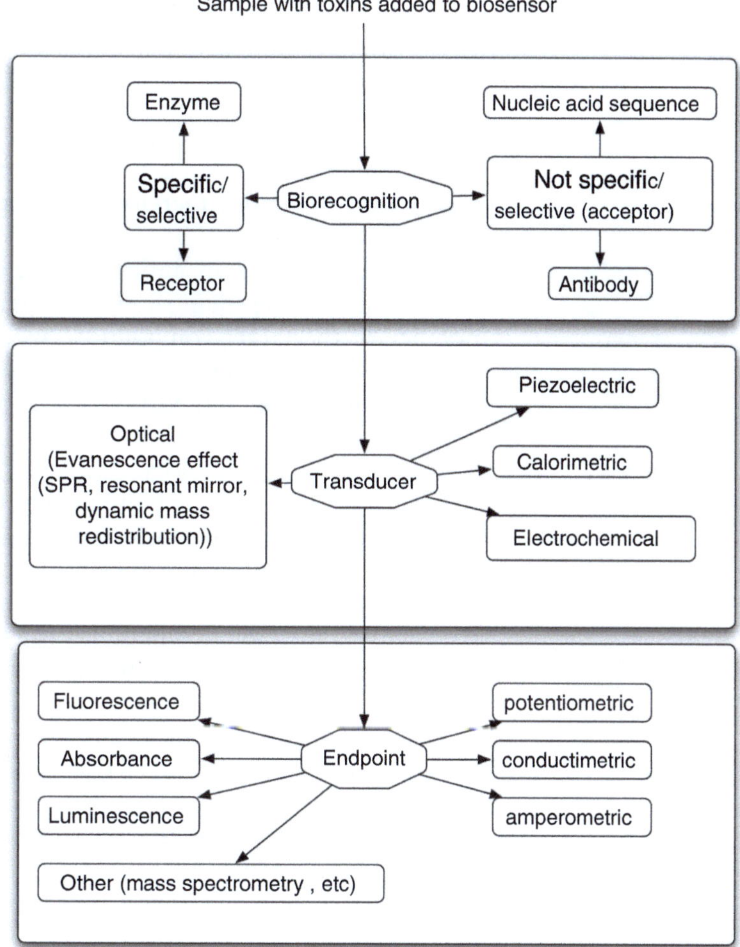

Fig. 1. Different approaches to biosensor detection.

Biorecognition can be achieved with antibodies developed against certain structures of the molecule. A similar approach can be taken by means of aptamers (68, 69), which are specific nucleic acid sequences (RNA or single-stranded DNA molecules) that bind target molecules with high affinity and specificity (70). Both aptamers and antibodies are able to detect very low amount of target compounds, but they are limited to their crossreactivity, and it is difficult to predict their detection capabilities with unknown or new substances that belong to a target group, and share the mechanism of action, but may not be recognized by the aptamer of the antibody (71, 72).

The basic difference with regard to recognition is that receptors or enzymes will bind to any compound sharing the mechanism of the group, while antibodies or aptamers will bind only those compounds they can recognize; hence, receptors and enzymes are dynamic in their recognition, while antibodies or aptamers are not. This is in fact the reason why so many antibody-based methods have failed to detect marine toxins, since they only recognize a few of the total number of compounds in a group. It is important to keep in mind that a toxin group is a set of compounds that share a mechanism of action; therefore, a receptor-based method is the best possible approach.

Recent advances in antibody development allow the use of selected antibodies that show cross-reactivities that match the potency of the target compounds (73), but still this approach is reliable with toxin groups with few components, such as the protein phosphatase inhibitor group of okadaic acid analogs (74).

1.2.2. Transducers and Endpoint

There are several technologies that may be used as transducers of the signal generated at the biorecognition to be viewed at the end-point. Calorimetric, piezoelectric, and electrochemical systems have not been successful or fully explored for marine toxins, since they require large amounts of sample, or do not provide enough selectivity to the target compounds (63). By far, the most successful approach is the optical technology. Optical transducers are label-free systems that monitor the binding of specific molecular recognition compounds by means of several technologies commercially available such as SPR (75), resonant mirror (76), or dynamic mass redistribution (1). They use evanescent electric fields to measure the change in local refractive index after binding between the target compound and a molecule that is bound to the sensor (75). This technology allows real-time interaction and kinetic analysis; hence, it is very versatile (77). Currently, SPR is the technology most widely used for the development of new methods, as this has been the most successful from a commercial point of view. Multisampling is possible now with multiplex analysis for large and small molecular weight compounds by means of multiple channel chips for SPR (78), microplate readers in the case of the dynamic mass redistribution using waveguide resonance (1), or by means of image analysis (75). Future developments intend to associate the SPR detection with mass spectrometry, Raman spectroscopy, imaging, etc. (79, 80).

Although there are many technological options for biosensing marine toxins, there is a problem, related to the food analysis in general, and it is the low profit margin of this sector. Therefore, competition is fierce, and here comes the main problem of biosensing technology, which is the cost. Either the chips or the equipment is rather expensive at the time, and mass spectrometry is now the preferred approach. On the other hand, rapid detection by automated immunoassays is inexpensive, and therefore, unless biosensors find a way through the low cost detection market, they will always be a marginal option for food safety analysis.

A good example of a marine toxin that has been deeply studied for its biosensor-detection options is yessotoxin. The mechanism of action of yessotoxin is not well defined, as several targets were proposed, such as cadherin (81) and phosphodiesterase (60). In addition, yessotoxin is a lipophilic toxin that is coextracted in the same fraction of other lipophilic toxins with totally different mechanisms of action, such as okadaic acid (phosphatase inhibitor (53, 82)), pectenotoxin (actin inhibitor (56, 83)), or azaspiracids (unknown target (84–88)). The fact that yessotoxin appears on the same extraction fraction of other toxins, but has a different potency and toxicity (89) and shows a totally different bioassay time response, has always been a management problem for regulators. Yessotoxin above legal limits shows a mouse death time of 6 h, while the other toxins above the legal threshold need 24 h to kill a mouse. Therefore, a selective extraction procedure for yessotoxin is needed, and this complicates notably its regulatory management (90). The use of isolated phosphodiesterases as biorecognition compounds for yessotoxin allowed the development of several rapid and biosensor-based approaches for detection. An indirect microplate fluorescent method was first developed based on the use of a fluorescent substrate of phosphodiesterases (anthranyloyl-cAMP) that would change fluorescence upon activation of phosphodiesterases by yessotoxin. This approach was further evolved to a direct assay and determined just the interaction of yessotoxin with the phosphodiesterase, by means of a resonant mirror biosensor (76) and a polarization fluorescence assay (91). The use of the biosensor allowed the study of the kinetic parameters of the binding of yessotoxin to phosphodiesterases (62, 77). The next step was the development of an SPR-based method for yessotoxin (92) and for ladder-shaped polyethers (93). Among all these methods, there is a philosophical difference with regard to binding and biorecognition. On the one hand, resonant mirror or polarization fluorescence are direct binding assays; hence, the more the binding, the more the signal. On the other hand, SPR are often inhibition assays, where the receptors or large molecules are not constrained on the sensor chip surface, and it is the small molecule, the toxin, that it is attached to the sensor. The inhibition assay reflects the inhibition of the binding of the enzyme to the attached toxin on the surface in the presence of competing toxin from the sample (71, 93). In addition to these biosensor methods, immunoassays for yessotoxin were also developed (94, 95), and the availability of antibodies allows the use of antibody-based biosensor assays.

1.3. Methods

For most of the marine toxin groups, there are biosensor methods available. Only those groups that have not receptor candidates, such as azarpiracids (34), pose problems to develop receptor-based biosensors, and antibodies are an alternative solution to the problem. On the other hand, toxins with very high potency are difficult to use as a source of antibodies, as they are lethal even at very high dilutions, and therefore only antibodies with low titles can be obtained (96, 97). A recent paper describes a direct SPR method for palytoxin using mouse antibodies and with a good sensitivity in the low to sub-ppb range (98).

1.3.1. Sodium Channel-Active Toxins (Saxitoxins, Brevetoxins, Ciguatoxins)

Paralytic shellfish toxins were the focus of a large European project (BIOCOP) that provided an SPR method based on the use of antibodies (71), the method was in house validated (99) and has been proven in an inter laboratory study (100) to be an empowering tool in the drive towards the reduction and replacement of the mouse bioassay (101) and as a support screening method for the reference HPLC method (102). With the exception of antibodies, receptor-based biosensors for PSP are rather difficult to develop since the sodium channel is unstable (103). A rather unique type of biosensors for PSP toxins are those based on the measurement of the electrical currents modulated by PSP, or tetrodotoxin for this same matter, as they share the same receptor. One approach has been described based on the measurements of the electrophysiological currents in cultured cells that express STX-sensitive skeletal muscle Na channel (104). A similar approach, that used cultured neuronal networks grown on integrated microelectrode arrays, has been developed for PSP and brevetoxins (105, 106). This neuronal network biosensor measures the bioelectrical activity of the network as extracellular action potentials and can monitor noninvasively the amount of neuroactive marine toxins. The method might be probably adapted to ciguatoxins as they are also active on sodium channels (44). The combined use of antibodies and screen-printed electrodes has been reported for okadaic acid, brevetoxin, domoic acid, and tetrodotoxin (107). This type of methods combine an ELISA with the use disposable carbon electrodes coupled to amperometric detection of *p*-aminophenol/alkaline phosphatase as endpoint. Although brevetoxins and ciguatoxins are difficult to detect by means of using sodium channels as the binder in biosensors, there are two methods that can screen them. One, mentioned above, is the SPR-phosphodiesterase sensor developed for ladder-shaped polyethers such as yessotoxin or brevetoxins (93); the second one, that has been described for ciguatoxins, but can be modified to brevetoxins, is the use of electrodes clamped to cerebellar neurons to measure the electrical current in voltage gated sodium or potassium channels (44). This last method can detect the toxins in the very low nanomolar range.

1.3.2. Phosphatase Inhibitors (Okadaic Acid and Analogs)

For okadaic acid and analogs, a pseudofunctional method that uses a monoclonal antibody that has been described, it shows a cross-reactivity that matches the relative toxicity of each of the analogs (73, 108, 109), and the results show a good correspondence with LC-MS equivalent data. To date, there is not an SPR-based phosphatase assay, but phosphatase has been used in an electrochemical method sensitive to dephosphorylation (110), a biosensor based on a screen-printed electrode with immobilized okadaic acid that binds to monoclonal antibodies and uses amperometric measurements as end point (111), and an electrochemical biosensor that detects H_2O_2 produced by oxidation of the glucose produced by the effect of phosphatase on glycogen (112). A different sensor design used anti-OA monoclonal antibodies labeled with horseradish peroxidase that compete with sample and OA-bovine serum albumin (BSA) bound to polyethersulfone membranes (113), chemiluminescence being the end point.

1.3.3. Yessotoxin

For yessotoxins, the several approaches available were described above, but since yessotoxin is a ladder-shaped polyether toxin, sharing similitudes with brevetoxins, there is a phosphodiesterase-based SPR method that can also detect brevetoxin 2 (93).

1.3.4. Domoic Acid

Since domoic acid receptor is very complex to handle, there is no method based on the use of the receptor. Domoic acid is the only marine toxin that has been always detected by HPLC with UV detection (114, 115); hence, this toxin group has not been a concern in terms of animal use. In addition, there is an ELISA available for domoic acid and legally endorsed in the EU (116). For domoic acid, an SPR inmunobiosensor with DA attached to the chip surface (117), that uses rabbit polyclonal antibody, has been developed and compared to the reference method with acceptable results. Screen-printed electrodes were used for detection of domoic acid (118); they combine antibody recognition with differential pulse voltammetry quantification, but in a similar fashion as for neurotoxins, the biosensor is only used for final quantification, being an ELISA the initial part of the method.

1.3.5. Cyclic Imines

There are several receptor-based methods for cyclic imines (52, 119), but since cholinergic receptors (49, 50) are stable only if used in vesicles, their use in biosensors is complex. Therefore, future SPR methods for these compounds will probably be based on antibodies rather than receptors.

1.3.6. Other

An SPR method has been mentioned above for palytoxin (98). For maitotoxin, azaspiracids, etc., there is no method yet, and this is attributable to the scarcity of the toxins to develop antibodies, or the difficulty in isolating a reliable functional receptor.

1.4. Validation

For any method, it is mandatory to validate it in an interlaboratory exercise, and following international guidelines (120–122). This is actually one of the major drawbacks of alternative methods, as validation is essential given the high toxicity of some of the marine toxins, and there is no dedicated international body to this task.

It is possible to run a validation study by means of AOAC that has a protocol for validation, or ECVAM (123). But in general, the cost of a validation is high and complex, and there are very few methods that were fully validated. At this time, only the HPLC method for PSP toxins (124) and domoic acid (125), and the ELISA method for domoic acid (126) were recognized as validated methods. There is no biosensor-based validated method, although the SPR method for PSP is the only biosensor method that has gone through a pilot interlaboratory validation study (100).

In summary, very little advance has been achieved in developing biosensors for marine toxins. Although SPR is by far the preferred choice of scientist, it is difficult to use receptors, and antibodies will always fall short in terms of cross-reactivity, specially for large toxin groups. The future of biosensors in marine toxin detection is unclear, as the acceptance of LC-MS as reference method will probably reduce the efforts to have alternative methods. Still, biosensors are a good screening option for marine toxins, once they are proven in interlaboratory validation exercises that they are reliable and provide good consumer protection.

2. Reagents and Equipment

2.1. Reagents for PSP Detection by SPR-Based Biosensors

2.1.1. Immobilization

Saxitoxin dihydrochloride

CM5 sensor chips from Biacore

Amine coupling kit (EDC, NHS and ethanolamine)

HBS-EP buffer (pH 7.4, 0.01 M HEPES, 0.15 M NaCl, 3 mM EDTA, 0.005 % polysorbate) (see Note 1)

Formaldehyde (37 %)

2,2-(Ethylenedioxy) bis-(ethylamine) (Jeffamine)

Borate buffer, 50 mM, pH 8.5

2.1.2. Sample Extraction Protocol

Glacial acetic acid

Sodium acetate minimum 99 %

2.1.3. Sample Analysis

Biacore Q SPR biosensor

HCl (1 M)

HBS-EP buffer (pH 7.4, 0.01 M HEPES, 0.15 M NaCl, 3 mM EDTA, 0.005 % polysorbate)

Saxitoxin dihydrochloride

Polyclonal antibody as saxitoxin binding protein

2.1.4. Evaluation of Results

BIAevaluation software version 4.1.

2.2. Reagents for YTX Detection: Resonant Mirror Optical Biosensor

2.2.1. Reagents

1. Iasys carboxymethyl dextran (CMD) cuvettes (two wells) (Lab systems, UK)
2. Coupling kit: 1-ethyl-3-(3-dimethylaminopropyl) carbodiimide (EDC), *N*-hydroxysuccinimide (NHS) and 1 M ethanolamine (Lab systems, UK)
3. Phosphodiesterase 3,5-cyclic-nucleotide-specific from bovine brain (PDEs) (Sigma Chemical Co, Madrid, Spain)
4. Sodium acetate (Sigma Chemical Co, Madrid, Spain)
5. Tween 20 (Sigma Chemical Co, Madrid, Spain)
6. Yessotoxin (CIFGA Laboratories (http://www.cifga.com), Spain)
7. Deionized water (Milli-Q)

2.2.2. Buffers

1. PBS/T (mM): Na^+ 145.2; K^+ 4.7, HPO_4^{2-} 8.2; $H_2PO_4^-$ 1.5; Cl^- 141.2; and 0.05 %v/v Tween 20
2. PB (mM): Na^+ 20; HPO_4^{2-} 10. pH 7.7

2.2.3. Equipment

1. Iasys Affinity Sensor (Lab system, UK) with simultaneous dual channel monitoring system to use dual cuvette surfaces.
2. Stirred cuvette system connected (to ensure minimal mass-transport effect): 85 rpm rate.
3. Data-sampling interval: 0.3 s.
4. Temperature: 22 °C.

3. Protocol

3.1. Protocol for PSP Detection by SPR-Based Biosensors

3.1.1. Immobilization of Saxitoxin on the Chip Surface

Saxitoxin is covalently bound to the chip surface by amino–amino coupling.

1. Apply a mixture (1:1, v/v) of 75 mg/ml 1-ethyl-3-(3-dimethylaminopropyl) carbodiimide hydrochloride (EDC) and 11.5 mg ml^{-1} NHS for 30 min to the surface of a CM5 chip.
2. Remove excess solution and apply 20 % jeffamine (2,2-ethylenedioxy)-bis (ethylamine) in borate buffer for 1 h.
3. Mix 36.6 µl of saxitoxin solution (100 µg in 0.1 M acetic acid), 64 µl of water and 46 µl of 37 % formaldehyde. Apply to the chip surface after removal of the solution added in step 2 and incubate for 3 days.
4. Remove solution and deactivate the chip surface by exposure to 1 M ethanolamine for 30 min.
5. The chip surface is then washed with a flow rate of 25 µL min^{-1} of HBS-EP buffer prior to initial use.

3.1.2. Sample Extraction

1. Removed shellfish samples from their shells and drain.
2. Homogenize 100 g of shellfish meat.
3. Weight 1 g of homogenate into centrifuge tubes and add 5 ml of sodium acetate buffer, 0.2 M, pH 5 (see Note 2).
4. Vortex for 10 s and roller mix for 30 min.
5. Centrifuge samples at 3,600 × *g* for 10 min at room temperature.
6. Collect the supernatant and dilute 1:20 in HBS-EP buffer.

3.1.3. Sample Analysis

1. Prepare saxitoxin calibrants at concentrations of 0, 1.0, 2.5, 5.0, 7.5, and 10.0 ng ml^{-1} in HBS-EP.
2. Dilute the saxitoxin binding protein (antibody) to the adequate concentration.
3. Load samples, calibrants, saxitoxin binding protein, and regeneration solution on the BiaCore Biosensor racks.
4. Start the BiaCore biosensor protocol for analysis. The procedure for the analysis are:
   a. Mix equal volumes of the binding protein and the saxitoxin calibrant (or sample).
   b. Inject the mixture over the STX sensor chip surface at a flow rate of 12 μL min^{-1} for 120 s.
   c. Regenerate the chip surface with 8 μL of 50 mM HCl at a flow rate of 12 μL min^{-1}.
   d. Mix and inject the next calibrant solution (or sample).

The duration of the procedure for one sample would be approximately 1 h for the extraction (ten samples can be easily managed simultaneously), 30 min for calibration curve and sample preparation and 7 min per cycle for biosensor detection (14 min for each sample in duplicates).

3.1.4. Evaluation of Results

Each binding cycle in the biosensor has a baseline, a binding phase, a dissociation phase and a regeneration step that yields the binding surface ready for the next binding cycle (Fig. 1). Report points for each cycle are taken 10 s before injection and 30 s after the injection is stopped and the dissociation phase is started.

The relative response is calculated by the difference between report points. The values of the relative response corresponding to the calibrant solution are used to generate a calibration curve with a four-parameter fit. The binding response for each sample is used to calculate toxin content in saxitoxin equivalents using the BIA evaluation software version 4.1 (Fig. 2).

The assay provides a dynamic range for saxitoxin from 2.8 to 7.8 ng ml^{-1} and the detection capability in shellfish extracts for this toxin has been estimated in 120 μg kg^{-1}. The estimation of PSP

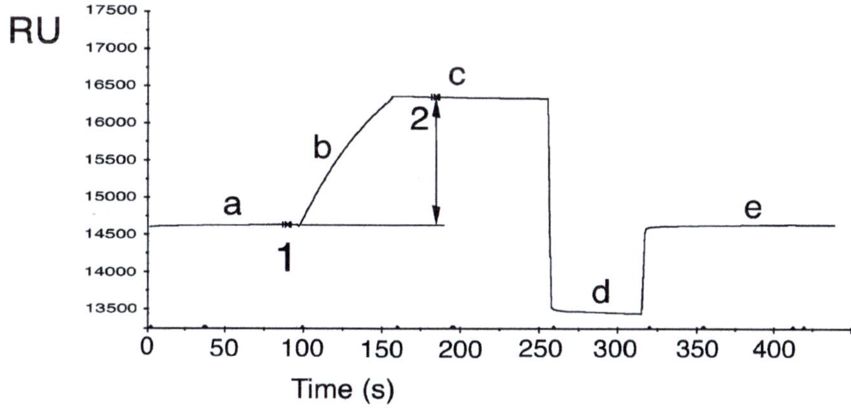

Fig. 2. Sensogram of one cycle of sample analysis in a BiaCore Q biosensor. The parts of the sensogram are identified as: (*a*) baseline, (*b*) binding (injection of binding protein and extract mixture), (*c*) dissociation (HBS injection), (*d*) regeneration (50 mM HCl injection), and (*e*) new baseline. Reporting points are marked as *1* (baseline) and *2* (response). *Arrow* indicates binding signal.

Table 1
Relative toxicity factors and R895 antibody cross-reactivity of PSP toxins

PSP toxin	Relative toxicity factor	R895 antibody cross-reactivity (%)
STXdiHCl	1.000	100
NEO	1.091	33
GTX 1/4	0.899	<0.7
GTX 2/3	0.601	39
dcSTX	0.745	142
dcNEO	0.701	6
dcGTX 2/3	0.398	13
C1/C2	0.075	20
GTX5	0.063	76
C3/C4	0.043	<7

content for each sample using this assay will depend on the cross-reactivity profile of the saxitoxin binding protein and the toxin profile of the sample. Table 1 shows the cross-reactivity profile of the polyclonal R895 antibody and the toxicity equivalency factors of PSP toxins.

This technique is very robust and no problems should be encountered other than the stability of reagents due to aging or inadequate storage. Each sensor chip has four cells and each can be used for at least 300 binding cycles (see Note 3). In any case, it is a good practice to monitor baseline and binding response values

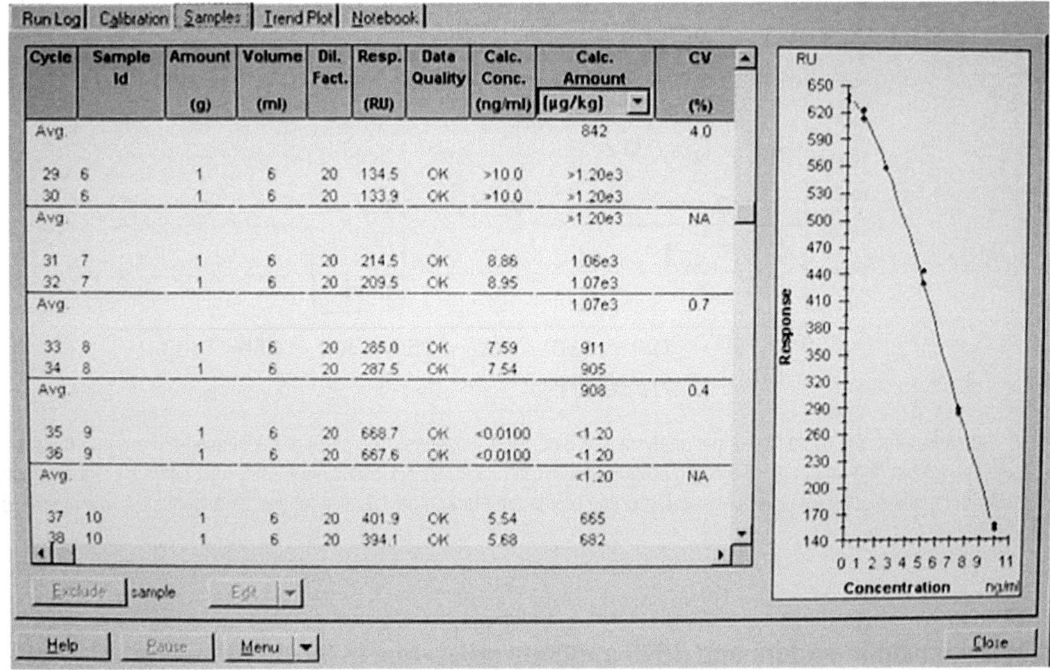

Cycle	Sample Id	Amount (g)	Volume (ml)	Dil. Fact.	Resp. (RU)	Data Quality	Calc. Conc. (ng/ml)	Calc. Amount [µg/kg]	CV (%)
Avg.								842	4.0
29	6	1	6	20	134.5	OK	>10.0	>1.20e3	
30	6	1	6	20	133.9	OK	>10.0	>1.20e3	
Avg.								>1.20e3	NA
31	7	1	6	20	214.5	OK	8.86	1.06e3	
32	7	1	6	20	209.5	OK	8.95	1.07e3	
Avg.								1.07e3	0.7
33	8	1	6	20	285.0	OK	7.59	911	
34	8	1	6	20	287.5	OK	7.54	905	
Avg.								908	0.4
35	9	1	6	20	668.7	OK	<0.0100	<1.20	
36	9	1	6	20	667.6	OK	<0.0100	<1.20	
Avg.								<1.20	NA
37	10	1	6	20	401.9	OK	5.54	665	
38	10	1	6	20	394.1	OK	5.68	682	

Fig. 3. Results displayed by the biosensor software after the PSP detection assay is finished.

for indications of surface or binding protein decay (see Note 4). The performance of periodic maintenance and cleaning procedures in the biosensor is essential to avoid problems with the microfluidic system (Fig. 3).

3.2. Protocol for YTX Detection: Resonant Mirror Optical Biosensor

Principle. Method based on the measurement of phosphodiesterases-yessotoxin association by using a resonant mirror biosensor.

3.2.1. CMD Surfaces Activation

1. Prepare activation solution following manufacturer's instructions (mix equal volume of 0.2 M EDC and 0.05 M NHS just before using) (see Note 5).
2. Add EDC/NHS solution in both wells. Incubate for 10 min.
3. Replace the activation solution by PBS/T. Incubate for 1 min.
4. Replace by 10 mM acetate buffer pH 5.0. Incubate for 1 min.

3.2.2. Ligand Immobilization

1. Add 0.1 mg ml^{-1} PDEs to the cuvette (PB solution). Incubate 5 min.
2. Wash with PB.
3. Add ethanolamine pH 8.5. Incubate for 3 min (nonspecific binding, NSB).
4. Replace the solution by PBS/T.

3.2.3. Toxin Detection by Toxin Association/ Dissociation and Sensor Regeneration

1. Prepare five YTX working standards: For example: 1, 3.75, 5, 7.5, and 10 μM YTX (PB, pH 7.7).
2. Add 10 μM YTX to one cuvette well.
3. Add the same amount of standard vehicle to the other cuvette well (control well).
4. Monitor association curves (response) for 20 min.
5. Wash the cuvette (both wells) with PB, recording dissociation phase for 5 min.
6. Cuvette regeneration: Add 0.01 M HCl and incubate for 2 min.
7. 5 min equilibration with PB.
8. Add the other YTX concentrations and repeat steps 3–7.
9. Add shellfish samples (extracted following [EC, 2005 #1]) and repeat steps 3–7 (see Note 6).
10. Add PBS/T to the cuvette and store at 4 °C. The cuvette (with the immobilized PDEs) is ready for a new experiment (see Note 7).

3.2.4. Data Analysis

1. Normalize data by subtracting control well to the response of YTX/sample well.
2. Calculate from each association curve the association rate constant k_{on} (s^{-1}) (62, 76, 77).

4. Typical Protocol Results

A typical experiment is shown in Fig. 4. As it can be observed, the activation and ligand immobilization is ready in less than 1 h. In these conditions, the cuvette can be stored at 4 °C, and it is ready for several experiments. The calibration curve can be done with four or five YTX concentrations. The extraction procedure of toxins from contaminated samples is fast, 30 min, and the toxin concentration is easily known. Several samples can be fast and selectively checked for the presence of YTX in a short period. Therefore, this is an accurate, feasible, and reliable method to detect YTX.

- Represent YTX concentration versus k_{on} (s^{-1}) obtained.

$$k_{on} = (k_{ass} \times [\text{YTX}] + k_{diss}).$$

(see Fig. 5).

- Transform shellfish k_{on} to YTX concentration (Fig. 6).

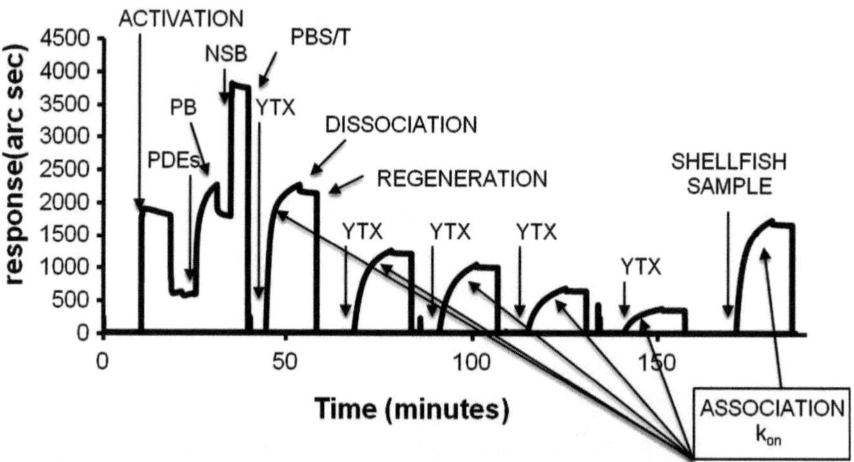

Fig. 4. Graphic profile of the steps described to be taken in this assay, from the surface activation to the addition of a test sample.

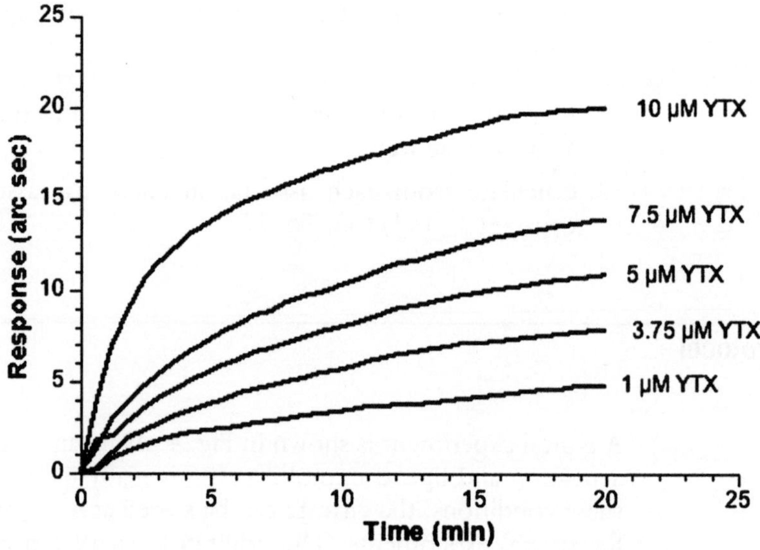

Fig. 5. Representative association curves (arc seconds response) after addition of different amounts of YTX over immobilized PDEs obtained from Fig. 4. From these curves, k_{on} (s^{-1}) for each YTX concentration is obtained.

5. Notes

Although for SPR the chip is very stable and reusable many times, the major problem concerning the resonant mirror method for yessotoxin is the stability of cuvettes once the immobilization is done. This item depends on storage conditions but also the purity of samples is important. Keeping at 4 °C the cuvettes with enough buffer, avoiding dryness, and contaminations, is essential to preserve

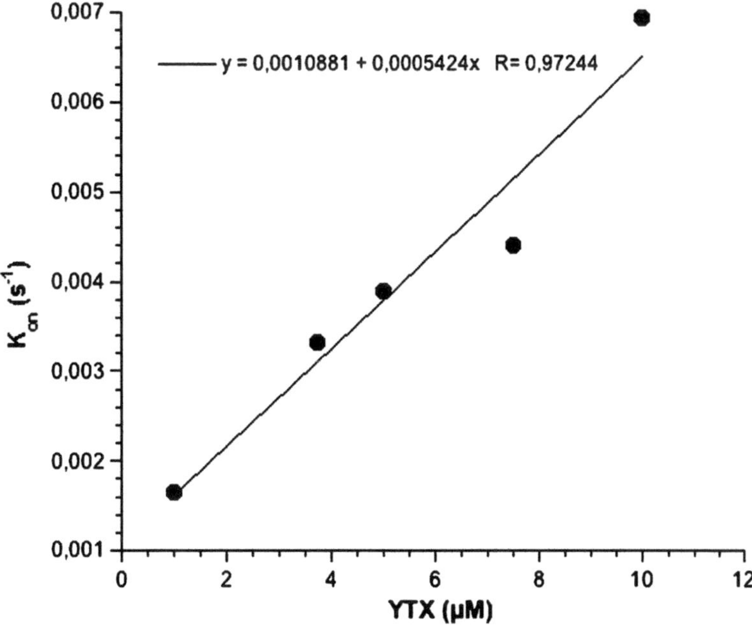

Fig. 6. Kinetic plot of apparent association constant k_{on} (s^{-1}) obtained from Fig. 5 plots versus YTX concentration. With this calibration curve the amount of YTX in a sample can be calculated.

an active surface. In addition, a filtration step trough 0.45 μm, before biosensor quantification, is sometimes recommended.

1. Degassed buffer should be used for good performance of the biosensor. One of the main problems of the technology is the presence of bubbles in the microfluidic system. The buffer should not be used for more than 2 days after opening the container.
2. Buffers should be prepared fresh.
3. Binding signals should not vary greatly among cells within one chip for good quality chips.
4. If binding signal diminishes significantly from initial values within one cell, that cell should be discarded for further use. Likewise if an increase of baseline values is observed, baseline signal for binding cycles may change slightly during the first cycles of one cell until it stabilizes.
5. Aminosilane cuvettes can also be used. In this case, the activation protocol is done with glutaraldehyde (4.2 %v/v).
6. The sample preparation includes the extraction of toxins from contaminated shellfish with acetone, three times, followed by three dichloromethane partitions. This process can be replaced by a methanol extraction to quantify YTX by other technologies. However when the resonant mirror optical biosensor is

used the organic partition is also necessary. In some cases, the cleaning procedure of samples includes a silica cartridge followed by a 10000 NMWL. These two steps can be replaced by a 0.45 filter even though the solid phase extraction improves sometimes the results.

7. Each activated cuvette can be used with 100 samples approximately.

6. Future Directions

The future of biosensors will be largely based on the two factors. One is the success of the mass spectrometry (LC-MS) methods, and their complexity. If LC-MS are a popular option, biosensors will no longer be pursued. On the other hand, the second factor to keep in mind is the need to develop antibodies with a cross-reactivity that mimics TEFs, and detect all the analogs required by the legislation. The realistic approach at this time is that there is no biosensor close to getting the status of usable, mostly because the equipments are rather expensive, and therefore downscale is very much needed, and because no method covers all the compounds required.

Another very important aspect is the need to validate the method once developed, and this will require a good usable method (not available yet) and a successful validation study. Therefore, although biosensors are promising, in the difficult arena of marine toxins, they are not ready yet.

Acknowledgements

This work was funded with the following FEDER cofunded-grants: from Ministerio de Ciencia y Tecnología, Spain: AGL2007-60946/ALI, SAF2009-12581 (subprograma NEF), AGL2009-13581-CO2-01, TRA2009-0189, AGL2010-17875. From Xunta de Galicia, Spain: GRC 2010/10, and PGIDT07CSA012261PR, PGDIT 07MMA006261PR, PGIDIT (INCITE) 09MMA003261PR, 2009/XA044, 2009/053 (Consell. Educación), 2008/CP389 (EPITOX, Consell. Innovación e Industria, programa IN.CI.TE.), 10PXIB261254 PR. From EU VIIth Frame Program: 211326 – CP (CONffIDENCE), 265896 BAMMBO, 265409 μAQUA, and 262649 BEADS. From the Atlantic Area Programme (Interreg IVB Trans-national): 2008-1/003 (Atlantox) and 2009-1/117 Pharmatlantic.

References

1. Noah JW (2010) New developments and emerging trends in high-throughput screening methods for lead compound identification. Int J High Throughput Screening 1:141–149
2. Mackenzie L (2008) Ecobiology of the brevetoxin, ciguatoxin and cyclic imine producers. In: Botana LM (ed) Seafood and freshwater toxins: pharmacology, physiology and detection, 2nd edn. CRC (Taylor and Francis Group), Boca Raton, FL, pp 433–476
3. Reguera B, Pizarro G (2008) Planktonic dinoflagellates that contain polyether toxins of the old "DSP complex". In: Botana LM (ed) Seafood and freshwater toxins: pharmacology, physiology and detection, 2nd edn. CRC (Taylor and Francis Group), Boca Raton, FL, pp 257–284
4. Aligizaki K, Katikou P, Nikolaidis G, Panou A (2008) First episode of shellfish contamination by palytoxin-like compounds from ostreopsis species (Aegean Sea, Greece). Toxicon 51:418–427
5. Katikou P (2008) Palytoxin and analogues: ecobiology and origin, chemistry, metabolism, and chemical analysis. In: Botana LM (ed) Seafood and freshwater toxins: pharmacology, physiology and detection, 2nd edn. CRC (Taylor and Francis Group), Boca Raton, FL, pp 631–664
6. Trevino CL, Escobar L, Vaca L, Morales V, Ocampo AY, Darszon A (2008) Maitotoxin: a unique pharmacological tool for elucidating Ca2+-dependent mechanisms. In: Botana LM (ed) Seafood and freshwater toxins: pharmacology, physiology and detection, 2nd edn. CRC (Taylor and Francis Group), Boca Raton, FL, pp 503–516
7. Botana LM (2008) The mouse bioassay as a universal detector. In: Botana LM (ed) Seafood and freshwater toxins: pharmacology, physiology and detection, 2nd edn. CRC (Taylor and Francis Group), Boca Raton, FL, pp 149–161
8. Communities E (1986) COUNCIL DIRECTIVE of 24 November 1986 on the approximation of laws, regulations and administrative provisions of the Member States regarding the protection of animals used for experimental and other scientific purposes (86/609/EEC). Off J Eur Commun, L358:1–29
9. Panel EC (2008) Influence of processing on the levels of lipophilic marine biotoxins in bivalve molluscs. Satement of the panel on contaminants in the food chain. The EFSA Journal 1016:1–10
10. Panel EC (2008) Scientific opinion on marine biotoxins in shellfish- yessotoxin group. EFSA panel on contaminants in the food chain (CONTAM). The EFSA Journal 907:1–62
11. Panel EC (2008) Opinion of the scientific panel on contaminants in the food chain on a request from the European commission on marine biotoxins in shellfish – azaspiracids. The EFSA Journal 723:1–52
12. Panel EC (2008) Opinion of the scientific panel on contaminants in the food chain on a request from the European commission on marine biotoxins in shellfish – okadaic acid and analogues. The EFSA Journal 589:1–62
13. Panel EC (2009) Scientific opinion on marine biotoxins in shellfish- pectenotoxin group. EFSA panel on contaminants in the food chain (CONTAM). The EFSA Journal 1109:1–47
14. Panel EC (2009) Scientific opinion on marine biotoxins in shellfish- domoic acid. EFSA panel on contaminants in the food chain (CONTAM). The EFSA Journal 1181:1–61
15. Panel EC (2009) Scientific opinion on marine biotoxins in shellfish- palytoxin group. EFSA panel on contaminants in the food chain (CONTAM). The EFSA Journal 7:1393 (1338 pp.)
16. Panel EC (2009) Opinion of the scientific panel on contaminants in the food chain on a request from the European commission on marine biotoxins in shellfish – saxitoxin group. The EFSA Journal 1019:1–76
17. Panel EC (2009) Marine toxins in shellfish. Summary on regulated marine biotoxins. Scientific opinion of the panel on contaminants in the food chain. The EFSA Journal 1306:1–23
18. Panel EC (2010) Statement on further elaboration of the consumption figure of 400 g shellfish meat on the basis of new consumption data. EFSA panel on contaminants in the food chain (CONTAM). The EFSA Journal 8:1706 (1720 pp.)
19. Panel EC (2010) Scientific opinion on marine biotoxins in shellfish- cyclic imines (spirolides, gymnodimines, pinnatoxins and pteriatoxins). EFSA panel on contaminants in the food chain (CONTAM). The EFSA Journal 8:1628 (1639 pp.)
20. Panel EC (2010) Scientific opinion on marine biotoxins in shellfish- emerging toxins: ciguatoxin-group toxins. EFSA panel on contaminants in the food chain. The EFSA Journal 8:1627 (1638 pp.)
21. REGULATION C (2011). COMMISSION REGULATION (EU) No 15/2011 of 10

January 2011 amending Regulation (EC) No 2074/2005 as regards recognised testing methods for detecting marine biotoxins in live bivalve molluscs. Off J Eur Commun, L:3–4

22. Botana LM, Vilariño N, Elliott CT, Campbell K, Alfonso A, Vale C, Louzao MC, Botana AM (2010) The problem of toxicity equivalent factors in developping alternative methods to animal bioassays for marine toxin detection. Trends Anal Chem 29:1316–1325
23. Botana LM, Alfonso A, Botana A, Vieytes MR, Vale C, Vilariño N, Louzao MC (2009) Functional assays for marine toxins as an alternative, high-throughput screening solution to animal tests. Trends Anal Chem 28:603–611
24. Botana LM, Louzao MC, Alfonso A, Botana AM, Vieytes MR, Vilariño N, Vale C (eds) (2011) Measurement of algal toxins in the environment. Willey, Chichester
25. Vale P (2008) Complex profiles of hydrophobic paralytic shellfish poisoning compounds in gymnodinium catenatum identified by liquid chromatography with fluorescence detection and mass spectrometry. J Chromatogr A 1195:85–93
26. Catterall WA (1979) Neurotoxins as allosteric modifiers of voltage-sensitive sodium channels. Adv Cytopharmacol 3:305–316
27. Gallacher S, Birkbeck TH (1993) Effect of phosphate concentration on production of tetrodotoxin by alteromonas tetraodonis. Appl Environ Microbiol 59:3981–3983
28. Noguchi T, Arakawa O, Takatani T (2006) TTX accumulation in pufferfish. Comp Biochem Physiol Part D Genomics Proteomics 1:145–152
29. Rodriguez P, Alfonso A, Vale C, Alfonso C, Vale P, Tellez A, Botana LM (2008) First toxicity report of tetrodotoxin and 5,6,11-trideoxyTTX in the trumpet shell charonia lampas lampas in Europe. Anal Chem 80: 5622–5629
30. Satake M, Ofuji K, Naoki H, James KJ, Furey A, McMahon T, Silke J, Yasumoto T (1998) Azaspiracid, a New marine toxin having unique Spiro ring assemblies, isolated from Irish mussels, mytilus edulis. J Am Chem Soc 120:9967–9968
31. James KJ, O′Driscoll D, García J, Furey A (2008) Azaspiracids: chemistry, bioconversion, and determination. In: Botana LM (ed) Seafood and freshwater toxins: pharmacology, physiology and detection, 2nd edn. CRC (Taylor and Francis Group), Boca Raton, FL, pp 763–773
32. Vale P, Bire R, Hess P (2008) Confirmation by LC-MS/MS of azaspiracids in shellfish from the Portuguese north-western coast. Toxicon 51:1449–1456
33. Vilarino N (2008) Marine toxins and the cytoskeleton: azaspiracids. FEBS J 275: 6075–6081
34. Vale C, Botana L (2010) Letter to the editor. The complexity of the cellular effects of azaspiracid prevents to highlight only one candidate as the target of the toxin. Toxicol Sci 115:611
35. Vale C, Nicolaou KC, Frederick MO, Vieytes MR, Botana LM (2010) Cell volume decrease as a link between azaspiracid-induced cytotoxicity and c-Jun-N-terminal kinase activation in cultured neurons. Toxicol Sci 113:158–168
36. Vale-Gonzalez C, Pazos MJ, Alfonso A, Vieytes MR, Botana LM (2007) Study of the neuronal effects of ouabain and palytoxin and their binding to Na, K-ATPases using an optical biosensor. Toxicon 50:541–552
37. Louzao MC, Ares IR, Cagide E, Espina B, Vilarino N, Alfonso A, Vieytes MR, Botana LM (2011) Palytoxins and cytoskeleton: an overview. Toxicon 57:460–469
38. Catterall WA, Gainer M (1985) Interaction of brevetoxin A with a new receptor site on the sodium channel. Toxicon 23:497–504
39. Furey A, García J, O'Callaghan K, Lehane M, Fernández M, James KJ (2007) Brevetoxins: structure, toxicology and origin. In: Botana LM (ed) Phytotoxins chemistry and biochemistry. Blackwell Publishing, Ames, Iowa, pp 19–46
40. Dickey RW (2008) Ciguatera toxins: chemistry, toxicology, and detection. In: Botana LM (ed) Seafood and freshwater toxins: pharmacology, physiology and detection, 2nd edn. CRC (Taylor and Francis Group), Boca Raton, FL, pp 479–500
41. Stewart I, Lewis RJ, Eaglesham GK, Graham GC, Poole S, Craig SB (2010) Emerging tropical diseases in Australia. Part 2. Ciguatera fish poisoning. Ann Trop Med Parasitol 104:557–571
42. Molgo J, Comella JX, Legrand AM (1990) Ciguatoxin enhances quantal transmitter release from frog motor nerve terminals. Br J Pharmacol 99:695–700
43. Lewis RJ (2000) Ion channel toxins and therapeutics: from cone snail venoms to ciguatera. Ther Drug Monit 22:61–64
44. Perez S, Vale C, Alonso E, Alfonso C, Rodriguez P, Otero P, Alfonso A, Vale P, Hirama M, Vieytes MR, Botana LM (2011) A comparative study of the effect of ciguatoxins on voltage-dependent Na(+) and K(+) channels in cerebellar neurons. Chem Res Toxicol 24:587–596

45. Debonnel G, Beauchesne L, de Montigny C (1989) Domoic acid, the alleged "mussel toxin," might produce its neurotoxic effect through kainate receptor activation: an electrophysiological study in the dorsal hippocampus. Can J Physiol Pharmacol 67:29–33
46. Larm JA, Beart PM, Cheung NS (1997) Neurotoxin domoic acid produces cytotoxicity via kainate- and AMPA-sensitive receptors in cultured cortical neurones. Neurochem Int 31:677–682
47. Doucette TA, Tasker RA (2008) Domoic acid: detection methods, pharmacology, and toxicology. In: Botana LM (ed) Seafood and freshwater toxins: pharmacology, physiology and detection, 2nd edn. CRC (Taylor and Francis Group), Boca Raton, FL, pp 397–429
48. Munday R (2008) Toxicology of cyclic imines: gymnodimine, spirolides, pinnatoxins, pteriatoxins, prorocentrolide, spiro-prorocentrimine, and symbioimines. In: Botana LM (ed) Seafood and freshwater toxins: pharmacology, physiology and detection, 2nd edn. CRC (Taylor and Francis Group), Boca Raton, FL, pp 581–594
49. Bourne Y, Radic Z, Araoz R, Talley TT, Benoit E, Servent D, Taylor P, Molgo J, Marchot P (2010) Structural determinants in phycotoxins and AChBP conferring high affinity binding and nicotinic AChR antagonism. Proc Natl Acad Sci U S A 107: 6076–6081
50. Wandscheer CB, Vilarino N, Espina B, Louzao MC, Botana LM (2010) Human muscarinic acetylcholine receptors are a target of the marine toxin 13-desmethyl C spirolide. Chem Res Toxicol 23:1753–1761
51. Janes RW (2005) Alpha-conotoxins as selective probes for nicotinic acetylcholine receptor subclasses. Curr Opin Pharmacol 5: 280–292
52. Vilarino N, Fonfria ES, Molgo J, Araoz R, Botana LM (2009) Detection of gymnodimine-A and 13-desmethyl C spirolide phycotoxins by fluorescence polarization. Anal Chem 81:2708–2714
53. Bialojan C, Takai A (1988) Inhibitory effect of a marine-sponge toxin, okadaic acid, on protein phosphatases. Specificity and kinetics. Biochem J 256:283–290
54. Vale C, Botana LM (2008) Marine toxins and the cytoskeleton: okadaic acid and dinophysistoxins. FEBS J 275:6060–6066
55. Allingham JS, Miles CO, Rayment I (2007) A structural basis for regulation of actin polymerization by pectenotoxins. J Mol Biol 371:959–970
56. Ares IR, Louzao MC, Espina B, Vieytes MR, Miles CO, Yasumoto T, Botana LM (2007) Lactone ring of pectenotoxins: a key factor for their activity on cytoskeletal dynamics. Cell Physiol Biochem 19:283–292
57. Espina B, Louzao MC, Ares IR, Cagide E, Vieytes MR, Vega FV, Rubiolo JA, Miles CO, Suzuki T, Yasumoto T, Botana LM (2008) Cytoskeletal toxicity of pectenotoxins in hepatic cells. Br J Pharmacol 155:934–944
58. Miles CO, Wilkins AL, Hawkes AD, Selwood AI, Jensen DJ, Munday R, Cooney JM, Beuzenberg V (2005) Polyhydroxylated amide analogs of yessotoxin from protoceratium reticulatum. Toxicon 45:61–71
59. Hess P, Aasen J (2007) Chemistry, origins and distribution of yessotoxin and its analogues. In: Botana LM (ed) Phytotoxins, chemistry and biochemistry. Blackwell Publishing, Ames, Iowa, pp 187–202
60. Alfonso A, de la Rosa L, Vieytes MR, Yasumoto T, Botana LM (2003) Yessotoxin, a novel phycotoxin, activates phosphodiesterase activity. Effect of yessotoxin on cAMP levels in human lymphocytes. Biochem Pharmacol 65:193–208
61. Leira F, Alvarez C, Cabado AG, Vieites JM, Vieytes MR, Botana LM (2003) Development of a F actin-based live-cell fluorimetric microplate assay for diarrhetic shellfish toxins. Anal Biochem 317:129–135
62. Pazos MJ, Alfonso A, Vieytes MR, Yasumoto T, Botana LM (2006) Study of the interaction between different phosphodiesterases and yessotoxin using a resonant mirror biosensor. Chem Res Toxicol 19:794–800
63. Luong JH, Bouvrette P, Male KB (1997) Developments and applications of biosensors in food analysis. Trends Biotechnol 15: 369–377
64. Vilarino N, Fonfría E, Louzao MC, Botana LM (2009) Use of biosensors as alternatives to current regulatory methods for marine biotoxins. Sensors 9:9414–9443
65. Samanta D, Sarkar A (2011) Immobilization of bio-macromolecules on self-assembled monolayers: methods and sensor applications. Chem Soc Rev 40:2567–2592
66. Krueger BK, Ratzlaff RW, Strichartz GR, Blaustein MP (1979) Saxitoxin binding to synaptosomes, membranes, and solubilized binding sites from rat brain. J Membr Biol 50:287–310
67. Vieytes MR, Cabado AG, Alfonso A, Louzao MC, Botana AM, Botana LM (1993) Solid-phase radioreceptor assay for paralytic shellfish toxins. Anal Biochem 211:87–93

68. Cella LN, Sanchez P, Zhong W, Myung NV, Chen W, Mulchandani A (2010) Nano aptasensor for protective antigen toxin of anthrax. Anal Chem 82:2042–2047
69. Yang C, Wang Y, Marty JL, Yang X (2011) Aptamer-based colorimetric biosensing of ochratoxin A using unmodified gold nanoparticles indicator. Biosens Bioelectron 26: 2724–2727
70. Ellington AD, Szostak JW (1990) In vitro selection of RNA molecules that bind specific ligands. Nature 346:818–822
71. Fonfria ES, Vilarino N, Campbell K, Elliott C, Haughey SA, Ben-Gigirey B, Vieites JM, Kawatsu K, Botana LM (2007) Paralytic shellfish poisoning detection by surface Plasmon resonance-based biosensors in shellfish matrixes. Anal Chem 79:6303–6311
72. Jing M, Bowser MT (2011) Methods for measuring aptamer-protein equilibria: a review. Anal Chim Acta 686:9–18
73. Stewart LD, Hess P, Connolly L, Elliott CT (2009) Development and single-laboratory validation of a pseudofunctional biosensor immunoassay for the detection of the okadaic acid group of toxins. Anal Chem 81: 10208–10214
74. Holmes CF, Luu HA, Carrier F, Schmitz FJ (1990) Inhibition of protein phosphatases-1 and -2A with acanthifolicin. Comparison with diarrhetic shellfish toxins and identification of a region on okadaic acid important for phosphatase inhibition. FEBS Lett 270:216–218
75. Kodoyianni V (2011) Label-free analysis of biomolecular interactions using SPR imaging. Biotechniques 50:32–40
76. Pazos MJ, Alfonso A, Vieytes MR, Yasumoto T, Vieites JM, Botana LM (2004) Resonant mirror biosensor detection method based on yessotoxin-phosphodiesterase interactions. Anal Biochem 335:112–118
77. Pazos MJ, Alfonso A, Vieytes MR, Yasumoto T, Botana LM (2005) Kinetic analysis of the interaction between yessotoxin and analogues and immobilized phosphodiesterases using a resonant mirror optical biosensor. Chem Res Toxicol 18:1155–1160
78. Campbell K, McGrath T, Sjolander S, Hanson T, Tidare M, Jansson O, Moberg A, Mooney M, Elliott C, Buijs J (2011) Use of a novel micro-fluidic device to create arrays for multiplex analysis of large and small molecular weight compounds by surface Plasmon resonance. Biosens Bioelectron 26:3029–3036
79. Abbas A, Linman MJ, Cheng Q (2011) Patterned resonance plasmonic microarrays for high-performance SPR imaging. Anal Chem 83:3147–3152
80. Abbas A, Linman MJ, Cheng Q (2011) New trends in instrumental design for surface Plasmon resonance-based biosensors. Biosens Bioelectron 26:1815–1824
81. Ronzitti G, Callegari F, Malaguti C, Rossini GP (2004) Selective disruption of the E-cadherin-catenin system by an algal toxin. Br J Cancer 90:1100–1107
82. Bagu JR, Sykes BD, Craig MM, Holmes CF (1997) A molecular basis for different interactions of marine toxins with protein phosphatase-1. Molecular models for bound motuporin, microcystins, okadaic acid, and calyculin A. J Biol Chem 272:5087–5097
83. Espina B, Rubiolo JA (2008) Marine toxins and the cytoskeleton: pectenotoxins, unusual macrolides that disrupt actin. FEBS J 275: 6082–6088
84. Roman Y, Alfonso A, Vieytes MR, Ofuji K, Satake M, Yasumoto T, Botana LM (2004) Effects of azaspiracids 2 and 3 on intracellular cAMP, [Ca2+], and pH. Chem Res Toxicol 17:1338–1349
85. Alfonso A, Roman Y, Vieytes MR, Ofuji K, Satake M, Yasumoto T, Botana LM (2005) Azaspiracid-4 inhibits Ca2+ entry by stored operated channels in human T lymphocytes. Biochem Pharmacol 69:1627–1636
86. Alfonso A, Vieytes MR, Ofuji K, Satake M, Nicolaou KC, Frederick MO, Botana LM (2006) Azaspiracids modulate intracellular pH levels in human lymphocytes. Biochem Biophys Res Commun 346:1091–1099
87. Vale C, Wandscheer C, Nicolaou KC, Frederick MO, Alfonso C, Vieytes MR, Botana LM (2008) Cytotoxic effect of azaspiracid-2 and azaspiracid-2-methyl ester in cultured neurons: involvement of the c-Jun N-terminal kinase. J Neurosci Res 86: 2952–2962
88. Vilarino N, Nicolaou KC, Frederick MO, Cagide E, Alfonso C, Alonso E, Vieytes MR, Botana LM (2008) Azaspiracid substituent at C1 is relevant to in vitro toxicity. Chem Res Toxicol 21:1823–1831
89. Munday R, Aune T, Rossini GP (2008) Toxicology of the yessotoxins. In: Botana LM (ed) Seafood and freshwater toxins: pharmacology, physiology and detection, 2nd edn. CRC (Taylor and Francis Group), Boca Raton, FL, pp 329–339
90. Ciminello P, Fattorusso E (2008) Yessotoxin: chemistry, metabolism, and chemical analysis. In: Botana LM (ed) Seafood and freshwater toxins: pharmacology, physiology and detection, 2nd edn. CRC (Taylor and Francis Group), Boca Raton, FL, pp 287–314

91. Alfonso C, Alfonso A, Vieytes MR, Yasumoto T, Botana LM (2005) Quantification of yessotoxin using the fluorescence polarization technique and study of the adequate extraction procedure. Anal Biochem 344:266–274
92. Fonfria ES, Vilarino N, Vieytes MR, Yasumoto T, Botana LM (2008) Feasibility of using a surface Plasmon resonance-based biosensor to detect and quantify yessotoxin. Anal Chim Acta 617:167–170
93. Mouri R, Oishi T, Torikai K, Ujihara S, Matsumori N, Murata M, Oshima Y (2009) Surface Plasmon resonance-based detection of ladder-shaped polyethers by inhibition detection method. Bioorg Med Chem Lett 19:2824–2828
94. Briggs LR, Miles CO, Fitzgerald JM, Ross KM, Garthwaite I, Towers NR (2004) Enzyme-linked immunosorbent assay for the detection of yessotoxin and its analogues. J Agric Food Chem 52:5836–5842
95. Samdal IA, Naustvoll LJ, Olseng CD, Briggs LR, Miles CO (2004) Use of ELISA to identify protoceratium reticulatum as a source of yessotoxin in Norway. Toxicon 44:75–82
96. Levine L, Fujiki H, Gjika HB, Van Vunakis H (1987) Production of antibodies to palytoxin: neutralization of several biological properties of palytoxin. Toxicon: official journal of the International Society on Toxinology 25: 1273–1282
97. Bignami GS, Raybould TJ, Sachinvala ND, Grothaus PG, Simpson SB, Lazo CB, Byrnes JB, Moore RE, Vann DC (1992) Monoclonal antibody-based enzyme-linked immunoassays for the measurement of palytoxin in biological samples. Toxicon 30:687–700
98. Yakes BJ, Degrasse SL, Poli M, Deeds JR (2011) Antibody characterization and immunoassays for palytoxin using an SPR biosensor. Anal Bioanal Chem 400:2865–2869
99. Campbell K, Haughey SA, van den Top H, van Egmond H, Vilarino N, Botana LM, Elliott CT (2010) Single laboratory validation of a surface Plasmon resonance biosensor screening method for paralytic shellfish poisoning toxins. Anal Chem 82:2977–2988
100. Van den Top H, Elliott C, Haughey SA, Vilarino N, Van Egmond HP, Botana LM, Campbell K (2011) Surface Plasmon resonance biosensor screening method for paralytic shellfish poisoning toxins: a pilot inter-laboratory study. Anal Chem 83: 4206–4213
101. Hollingworth T, Wekell MM (1990) Fish and other marine products, 959.08. Paralytic shellfish poison biological method, final action. In: Hellrich K (ed) Official methods of analysis of the AOAC, 15th edn. Arlington, AOAC, pp 881–882
102. Lawrence JF, Niedzwiadek B, Menard C (2005) Quantitative determination of paralytic shellfish poisoning toxins in shellfish using prechromatographic oxidation and liquid chromatography with fluorescence detection: collaborative study. J AOAC Int 88:1714–1732
103. Hartshorne RP, Catterall WA (1984) The sodium channel from rat brain. Purification and subunit composition. J Biol Chem 259:1667–1675
104. Velez P, Sierralta J, Alcayaga C, Fonseca M, Loyola H, Johns DC, Tomaselli GF, Marban E, Suarez-Isla BA (2001) A functional assay for paralytic shellfish toxins that uses recombinant sodium channels. Toxicon 39: 929–935
105. Kulagina NV, O'Shaughnessy TJ, Ma W, Ramsdell JS, Pancrazio JJ (2004) Pharmacological effects of the marine toxins, brevetoxin and saxitoxin, on murine frontal cortex neuronal networks. Toxicon 44: 669–676
106. Kulagina NV, Mikulski CM, Gray S, Ma W, Doucette GJ, Ramsdell JS, Pancrazio JJ (2006) Detection of marine toxins, brevetoxin-3 and saxitoxin, in seawater using neuronal networks. Environ Sci Technol 40: 578–583
107. Kreuzer MP, Pravda M, O'Sullivan CK, Guilbault GG (2002) Novel electrochemical immunosensors for seafood toxin analysis. Toxicon 40:1267–1274
108. Llamas NM, Stewart L, Fodey T, Higgins HC, Velasco ML, Botana LM, Elliott CT (2007) Development of a novel immunobiosensor method for the rapid detection of okadaic acid contamination in shellfish extracts. Anal Bioanal Chem 389:581–587
109. Stewart L, Elliott C, Walker A, Curran R, Connolly L (2009) Development of a monoclonal antibody binding okadaic acid and dinophysistoxins-1, -2 in proportion to their toxicity equivalence factors. Toxicon 54: 491–498
110. Campas M, Marty JL (2007) Enzyme sensor for the electrochemical detection of the marine toxin okadaic acid. Anal Chim Acta 605:87–93
111. Campas M, de la Iglesia P, Le Berre M, Kane M, Diogene J, Marty JL (2008) Enzymatic recycling-based amperometric immunosensor for the ultrasensitive detection of okadaic acid in shellfish. Biosens Bioelectron 24:716–722
112. Volpe G, Cotroneo E, Moscone D, Croci L, Cozzi L, Ciccaglioni G, Palleschi G (2009)

A bienzyme electrochemical probe for flow injection analysis of okadaic acid based on protein phosphatase-2A inhibition: an optimization study. Anal Biochem 385:50–56

113. Marquette CA, Coulet PR, Blum LJ (1999) Semi-automated membrane based chemiluminescent immunosensor for flow injection analysis of okadaic acid in mussels. Anal Chim Acta 398:173–182

114. Quilliam MA, Xie M, Hardstaff WR (1995) Rapid extraction and cleanup for liquid chromatography determination of domoic acid in unsalted food. J Assoc Off Anal Chem Int 78:543–554

115. AOAC (2005) Domoic acid in mussels, liquid chromatographic method. Official methods of analysis of the AOAC (Association of Official Analytical Chemists). Method 991.926

116. Commission E (2007) Commission regulation (EC) No 1244/2007 of 24 October 2007 amending regulation (EC) No 2074/2005 as regards implementing measures for certain products of animal origin intended for human consumption and laying down specific rules on official controls for the inspection area. Off J Eur Union L 281:12–18

117. Traynor IM, Plumpton L, Fodey TL, Higgins C, Elliott CT (2006) Immunobiosensor detection of domoic acid as a screening test in bivalve molluscs: comparison with liquid chromatography-based analysis. J AOAC Int 89:868–872

118. Micheli L, Radoi A, Guarrina R, Massaud R, Bala C, Moscone D, Palleschi G (2004) Disposable immunosensor for the determination of domoic acid in shellfish. Biosens Bioelectron 20:190–196

119. Fonfria ES, Vilarino N, Espina B, Louzao MC, Alvarez M, Molgo J, Araoz R, Botana LM (2010) Feasibility of gymnodimine and 13-desmethyl C spirolide detection by fluorescence polarization using a receptor-based assay in shellfish matrixes. Anal Chim Acta 657:75–82

120. Balls M, Karcher W (1995) The validation of alternative test methods. ATLA 23:884–886

121. Huber L (1998) Validation of analytical methods: review and strategy. LC-GC Int 11:96–105

122. Thompsom M, Ellison S, Wood R (2002) Harmonized guidelines for single-laboratory validation of methods of analysis. Pure Appl Chem 74:835–855

123. Hess P, Grune B, Anderson DB, Aune T, Botana LM, Caricato P, van Egmond HP, Halder M, Hall S, Lawrence JF, Moffat C, Poletti R, Richmond J, Rossini GP, Seamer C, Vilageliu JS (2006) Three Rs approaches in marine biotoxin testing. The report and recommendations of a joint ECVAM/DG SANCO workshop (ECVAM workshop 54). Altern Lab Anim 34:193–224

124. AOAC (2005) Method 2005.06: paralytic shellfish poisoning toxins in shellfish. Prechromatographic oxidation and liquid chromatography with fluorescence detection. Official methods of analysis of the Association of Official Analytical Chemists. Method 2005.06, First Action

125. International A (2000) AOAC official method 991.26. Domoic acid in mussels, liquid chromatographic method. In: Horowitz (ed) Official methods of analysis of the AOAC. Gaithersburg, ML: AOAC

126. Kleivdal H, Kristiansen SI, Nilsen MV, Goksoyr A, Briggs L, Holland P, McNabb P (2007) Determination of domoic acid toxins in shellfish by biosense ASP ELISA–a direct competitive enzyme-linked immunosorbent assay: collaborative study. J AOAC Int 90:1011–1027

Chapter 12

Electrochemical Detection of Harmful Algae by Means of a Sandwich Hybridization Assay on an Electrode Surface

Jahir Orozco and Linda K. Medlin

Abstract

We report on the development of a sandwich hybridization assay on an electrode surface-based sensor device for the detection of toxic algae. This DNA (rRNA)-based sensor uses an electrochemical detection of a target sequence by means of its hybridization to a capture probe, with an attachment label at its 5′ end that binds it to the electrode surface, and a signal probe with a DIG detection label at the 3′ end. The thiol (biotin)-labelled capture probe is immobilized onto gold (carbon, graphite or carbon nanotubes) screen printed electrodes. Synthetic positive control DNA or real rRNA isolated from algal cells is applied to the sensor and allowed to hybridize to the capture probe. A signal probe with a digoxigenin label is then hybridized to this construct, followed by an anti DIG antibody coupled to horseradish peroxidase (HRP) and a substrate. The electrical signal obtained from the redox reaction is proportional to the amount of DNA (rRNA) applied to the biosensor, which is in turn proportional to the number of cells harvested when applied to real samples. The optimized hybridization conditions and steps of the fabrication process are presented here. The biosensor has been used to detect different algal species, such as *Prymnesium parvum*, *Gymnodinium catenatum*, *Pseudo-nitzschia australis*, *Alexandrium ostenfeldii* and *Alexandrium minutum*, by immobilizing their genetic material over the different transducer platforms.

Key words: Sandwich hybridization assay, DNA sensor, Harmful algae, Screen-printed electrodes, Environmental monitoring, *Prymnesium parvum*, *Gymnodinium catenatum*, *Pseudo-nitzschia australis*, *Alexandrium ostenfeldii*, *Alexandrium minutum*

1. Introduction

Monitoring of aquatic pathogens is an essential requirement to guarantee safety of water for humans. Identification and detection of harmful algae, as hazardous aquatic pathogens, are of special interest because they can produce powerful toxins to fish and shellfish (e.g. oysters, mussels) and cause amnesic, paralytic or diarrhetic shellfish poisoning (1–3) when the toxic animals are consumed by humans. To mitigate their negative effects not only to fisheries,

Sonia M. Tiquia-Arashiro (ed.), *Molecular Biological Technologies for Ocean Sensing*, Springer Protocols Handbooks, DOI 10.1007/978-1-61779-915-0_12, © Springer Science+Business Media New York 2012

aquaculture and tourism, but also to local economies and to health problems for humans, the development of rapid, accurate and cost-effective detection systems for toxic algal identification, and detection in aquatic environments is urgently required. In the past, monitoring and mitigation efforts have relied on visual confirmation of water discoloration, fish kills and laborious cell counts. These practices are very time consuming, require specialized or trained personnel, expensive equipment, then being inappropriate when many samples have to be routinely analyzed. Conventional methods for species detection often require transportation of samples to specialized laboratories and identification of species, which can take several days. Currently, using conventional methods by light microscopy, it is typical for up to 5 working days to elapse between specimen collection and the delivery of a diagnostic report of the species present in a sample. Molecular methods that are potentially faster and more accurate than traditional light microscopic methods have been used for the identification of phytoplankton, not just those that are toxic (4–11). The simplicity and low power requirements of electrochemical biosensors as well as their speed, sensitivity, accuracy and versatility have made them attractive candidates to overcome traditional detection and quantification limitations and have allowed the fabrication of compact and inexpensive devices (12, 13). Moreover, the ability of electrochemical sensors to identify directly nucleic acids in complex samples is a valuable advantage over other approaches, such as polymerase chain reactions (PCR) that require target purification and amplification (14).

Biosensors are powerful tools for species detection. Among them, those based on the direct electrochemical detection of nucleic acid target molecules have successfully been demonstrated in a variety of approaches by linking DNA or RNA hybridization events to an oligonucleotide-modified electrode surface (15, 16). The study of toxic algal biodiversity and distribution is greatly facilitated by the application of rRNA probes. The detection strategy is an electrochemical sandwich hybridization assay (SHA) in which target DNA or RNA is bound by both a capture and a signal probe (17, 18). Only one of the two probes needs to be specific for the target species. A capture probe is immobilized on the electrode. If the target sequence binds to the immobilized capture probe in the first hybridization event, thus its detection takes place via a second hybridization event with a signal probe linked to a recorder molecule (2), such as a fluorochrome or digoxigenin. An antibody to the recorder molecule is coupled to a horseradish peroxidase (HRP) enzyme for electrochemical signal amplification and forms the final complex. HRP converts electrochemically inactive substrates to an electroactive product that can be detected amperometrically, in which the current signal is proportional to the amount of the bound enzyme (and hence to the analyte concentration in a sample). Oligonucleotide probes detection assays involving the amplification of hybridization signals through enzyme tracer molecules have the advantage of being potential ultrasensitive (19). Besides, the assay format maximizes discrimination of the target sequences and purification of rRNA is not required. The reactions are rapid, easy to execute and amenable to automation. Quantification of the target species can be performed by using smaller, portable and less expensive instrumentation.

Preliminary data from an innovative approach to harmful algal cell enumeration based on fibre optic genosensors for *A. tamarense* (North American clade), *P. australis* and *A. ostenfeldii* have been introduced (20). A faster, cheaper, and more user friendly and reliable biosensor method, together with a handheld device for the detection and identification of the toxic dinoflagellate *A. ostenfeldii* and *A. minutum* have been reported (13). Development and adaptation of a multiprobe biosensor for the use in a semi-automated device for the simultaneous detection of 14 target species of toxic algae has been also recently devised (12). To test the biosensor platform format that was initially developed for sensing *Alexandrium* species (6, 13) the protocol was assessed using probes for *G. catenatum* and *P. parvum* developed by (21) in a microtiter plate assay following the same SHA format, detection protocol and conditions. This study has indicated that the electrochemical sensor approach has the potential for species-specific detection of toxic algae. More recently, elucidation of the different steps of the biosensor fabrication process from the electrochemical point of view, proof of concept with different algal species, and the evaluation of the influence of the transducer platform geometry and material in the biosensor analytical performance have been recently published by us (22, 23). However, a multianalyte system not only to identify a broad spectrum of toxic algal species, but also to quantify very low concentrations of them without filtration of large volume of water sample is still an object of study.

The protocol reported here concern the validated SHA format for the sensing of different toxic algae over different transducer platform sizes and materials. It includes not only the optimized hybridization conditions, but also the different steps of the fabrication process. In this context, the formation of a mixed self-assembled monolayer (SAM) and the use of casein and bovine serum albumin (BSA) as blockage agents were included in the fabrication protocol. The best orientation of the probes and the best mediator/substrate concentration ratio was selected. The entire protocol and detection system improve the biosensor performance in terms of higher sensitivity and enhanced limits of detection when testing real samples. Proof of concept with different algal species demonstrated the feasibility of their use for the detection of *P. parvum*, *G. catenatum*, *P. australis*, *A. ostenfeldii* and *A. minutum*.

2. Reagents and Equipments

2.1. Apparatus and Materials

1. An EmStat (Electrochemical Sensor Interface), using the PSTrace software from Palm Instruments BV (The Netherlands) and a boxer connector DSC from Dropsens (Spain) are used for Voltamperometric measurements (see Fig. 1a).
2. Measurements are carried out with a three electrode cell printed over ceramic substrates, whose total volume is 50 μl. This cell consists of a DNA-modified working electrode and an on-chip counter and pseudoreference electrode of different materials (see Fig. 1b). All characteristics are summarized in Table 1.

2.2. Reagents and Solutions

2.2.1. Probes

1. Synthetic oligonucleotides probes (Thermo Scientific, Germany) (see Note 1), are listed in Table 2.

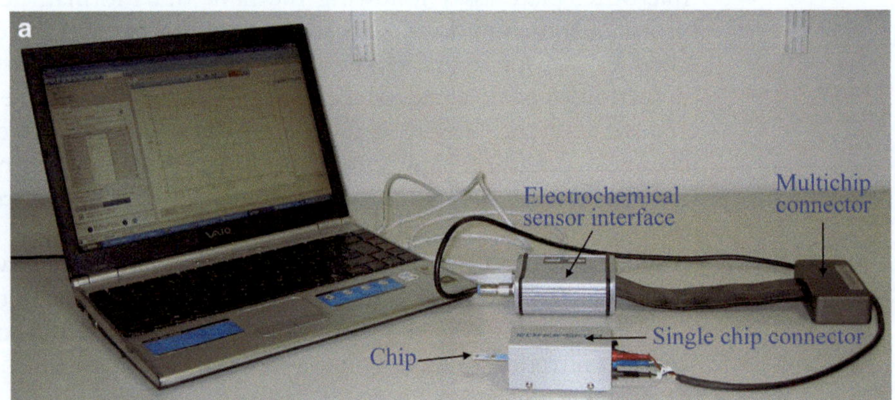

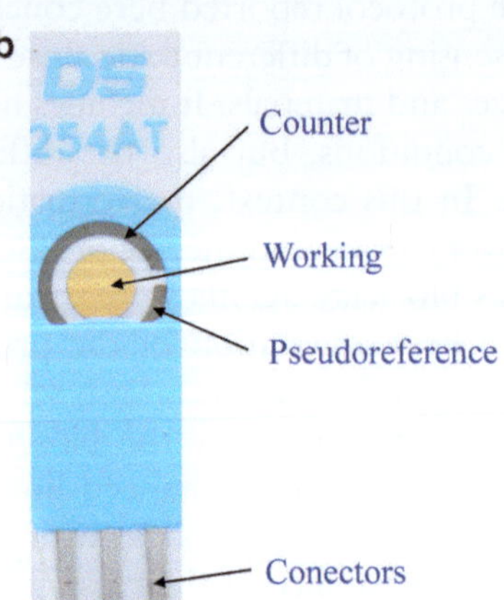

Fig. 1. Experimental set-up. (**a**) Electrochemical sensor interface, chip connector and chip; (**b**) three electrode cell printed on a ceramic substrate (SPAuE from Dropsens).

Table 1
Characteristics of the screen printed gold (SPAuE-4, SPAuE-1), graphite (SPGE-1) electrodes and single wall carbon nanotubes electrodes (SWCNT-4)

Abbreviated name	Supplier	Area (mm^2)	Electrode: Working	Counter	Reference
SPAuE-4	Dropsens (Spain)	12.57	Au	Pt	Ag/AgCl
SWCNT-4		12.57	Carbon nanotubes	C	Ag
SPAuE-1	BVT technologies (Czech Republic)	0.79	Au	Au	Ag/AgCl
SPGE-1		0.79	Graphite	Au	Ag/AgCl

2.2.2. Buffers

1. 50 mM $NaHCO_3$ buffer solution (pH 9.6).
2. 10× PBS (0.5 M $NaH_2PO_4 \cdot H_2O$ buffer solution containing 1.54 M NaCl, pH 7.4).
3. Bead buffer (0.1 M Tris solution containing 0.3 M NaCl, pH 7.6).
4. 4× Hybridization buffer (0.08 M Tris solution containing 0.3 M NaCl and 0.04 % sodium dodecyl sulphate (SDS), pH 8.0).
5. 10× POP buffer (0.5 M $NaH_2PO_4 \cdot H_2O$ buffer solution containing 1.0 M NaCl, pH 6.45).
6. PBS-BT (1.0 M PBS containing 0.1 % BSA and 0.05 % Tween 20, pH 7.4).

2.2.3. Other Solutions

1. 0.5 M sulphuric acid solution, containing 0.01 M KCl.
2. 1 mM potassium hexacianoferrate solution, containing 1 M KCl.
3. 1 mmol l^{-1} 6-mercapto-1-hexanol aqueous solution.
4. 0.5 mg ml^{-1} neutravidin solution (see Note 2).
5. 3 % casein solution in 1× PBS
6. 7.5 U ml^{-1} anti-digoxigenin-HRP fragments (Roche, France) (see Note 3).
7. 3.48 μg μl^{-1} salmon sperm DNA solution was also used as blocking reagent in the hybridization mixture (see Note 4).
8. 100 mM hydrogen peroxide solution (see Note 5).
9. 0.4 mM *N*-phenyl-*p*-phenylenediamine monohydrochloride 99 % (ADPA) (Acros Organics, Belgium) (see Note 6).

Table 2
DNA probe sets designed with the ARB program (21, 22)

Abbreviated name	Sequence (5′–3′direction)	Modifications
B-Prym	CAG CCG ACG CCG AGC GCG	5′ Biotin
T-Prym	CAG CCG ACG CCG AGC GCG	5′ Thiol
B-Gym	TTT TTA AAA GAT TAC CCA ATC	5′ Biotin
T-Gym	TTT TTA AAA GAT TAC CCA ATC	5′ Thiol
T-PsAus	AAC GTC GTT CCG CCA AT	5′ Thiol
B-AlOst	CAA CCC TTC CCA ATA GTC AGG T	5′ Biotin
T-AlOst	CAA CCC TTC CCA ATA GTC AGG T	5′ Thiol
B-Almin	GAA GTC AGG TTT GGA TGC	5′ Biotin
T-Almin	GAA GTC AGG TTT GGA TGC	5′ Thiol
D-Prymnext	CGC CAT CCT ACC AGG CTC	3′ Digoxigenin
D-Gymnext	CTG TCG GAC AAG GTC GTA	3′ Digoxigenin
D-PsAusnext	CAA GGT GCT GAC GGA GAC GT	3′ Digoxigenin
D-AlOstnext	GAA TCA CCA AGG TTC CAA GCA G	3′ Digoxigenin
D-Alminnext	TAA TGA CCA CAA CCC TTC C	3′ Digoxigenin
Prymnesium parvum[a,b]	GAG CCT GG**T** **A**GG ATG GCG CGC GCT CGG CGT CGG CTG[b]	
Gymnodinium catenatum[a]	TAC GAC CTT GTC CGA CAG GAT TGG GTA ATC TTT TAA AAA	
Prymnesium parvum differently oriented[c]	CGC GCT CGG CGT CGG CTG GAG CCT GGT AGG ATG GCG	
Pseudo-nitzschia australis[c]	A CGT CTC CGT CAG CAC CTT G ATT GGC GGA ACG ACG TT	
Alexandrium ostenfeldii	GTT GGG AAG GGT TAT CAG TCC ACT TAG TGG TTC CAA GGT TCG TC	
Alexandrium minutum	GCA TCC AAA CCT GAC TTC GGA AGG GTT GTG GTC ATT A	
Negative control	ATA TAG ATT ATC ACC AGT TGT AAG AAC CTT CCT TAA	

[a]The regions specific for the capture and signal probes are such that their labels are distal to each other.
[b]The original positive control used by (21) contained an error and the correct nucleotides are indicated in bold.
[c]The regions specific for the capture and signal probes are inversed, thus their labels are proximal each other.

3. Protocols

3.1. Probe Design

Identification of toxic algae using oligonucleotide probes are usually carried out with probes designed against ribosomal RNA. Targets for the probes are the small and large subunit rRNA genes in the ribosomes of the cells. Their conserved and variable regions make it possible to develop probes specific for different taxonomic levels (24), the so-called hierarchical probe approach. Probes are designed with the probe match function within the ARB (latin, "arbor" = tree) software package (25), in which mismatches to non-targeted are positioned in the middle of the oligonucleotide (as close to the centre of the probe as possible to destabilize the helix formation with non-target DNA) (21). Only one of the two probes has to be specific for the target species. Theoretical probe specificity is dependent on the number of sequences of the targeted gene available in the databases. If molecular probes are designed from only a few sequences, then a danger of cross-hybridization to non-targeted species and organisms occurs, whose sequences are unknown and not yet in the database. Prior to the analysis of field samples, molecular probes are tested for specificity with cultivated target species and closely related species because in silico (in or by means of a computer simulation) and in situ results can show different specificity signals. In the in silico testing, programs (e.g. Oligo) are used to determine if the probes can bind to each other, fold back on themselves and bind. The melting temperature of the probe from its target is determined. Hybridization temperatures, usually two degrees below the melting temperature, are selected for initial testing.

3.2. Pretreatment of Gold Electrodes

A gold-activated surface has a remarkable positive effect over the intensity of the further amperometric signal. To pretreat the SPAuEs:

1. Place a 50 μl drop of a 0.5 M H_2SO_4 solution containing 0.01 M KCl on their surface.
2. Cycle the potential between 0.0 and +1.2 V, around 30 times, at a scan rate of 0.1 V s^{-1}.
3. Rinse with water and dry with a vacuum pump.
4. Evaluate the electrochemical behaviour in the sulphuric solution (see Note 7).

3.3. Characterization of all the Transducer Platforms

Before the development of the biosensor, it is very important to be sure of the transducer platform suitable stage. The performance of the biosensor, among other factors, depends on a straight electron transference process taking place on the electrode surface. To characterize the electrodes made of any material mentioned here:

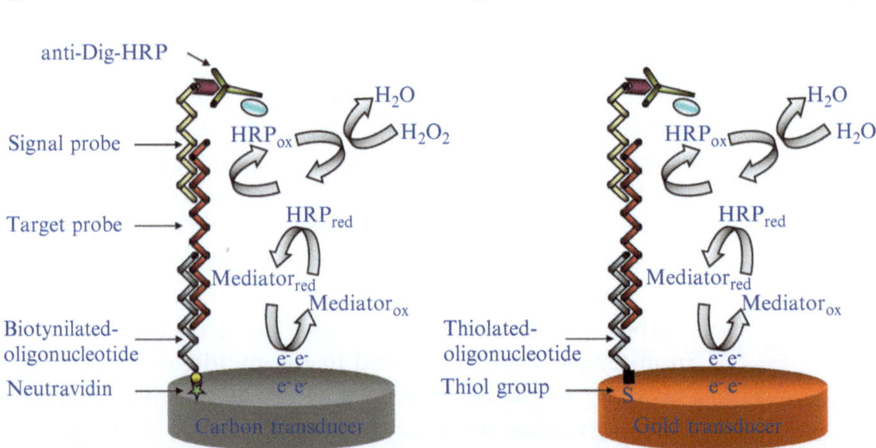

Fig. 2. Sandwich hybridization assay, where a target probe is immobilized on an electrode surface between a capture and a signal probe. (**a**) Carbon and (**b**) Gold surface.

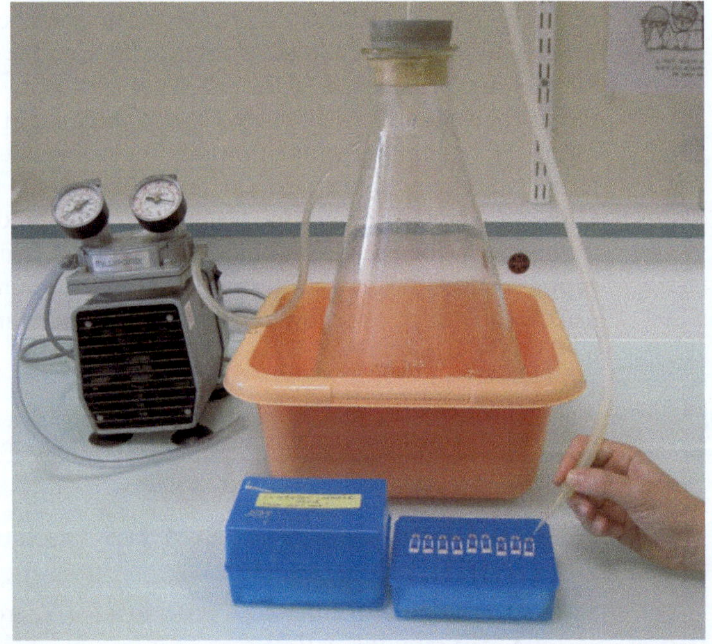

Fig. 3. Incubation chamber and drying vacuum pump.

1. Place a 50 μl drop of 1 M KCl solution containing 1 mM $K_3Fe(CN)_6$.
2. Cycle the potential between −0.25 to +0.5 V, around five times, at a scan rate of 0.50 $V s^{-1}$.
3. Rinse with water and dry with a vacuum pump.
4. Evaluate the electrochemical information from the voltamogramm obtained (see Note 8).

3.4. Development of the Biosensor Device

A general approach for species-specific identification and quantification of toxic algae using a DNA (rRNA)-based biosensor involves an SHA, in which the target sequence is immobilized between single-stranded oligonucleotide capture and signal probes (17, 18) (see Fig. 2).

3.4.1. General Considerations

1. All the incubations are made in a closed, incubation chamber (see Fig. 3) with wet Whatman filters to avoid evaporation, at room temperature (unless otherwise stated).
2. For electrodes with 4 and 1 mm diameter incubations are made by using 7 μl and 2 μl of the corresponding incubation solutions, respectively.
3. The rinsing of electrodes was performed three times by using ca. 40 μl of the corresponding solution (as it is indicated) and intercalating with drying with the vacuum pump as shown in Fig. 3.

3.4.2. Immobilization of the Capture Probe

Overall, different sizes and materials can be used as transducer platform (see Table 1). Based on the type of material, to explain better the fabrication process, they can be set in two groups—gold (SPAuE-4 and SPAuE-1) and carbon (SWCNT-4 and SPGE-1) electrodes, thus varying the way of capture probe immobilization.

For Gold Electrodes

To anchor the capture specific probe onto SPAuEs:

1. Incubate in 10 μM T-probe, overnight (see Note 9).
2. Rinse with 1× PBS and dry with a vacuum pump.
3. Post-treat with 1 mmol l^{-1} 6-mercapto-1-hexanol, for 1 h (see Note 10).
4. Rinse with 1× PBS and dry with a vacuum pump.

For Carbon Electrodes

To anchor the capture specific probe onto the SWCNT-4 and SPGE-1:

1. Incubate overnight in 0.5 mg ml^{-1} neutravidine solution, at 4 °C (see Note 11).
2. Remove excess of neutravidin by washing the chips in 1× PBS and dry with a vacuum pump.
3. Block with 3 % casein, for 1 h (see Note 12).
4. Rinse excess of casein with 1× PBS and dry with a vacuum pump.
5. Incubate in 10 μM B-capture probe, for 30 min.
6. Rinse excess of B-capture probe with 1× PBS and dry with a vacuum pump.

3.4.3. Hybridization of Target and Signal Probe

If the target sequence binds to the immobilized capture probe in a first hybridization event, then its detection takes place via a second hybridization event with a signal probe linked to digoxigenin (2), and later an antibody to the DIG coupled to an HRP enzyme is applied for electrochemical signal amplification. This process is common for both SPAuE and SWCNT.

Assemble target and signal probes over the capture-modified chips in a unique step by incubating in a hybridization mixture (master mix) containing:

1. 1 μl DIG-labelled DNA probe (1.4 μM).
2. 3.5 μl 4× hybridization buffer.
3. 1 μl salmon sperm DNA.
4. 7.5 μl milliQ water.
5. Either 1 μl target sequence (1.4 μM) for the positive control or 1 μl non-complementary sequence for the negative control.
6. 1 μl, 1 % BSA (see Note 13).
7. Incubate the chips in the master mix at 46 °C for 30 min (see Note 14).
8. Allow them to cool for 10 min.
9. Rinse excess of master mix with POP and dry with a vacuum pump.
10. Incubate in anti-DIG-HRP enzyme for 30 min to complete the biosensor development.

3.4.4. Electrochemical Detection

HRP electrochemically converts an inactive substrate to an electro-active product that can be detected amperometrically. The current signal is proportional to the amount of the bound enzyme (and hence to the analyte concentration in a sample).

1. Place the chips in the connector as shown in Fig. 1.
2. Pipette 45 μl of ADPA to the chip.
3. Apply a fix potential of −0.15 V between the working and reference electrode (by means of the electrochemical sensor interface software) and register the current for 60 s.
4. Add 5 μl of 100 mM H_2O_2 solution and continue registering the current 60 s more.

4. Typical Protocol Results

4.1. Synthetic Target

This protocol has been successfully tested for the electrochemical detection of synthetic *P. parvum* sequence by using a sandwich hybridization assay on an electrode surface of different sizes and

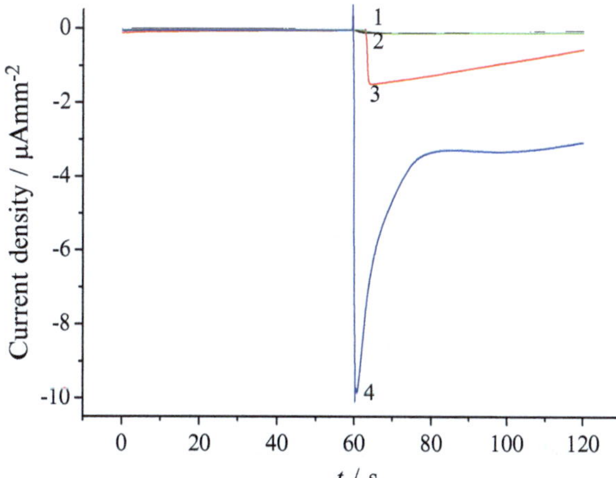

Fig. 4. Analytical signal recorded at a fix potential of −0.15 V. (*1*) POP background, (*2*) negative control, (*3*) positive control using an SWCNT-4 mm electrode, (*4*) positive control using the SPGE-1.

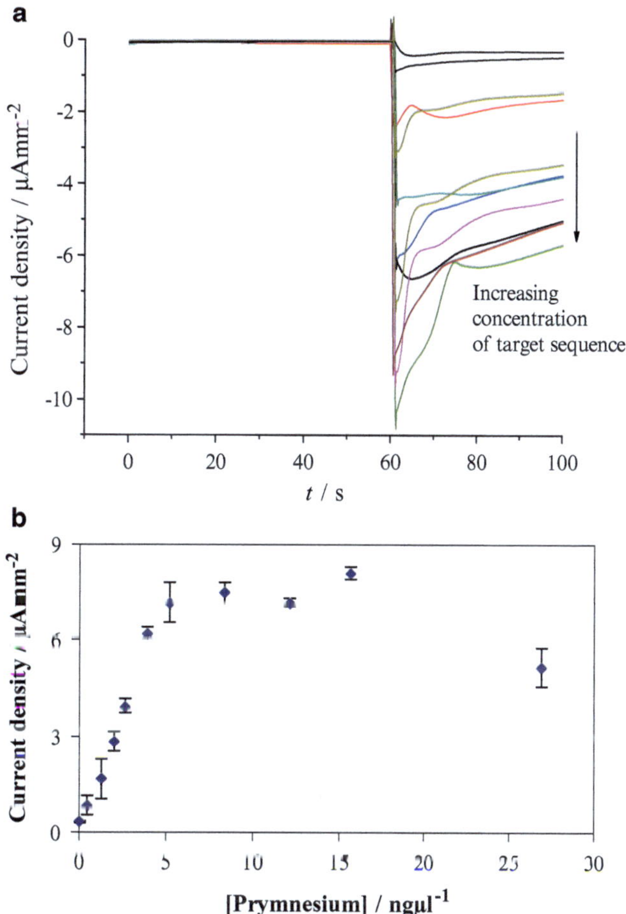

Fig. 5. (**a**) Example of the biosensor amperometric response recorded when different concentrations of *Prymnesium parvum* were set to react with the capture and signal probes. (**b**) The corresponding calibration curve.

materials. Figure 4 shows the typical amperometric signal expected for two kinds of electrodes (SPGE and SWCNT) (see Note 15). Figure 5a depicts the signal for increasing concentrations of target sequence immobilized over the SPGE from 0.0 to 5.2 ng μl^{-1} of *P. parvum* (see Note 16), and Fig. 5b the corresponding calibration curve.

The protocol has been also interrogated with other algal species, such as *G. catenatum, P. australis, A. ostenfeldii* and *A. minutum*, by hybridizing the synthetic sequence between the respective capture and signal probes (according to Table 2) over the different transducer platforms.

4.2. Real Targets

When the protocol is used for real sample analysis, the following steps have to be performed to preserve and store the samples and isolate the target sequence from the cells before its hybridization to the signal probe (Sect. 3.4.3).

4.2.1. Preservation and Storage

After collecting field samples, the algal cells can be stored at room temperature for several days by using RNALater from Ambion, Huntingdon, UK for a later RNA isolation. The cells can be frozen for long-term storage by flash-freezing in liquid nitrogen and an immediate transfer to –70 °C. Alternatively, the cells can be stored for a longer term if placed immediately in TRI Reagent (Sigma) for RNA isolation, but cells should be disrupted for long-term storage so that the RNA is released from the cells (see Note 17). Cell disruption is best done in a bead beater and the time and duration of the disruption should be visually verified to ensure that all cells are broken in the field sample, which will undoubtedly contain a variety of cell walls.

4.2.2. RNA Isolation

Isolation of RNA can be performed by using one of the two methods explained as follow (see Note 18). Although in general terms, both methods are user sensitive, some practice must be made to ensure consistent extraction efficiency.

RNA-Isolation with RNeasy Plant Mini Kit (QIAGEN)

1. Add 450 μl buffer RLT with ß-mercaptoethanol (ß-ME) to the cells.
2. Pipette the lysate into the glass beads and disrupt the lysate in a bead beater for 2 × 20 s.
3. Pipette the lysate directly onto a QIAshredder spin column (lilac) placed in a 2 ml collection tube and centrifuge for 15 min at maximum speed. Carefully transfer the supernatant of the flow-through fraction to a new microcentrifuge tube without disturbing the cell debris pellet in the collection tube. Use only this supernatant in subsequent steps.
4. Add half of volume (usually 225 μl) of ethanol (96–100 %) to the cleared lysate and mix immediately by pipetting. Do not centrifuge. Continue without delay.

5. Apply sample (usually 650 μl), including any precipitate that may have formed, into an RNeasy mini column (pink) placed in a 2 ml collection tube. Close the tube gently and centrifuge for 15 s at 8,000 × *g*. Discard the flow-through. Reuse the collection tube in the next step.
6. Add 700 μl buffer RW1 to the RNeasy column. Close the tube gently and wait for ca. 45 s, then centrifuge for 15 s at ≥8,000 × *g* to wash the column. Discard the flow-through and collection tube.
7. Repeat step 6.
8. Transfer the RNeasy column into a new 2 ml collection tube (supplied). Pipette 500 μl buffer RPE onto the RNeasy column. Close the tube gently, and centrifuge for 15 s at 8,000 × *g* to wash the column. Discard the flow-through. Reuse the collection tube in step 9.
9. Repeat step 8.
10. Add another 500 μl buffer RPE to the RNeasy column. Close the tube gently and centrifuge for 2 min at 8,000 × *g* to dry the RNeasy silica-gel membrane.
11. To elute, transfer the RNeasy column to a new 1.5 ml collection tube. Pipette 30–50 μl RNase-free water directly onto the RNeasy silica-gel membrane. Close the tube gently and centrifuge for 1 min at 8,000 × *g* to elute.
12. To obtain a higher total RNA concentration, a second elution step may be performed by using the first eluate (from step 11).
13. Measure the RNA concentration by using a spectrophotometer, e.g. Nanodrop spectrophotometer.

RNA-Isolation with TRI Reagent (Sigma)

1. Homogenize tissue samples in 10–20 volumes TRI Reagent solution.
2. Incubate the homogenate for 5 min at room temperature.
3. (Optional) Centrifuge at 12,000 × *g* for 10 min at 4 °C and transfer the supernatant to a fresh tube to remove any particles from the cell lysate.
4. Add 100 μl 1 bromo 3 chloropropane (BCP) per 1 ml of TRI Reagent solution, mix well, and incubate at room temperature for 5–15 min.
5. Centrifuge at 12,000 × *g* for 10–15 min at 4 °C. Then transfer the aqueous phase to a fresh tube.
6. Add 500 μl of isopropanol per 1 ml of TRI Reagent solution, vortex for 5–10 s, and incubate at room temperature for 5–10 min.
7. Centrifuge at 12,000 × *g* for 8 min at 4–25 °C, and discard the supernatant.

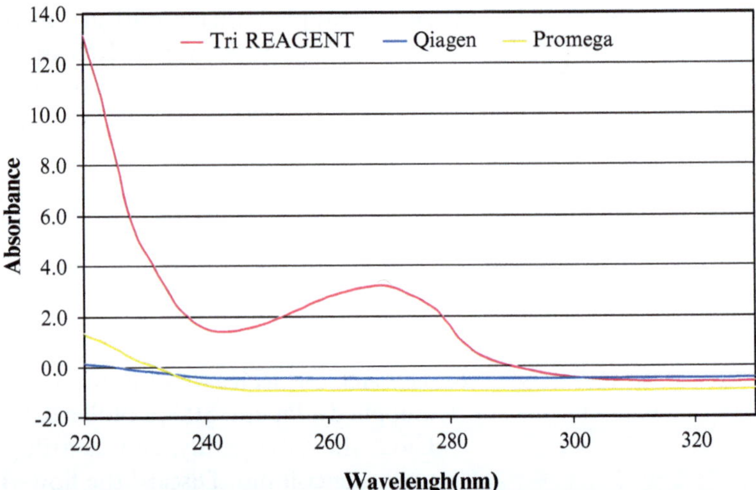

Fig. 6. Comparison among three RNA-Isolation methods: *Neasy Plant Mini Kit (QIAGEN)*, TRI Reagent (Sigma) and the Promega commercial kit.

8. Add 1 ml of 75 % ethanol per 1 ml of TRI Reagent solution.
9. Centrifuge at 7,500 × g for 5 min, remove the ethanol, and briefly air dry the RNA pellet.
10. Dissolve RNA in the buffer of choice.

Target sequence of *A. ostenfeldii* and *A. minutum* has been isolated from the respective algal growing using both protocols described above. However, Tri Reagent has empirically shown to be much more efficient for rRNA extraction (Fig. 6), thus is recommended for this purpose.

Once isolated the target sequence, it has to be hybridized to the signal probe as described in Sect. 3.4.3, and the protocol follows as explained for the synthetic target.

Finally, for both synthetic and real target, a calibration curve has to be determined for each probe. For each target species of real rRNA, the current density (nA mm^{-2}) for 1 ng rRNA, and the RNA concentration per cell have to be founded. Subsequently, the cell concentration of the target species in a sample can be calculated from the electrochemical signals (see Note 19).

5. Notes

1. Synthetic oligonucleotides probes are diluted in 18 MΩ cm ultrapure deionized water (free of DNA and RNA) to have 100 μM stock solutions.
2. A stock solution of neutravidine is prepared by diluting the entire content of 1 mg flask (Ref 40945, Sigma) in 500 μl of 50 mM

$NaHCO_3$ buffer solution (pH 9.6) and stored at 4 °C when not in use. A working solution is prepared by diluting the stock solution with the appropriate volume of 50 mM $NaHCO_3$ buffer solution (pH 9.6) to get a 0.5 mg ml^{-1} solution, prepared daily.

3. A stock solution of anti-DIG HRP fragments is prepared by diluting the entire content of a 150 U flask (Ref 11207733910, Roche) in 1 ml of 1.0 M PBS-BT and stored at 4 °C when not in use. A working solution is prepared by diluting 50 μl of the stock solution with 950 μl of PBS-BT buffer solution (pH 7.4) to get a 7.5 mg ml^{-1} solution and stored at 4 °C when not in use.
4. All reagents are prepared by diluting analytical grade reagents (Sigma-Aldrich, France) in 18 MΩ cm deionized water and stored at 4 °C when not in use, unless otherwise stated.
5. The solution is prepared by dilution of analytical grade reagent in 1× POP buffer solution.
6. ADPA is first dissolved in around 100 μl of 96 % ethanol and then diluted in 1× POP buffer solution. To avoid degradation of the solution, it has to be prepared before use.
7. SPAuEs should have an electrochemical behaviour characteristic of a gold electrode. This is the characteristic and stable gold oxide formation and reduction peaks.
8. Voltammograms should show a well-defined reversible oxidation and reduction peaks with an experimental current intensity close to the theoretical one (that expected based on the geometrical area). The theoretical current can be calculated with the Randles-Sevcik equation:

$$i_p = (2.687 \times 10^5) n A C \sqrt{n v D}$$

where: i_p, intensity of the peak (μA); n, # electrons involved in the semi-reaction for the redox couple (in this case = 1); A, geometrical area of the electrode (cm^2); C, concentration of the analyte (mol cm^{-3}) (in this case = 1.0×10^{-6}); v, scan rate (V s^{-1}) (in this case = 50); D, coefficient diffusion of the analyte (cm^2 s^{-1}) (in this case = 6.07×10^{-6} for the $K_3Fe(CN)_6$).

9. B-Prym, T-Prym, *P. parvum* and D-Prymnext are used as examples. However, the concept was also extended to other different species in Table 2. The reader can use the suitable coupling of previously designed capture and signal probes for detecting certain target sequences.
10. To enhance the accessibility of the DNA (rRNA) probes to the target, the surface was exposed to MCH after DNA (rRNA) immobilization. By this treatment, the thiol group of MCH rapidly displaces the weaker adsorptive contacts between the nucleotide chain and gold, covalently attaching to the interstitial regions between ssDNA (rRNA) chains. The post-treatment

ensures that the DNA probes are only attached to the gold surface through the terminal sulphur atom of the thiol linker because the DNA negligibly interacts with the hydroxyl end of MCH.

11. It is also necessary to note that the effect of capture probe density on signal level and hybridization efficiency has been empirically tested, and optimal conditions are proposed here.
12. Casein is commonly used as a blocking agent to minimize unspecific adsorption (some capture probe that can physically adsorb to the part of electrode uncovered by neutravidin).
13. When the SHA is performed over SPAuE, 1 % BSA is added to the master mix to minimize the non-specific interaction between the gold surface and the DNA probes.
14. Hybridization temperature is specific for the target sequence interrogated. It was previously selected and validated based on the ARB program.
15. Although the amperometric response should be independent of the material, it turned out to be highly dependent on the electrode size. One millimetre diameter electrode led to seven-fold higher current density compared to 4 mm diameter one.
16. By using the amperometric technique, a simply current reading in the display of the electrochemical sensor interface can be related to the concentration of a target species even by any inexperienced analyst.
17. TRI Reagent solution is a complete and ready-to-use reagent for the isolation of total RNA or the simultaneous isolation of RNA, DNA, and protein from diverse biological material, which can be completed in about 1 h.
18. Ribonucleases (RNases) are very stable, active enzymes that are difficult to inactivate; even minute amounts are sufficient to destroy RNA. Use only plasticware or glassware that has been chemically or heat treated to destroy RNases. Glassware should be cleaned with a detergent, thoroughly rinsed and baked at 180 °C for four or more hours before use. Always wear gloves while handling reagents and RNA samples to prevent RNase contamination from the surface of the skin or from dusty laboratory equipment, which are the most common sources of contamination. Change gloves frequently and keep tubes closed whenever possible. Keep isolated RNA on ice when aliquots are pipetted for downstream applications.
19. Probe signal (nA mm^{-2}) = Total RNA present in the sample (ng). Then, Cell number = Probe signal (ng)/per cell (ngRNA).

6. Future Directions

A prototype of a portable semi-automated electrochemical biosensor system has been developed to facilitate the detection of toxic algae in the field (12). This device enables the electrochemical detection of microalgae from water samples in less than 2 h, without the need of expensive equipment, but is still not available for general purchase. Nevertheless, probes for only a limited number of phytoplankton exist, they must be validated for each region where they are and high sample volume is required if the cell densities are expected to be relatively low. The manual isolation of RNA is currently the limiting factor of all systems. The concentration and quality of the RNA required is high so that large amounts of water have to be filtered and RNA isolation should be done by a trained molecular scientist. Separate users can isolate different qualities of rRNA from the same sample with an equal number of algae cells. The resulting signal intensities cannot be compared to cell counts determined using another enumeration technique, thus calibration curves are necessary. The validation of probe signals against total rRNA over the growth cycle of the target microalgae under different environmental conditions has to be conducted to verify the calibration curves. This will allow the extrapolation of the electrochemical readings into more accurate values of cells per litre. Future research is directed to overcome these difficulties and so an autonomous biosensor can be combined with the state-of-the-art in situ measurement systems for the reliable and high resolution monitoring of marine phytoplankton in order to evaluate consequences of environmental change in the oceans.

First attempts to make an automated device come from the labs of Chris Scholin in California (http://www.mbari.org/canon/ESP.htm). Their environmental process sampler is in a prototype stage of development and has been field tested. This instrument uses an SHA with chemiluminescent detection and exchangeable pucks with membrane filters for the dot blot hybridization. Sufficient pucks for about 30 days operation are in the present prototype. Ideally, an automatic system that can be recharged, such as the electrochemical sensor presented here, would ensure that the automatic deployment can be unlimited in time. Another system, using a submersible, with detection by surface plasmon resonance, has been developed by IFREMER but it has only been lab tested and so far can only analyze one probe at a time. The Autonomous Microbial Genosensor (AMG) USA uses a PCR amplification of the RNA present in the sample and is not available as a laboratory instrument but only as a buoy. Using a PCR step as part of the analysis is always open to biases. Clearly, there is a need for automated systems but the perfect one has yet to be designed.

Acknowledgments

We acknowledge to Dr. Julia Baudart and Sebastien Peuchet from the Observatoire Oceanologique de Banyuls Sur Mer for their help in the discussion of the results and to the latter for the results reported in Fig. 6.

References

1. Maso M, Garces E (2006) Harmful microalgae blooms (HAB); problematic and conditions that induce them. Mar Pollut Bull 53(10–12):620–630
2. Penna A, Magnani M (1999) Identification of *Alexandrium* (Dinophyceae) species using PCR and rDNA-targeted probes. J Phycol 35(3):615–621
3. Scholin CA, Marin R, Miller PE et al (1999) DNA probes and a receptor-binding assay for detection of *Pseudo-nitzschia* (Bacillariophyceae) species and domoic acid activity in cultured and natural samples. J Phycol 35(6):1356–1367
4. Ayers K, Rhodes LL, Tyrrell J, Gladstone M, Scholin C (2005) International accreditation of sandwich hybridization assay format DNA probes for micro-algae. N Z J Mar Freshw Res 39(6):1225–1231
5. Diercks S, Gescher C, Metfies K, Medlin LK (2009) Evaluation of locked nucleic acids for signal enhancement of oligonucleotide probes for microalgae immobilised on solid surfaces. J Appl Phycol 21(6):657–668
6. Diercks S, Medlin LK, Metfies K (2008) Colorimetric detection of the toxic dinoflagellate Alexandrium minutum using sandwich hybridization in a microtiter plate assay. Harmful Algae 7(2):137–145
7. Gescher C, Metfies K, Medlin LK (2008) The ALEX CHIP - Development of a DNA chip for identification and monitoring of *Alexandrium*. Harmful Algae 7(4):485–494
8. Greenfield DI, Marin R, Jensen S et al (2006) Application of environmental sample processor (ESP) methodology for quantifying *Pseudo-nitzschia australis* using ribosomal RNA-targeted probes in sandwich and fluorescent in situ hybridization formats. Limnol Oceanogr Methods 4:426–435
9. Haywood AJ, Scholin CA, Marin R, Steidinger KA, Heil C, Ray J (2007) Molecular detection of the brevetoxin-producing dinoflagellate Karenia brevis and closely related species using rRNA-targeted probes and a semiautomated sandwich hybridization assay. J Phycol 43(6):1271–1286
10. O'Halloran C, Silver MW, Holman TR, Scholin CA (2006) Heterosigma akashiwo in central California waters. Harmful Algae 5(2):124–132
11. Tyrrell JV, Connell LB, Sholin CA (2002) Monitoring for Heterosigma akashiwo using a sandwich hybridization assay. Harmful Algae 1(2):205–214
12. Diercks S, Metfies K, Medlin LK (2008) Development and adaptation of a multiprobe biosensor for the use in a semi-automated device for the detection of toxic algae. Biosens Bioelectron 23(10):1527–1533
13. Metfies K, Huljic S, Lange M, Medlin LK (2005) Electrochemical detection of the toxic dinoflagellate Alexandrium ostenfeldii with a DNA-biosensor. Biosens Bioelectron 20(7):1349–1357
14. Liao JC, Mastali M, Li Y et al (2007) Development of an advanced electrochemical DNA biosensor for bacterial pathogen detection. J Mol Diagn 9(2):158–168
15. Drummond TG, Hill MG, Barton JK (2003) Electrochemical DNA sensors. Nat Biotechnol 21(10):1192–1199
16. Wang J (2006) Electrochemical biosensors: towards point-of-care cancer diagnostics. Biosens Bioelectron 21(10):1887–1892
17. Rautio J, Barken KB, Lahdenpera J, Breinstein A, Molin S, Neubaure P (2003) Sandwich hybridisation assay for quantitative detection of yeast RNAs in crude cell lysates. Microb Cell Fact 2(1):4
18. Zammatteo N, Moris P, Alexandre I, Vaira D, Piette J, Remacle J (1995) DNA-probe hybridization in microwells using a new bioluminiscent system for the detection of PCR-amplified HIV-1 proviral DNA. J Virol Methods 55(2):185–197
19. Ronkainen-Matsuno NJ, Thomas JH, Halsall HB, Heinemann WR (2002) Electrochemical immunoassay moving into the fast lane. Trac-Trends Anal Chem 21(4):213–225
20. Anderson DM, Kulis D, Erdner D, Ahn S, Walt D (2006) Fibre optic microarrays for the detec-

tion and enumeration of harmful algal bloom species. Afr J Mar Sci 28(2):231–235

21. Diercks S, Metfies K, Medlin LK (2008) Molecular probe sets for the detection of toxic algae for use in sandwich hybridization formats. J Plankton Res 30(4):439–448
22. Orozco J, Medlin L (2010) Electrochemical performance of a DNA-based sensor device for detecting toxic algae. Sens Actuators B Chem 153:71–77
23. Orozco J, Baudart J, Medlin L (2011) Evaluation of probe orientation and effect of the digoxigenin-enzymatic label in a sandwich hybridization format to develop toxic algae biosensors. Harmful Algae 10:489–494
24. Groben R, John U, Eller G, Lange M, Medlin LK (2004) Using fluorescently labelled rRNA probes for hierarchical estimation of phytoplankton diversity. Nova Hedw 79:313–320
25. Ludwig W, Strunk O, Westram R, Richter L, Meier H, Yadhukumar, Buchner A, Lai T, Steppi S, Jobb G, Förster W, Brettske I, Gerber S, Ginhart AW, Gross O, Grumann S, Hermann S, Jost R, König A, Liss T, Lüßmann R, May M, Nonhoff B, Reichel B, Strehlow R, Stamatakis A, Stuckmann N, Vilbig A, Lenke M, Ludwig T, Bode A, Schleifer K-H (2004) ARB, a software environment for sequence data. Nucl Acids Res;32: 1363–1371

Chapter 13

Waterborne Pathogen Detection Using a Magnetoresistive Immuno-Chip

Sofia S.A. Martins*, Verónica C. Martins*, Filipe A. Cardoso, Paulo P. Freitas, and Luís P. Fonseca

Abstract

Magnetoresistive bioassays, in which the traditional optical labels are replaced by magnetic labels, hold the promise of increased response speed, sensitivity, and portability in the detection system. These properties make these systems ideal for the monitoring of microbiological quality of drinking water or hydric resources. In this chapter, we demonstrate the applicability of magnetoresistive biosensors for the detection of the environmental pathogen *Salmonella typhimurium*. The approach comprises the coating of magnetic nanoparticles with polyclonal antibodies to *Salmonella* spp. in order to specifically capture and concentrate the *Salmonella* cells from solution. Once captured, the magnetically labeled cells are recognized by a second immuno-recognition on top of the sensor's surface through the use of specific monoclonal antibodies. Quantitative data are then obtained using an electronic platform to measure the resistance change of the sensor due to presence of the magnetic particles.

Key words: Pathogen detection, Immunoassay, Chemical functionalization, Magnetoresistive biosensors

1. Introduction

Immunoassays are based on the unique ability of antibodies to bind with high specificity to a particular molecule (antigen), through particular structures on the molecule (epitopes). The enzyme-linked immunosorbent assay (ELISA) is the most commonly used immunoassay technique for the detection of a wide range of analytes, in particular whole microbial cells. In the last decade, ELISA has been combined with biosensors and biochips to create immuno-sensing devices with improved detection

* Sofia S.A. Martins and Verónica C. Martins contributed equally for this work.

Sonia M. Tiquia-Arashiro (ed.), *Molecular Biological Technologies for Ocean Sensing*, Springer Protocols Handbooks, DOI 10.1007/978-1-61779-915-0_13,

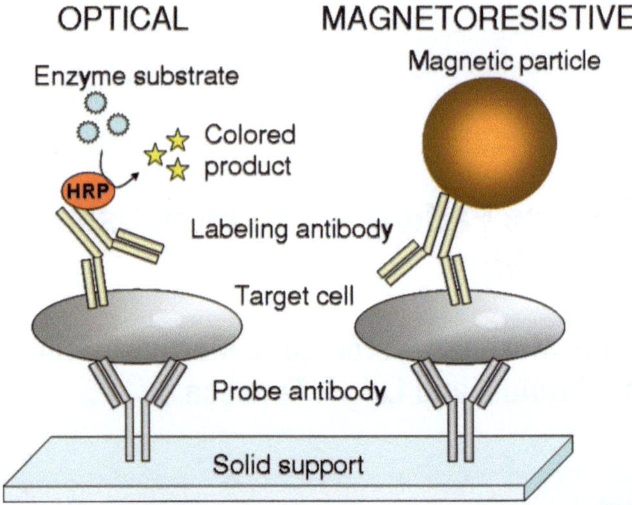

Fig. 1. Schematic representation of a "sandwich" enzyme-based immunoassay (ELISA) in analogy with a magnetoresistive immunoassay.

performances with respect to speed and sensitivity (1). Nevertheless, more advanced immuno-sensors are being developed based on the ELISA principle but avoiding the enzymatic label. Examples are label-free technologies that use direct detection of antigen bound to the antibody on the surface of the biosensor (e.g., SPR, piezoelectric, FET-based biosensors) (2–4) while others replace the reporter enzyme by a different label (e.g., magnetic particles, gold nanoparticles, or quantum dots) (5–8).

Magnetoresistive sensors may be used in association with the ELISA principle to perform immunoassays. These sensors replace the traditional optical detection system, since the labeling is performed by a magnetic nanoparticle rather than a colored, fluorescent, or bioluminescent label (Fig. 1). The construction of the magnetoresistive bioassay may be generally described in four steps: (1) antibodies (poly- or monoclonal) specific to the microbial cell of interest are immobilized (e.g., physically adsorbed, chemisorbed, or chemically bound) on the chip surface; (2) a second specific antibody is immobilized on the surface of the magnetic labels; (3) the target sample is brought into contact with the biochip surface, and the probe antibodies specifically recognize the target microbial cell; (4) target cells are magnetically labeled, either prior to or after surface biorecognition, by the second antibody conjugated with the magnetic particles. The assembly formed by probe antibody/target cell/second antibody/magnetic particles is the so-called *sandwich* assay format (9).

This chapter outlines the major strategies for the immunomagnetic capture and magnetoresistive detection of *Salmonella typhimurium* cells from solution. The main protocol considerations are therefore related to the type of immobilization chemistry required for antibody's immobilization on both the surface of magnetic particles and the surface of the chip, aiming at optimal specificity and sensitivity for cell capture and detection.

1.1. Immobilization of Antibodies on Solid Surfaces

In these bioassays the selection of the antibodies is very important in order to assure specificity. Antibodies (most commonly immunoglobulins G, IgG) are known to have Y-shape and to consist of one Fc and two Fab fragments (Fig. 2). Fab fragments are the variable active sites of the molecule which react with antigens to form immune complexes. Therefore, their spatial accessibility is an important factor in ensuring the immunosorbent activity. These molecules are formed by four smaller protein units (two heavy and two light chains) that are linked to each other by disulfide bonds (Fig. 2). This is an important aspect to consider in some immobilization approaches that use thiol functional groups.

The process to chemically attach the probe antibody to the chip surface is known to highly influence the performance of the system (10). Depending on the immobilized molecule orientation on the surface, a higher or lower amount of Fab fragments appear to be available for the interaction with the antigen present in solution. The manner of antibody immobilization determines its orientation, and, as a result, its activity in the immunoassay can vary. Random antibody immobilization may result in inaccessibility of the active binding site to the antigen molecule or originate spatial hindrances or even structural deformation of the molecule. This is likely to occur in the case of molecules' adsorption, either by hydrophobic or electrostatic interactions, or random covalent binding using bifunctional cross-linkers such as Sulfosuccinimidyl

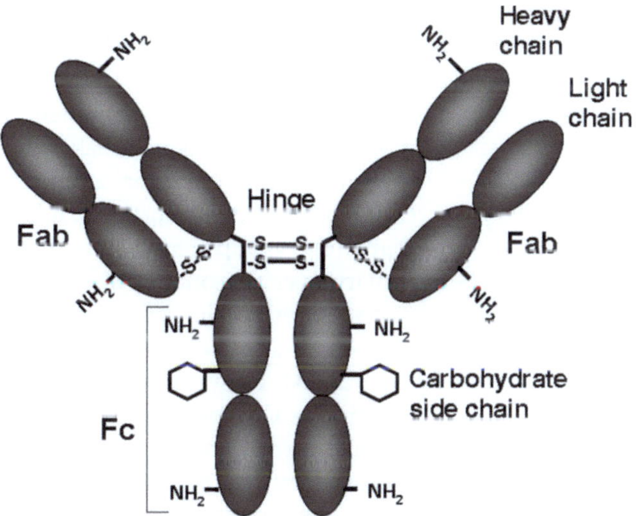

Fig. 2. Schematic representation of an antibody molecule depicting the main functional groups involved in immobilization reactions.

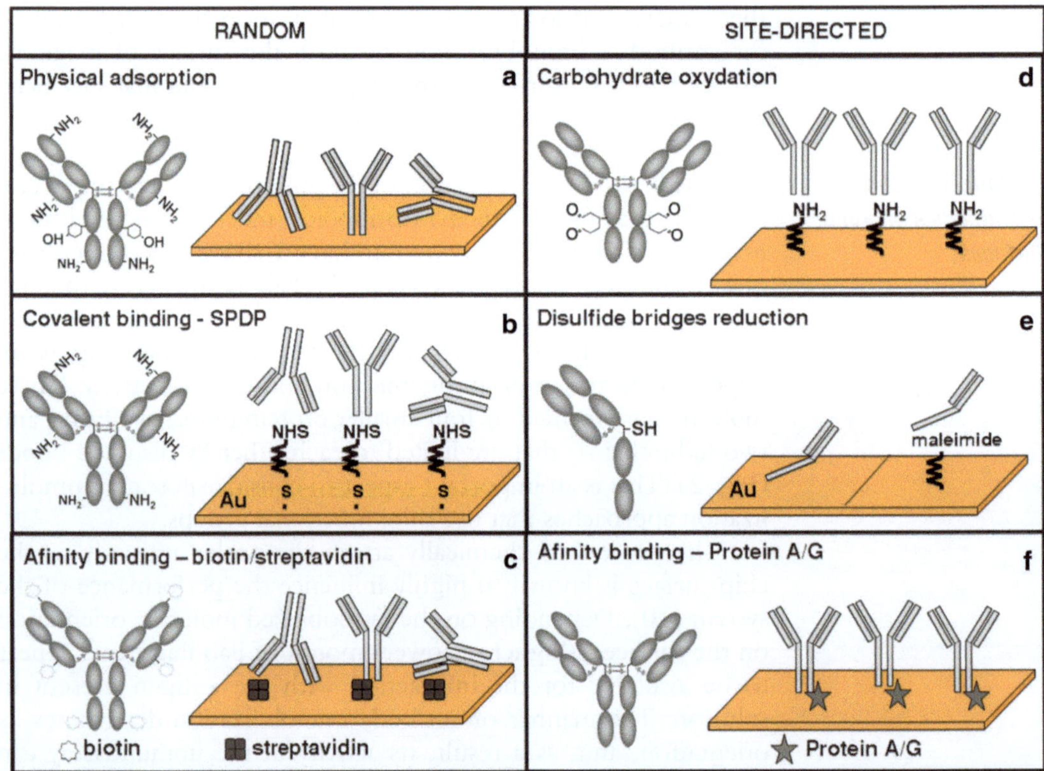

Fig. 3. Schematic representations of antibody immobilization on solid surfaces (gold in particular when stated (Au) and other functional surfaces in general) by random and site-directed chemistries. The main functional groups of the molecule involved in the immobilization process are represented on the *left-hand side* of each diagram. (**a**) Random physical adsorption through hydrophobic or electrostatic interactions. (**b**) Random covalent binding of antibody amine groups to N-hydroxysuccinimide (NHS) groups of SPDP molecules. (**c**) Random biotinylated antibody immobilized by biotin–streptavidin affinity interaction. (**d**) Site-directed immobilization through oxidized carbohydrate side chains of antibodies over amine, hydrazide, or hydrazine functional surfaces. (**e**) Thiolated antibody fragments by specific reduction of disulfide bridges on the hinge region, immobilized directly over gold surfaces or through a maleimido functional linker. (**f**) Site-directed immobilization through affinity binding of immobilized Protein A or G towards antibody Fc region (see Note 8).

6-[3′(2-pyridyldithio)-propionamido] hexanoate (Sulfo-LC-SPDP) or glutaraldehyde, which makes the bridge between the active solid surface and the functional amine groups of the antibody molecules (Fig. 3a–c). Furthermore, results were published demonstrating that amine groups of the antigen-binding site show higher chemical activity over the ones within the immunoglobulin molecule (11). Consequently, oriented antibody immobilizations are reported to be the best approach in order to retain all the immunological activity (12, 13). However a limited number of oriented alternatives are described. The most common includes oxidation of carbohydrate residues mostly located in the Fc fragment of the immunoglobulin, followed by the reaction of the resulting aldehyde groups with amine, hydrazine, or hydrazide groups on the support (Fig. 3d) (14, 15). Alternatively, disulfide bridges in the hinge region of the Fab fragments may be reduced to originate active thiol groups, allowing the direct chemisorption of the fragments to a gold surface (16) or the use of linkers with maleimido moieties (Fig. 3e). In this way the antigenic binding sites are available for antigen capture. A third option is the use of antibody-binding proteins such as Protein G and Protein A, which specifically target the Fc region on the antibody (Fig. 3f). In this case there is no need for antibody chemical modification avoiding the usually associated lost of activity. Other strategies are related with the recent advances on protein engineering, where recombinant antibodies are immobilized employing genetic fusions such as histidine or cysteine tags (17).

1.2. Target Pathogen Cells' Capture and Labeling

Following antibody immobilization target cell recognition and labeling is another relevant topic in the success of the analytical device. In particular for magnetoresistive biochips the target analyte has to be labeled with a magnetic particle (MP) (Fig. 4a, b). Then, the magnetoresistive sensors incorporated in the biochip are able to detect the small magnetic fringe fields coming from the labels (18).

The most commonly used labeling strategy makes use of a second antibody, also specific to the target analyte but preferentially recognizing a different epitope (Fig. 4d). By conjugating the second antibody with MPs through one of the immobilization chemistries previously mentioned, it is possible to use the immunoconjugate for cell capture, purification, concentration, and/or transportation, besides magnetic labeling (Fig. 4c) (19–23). However these particles were originally designed and produced to be used freely in solution, presenting some constraints to their application as magnetic labels over solid surfaces, namely nonspecific adsorption. To minimize the nonspecific interactions, a blocking step of the particle surface can be carried out using blocking agents,

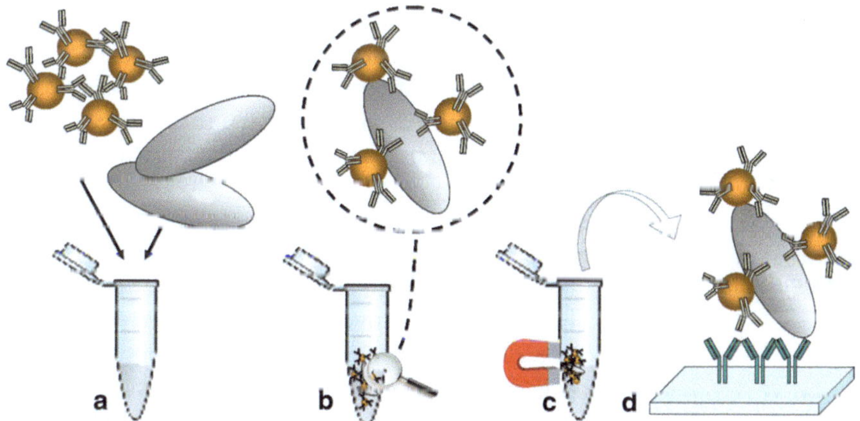

Fig. 4. Sketch out of the protocol steps for pathogen cell magnetoresistive detection. (**a**) Mixture of the *Salmonella typhimurium* cells with antibody-modified 250 nm magnetic particles. (**b**) Magnetic labeling of the target cells. (**c**) Magnetic capture of the magnetically labeled cells by a permanent magnet. (**d**) Loading and detection of the labeled sample over a magnetoresistive immuno-chip.

such as polyethylene glycol (PEG) or bovine serum albumin (BSA), in association with a surfactant in order to promote biological stabilization and prevent particles' aggregation and adhesion. Conversely, their remarkable magnetic properties associated to easiness and low-cost production as well as broad commercial availability make these labels a valuable choice (24). Additionally, because magnetic moments of magnetic particles exceed by far those of biological materials, background problems are eliminated and magnetic detection has little interferences even in samples with a complex biochemical composition.

The requisite for non-clustering, disperse MPs in homogeneous suspensions makes superparamagnetic particles a preferential choice. Superparamagnetic particles can quickly be attracted and agglomerate under an external magnetic field, but also promptly resuspended in a fluid upon removal of that field, exhibiting zero remanence. Therefore a simple permanent magnet can be easily used to assist on MPs' biochemical modification protocols. Presently, a variety of such particles are commercially available with a wide range of alternatives in terms of magnetic properties, types of coating (e.g., polymeric, silica, gold, latex) (25), sizes (from several micrometers down to few nanometers) (26), shapes (e.g., spherical, cluster-type, rods), and surface functionalization.

The nature of the functional groups (e.g., $-NH_2$, $-COOH$, $-SH$) or molecules (e.g., streptavidin, Protein A, or antibodies) on the particle surface will dictate the immobilization strategy as well as the labeling performance. Also the adopted size of MPs depends on the target analyte dimensions as well as on the number of target labeling sites available. For example on microbial cells the number of surface epitopes limits the number of labeling sites and therefore magnetic labels. Larger labels imply a lower labeling density and are more prone to leakage during protocol performance (e.g., washing steps). On the other hand smaller labels may reach higher labeling densities, which may hinder further biomolecular recognition events.

Strategies to estimate the density of antibodies on the magnetic particles' surface and further the magnetic labels' density over the target analyte are important for the design and optimization of the detection system.

1.3. Magnetoresistive Detection of Pathogen Cells

After successful magnetic labeling and capture of the target cells, on-chip biomolecular recognition and further magnetic detection occur. The magnetoresistive chip may be designed with different architectures. Important parameters in this design are the number of sensing sites and biomolecular active regions. Presently, the chip in use comprises up to 32 sensing sites of U-shaped spin-valve sensors. Four of those are reference sensors biologically inert to the target entities. The remaining sensors are covered with a bioactive region defined by a gold thin film pad of dimensions $13 \times 40\ \mu m^2$. The gold pad defines the probe antibody standing area through thiol–gold immobilization chemistry (Fig. 5).

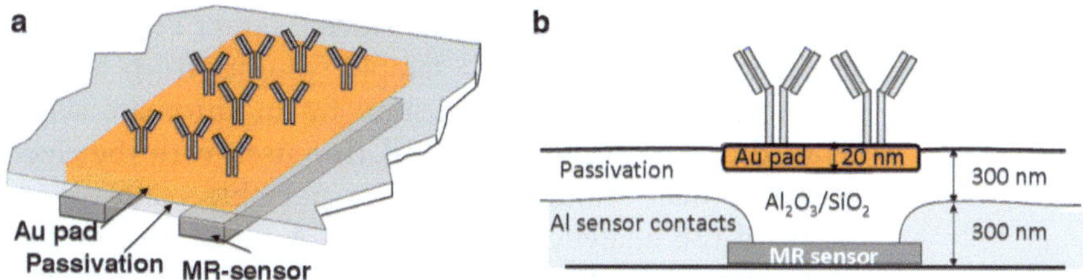

Fig. 5. Overview (**a**) and transversal cross section (**b**) of the magnetoresistive chip depicting the multilayer thicknesses.

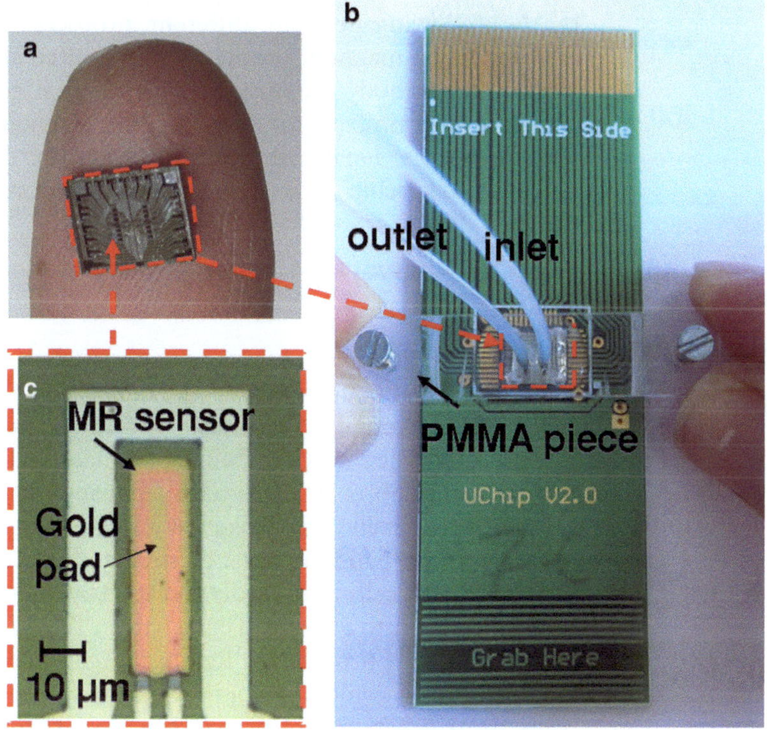

Fig. 6. Pictures from the magnetoresistive chip. (**a**) Chip on-finger. (**b**) Chip wirebonded on a chip carrier and integrated with a microfluidic system forming the assembled device. (**c**) Microscopic picture (800× magnification) of a sensing unit of the chip.

The chip (Fig. 6a) is produced by a standard microfabrication process taking place inside a clean-room environment (27). After microfabrication, the chip is wire bonded on a chip carrier to make the electrical contact that will allow the sensor readout by an electronic measurement setup (Fig. 6b) (28). The biological sample loading over the previously bio-functionalized and electronically connect chip is performed through a microfluidic system. The fluidic system consists of a U shaped channel printed in a polydimethylsiloxane (PDMS) block. The channel is connected with metallic adapters to tubing for the inlet and outlet (Fig. 6b). The sealing of the U-shaped channel to the chip is achieved by applying pressure over the PDMS, guaranteeing at the same time a correct alignment between the channel and the sensors (Fig. 6c) (29).

Before sample loading the labeled target cells are magnetically concentrated in a reduced sample volume (from 5 to 10 μl) and injected inside the fluidic system by an automate syringe pump. Thus the assemblage of the MR chip itself with the electronic system and microfluidics provides a portable platform suitable for point-of-use analytical applications. A full description of the MR chip microfabrication process as well as the electronic platform for sensor signal readout is out of the scope of the current chapter. Nevertheless, the reader is invited to consult several recent published works on this matter (27–29).

In the following sections different protocols for antibody immobilization on magnetic particles are presented and optimized for maximum antibody loading. *S. typhimurium* cell capture is performed and optimized for the different magnetically labeled antibody protocols. Finally, the captured cells are detected by a magnetoresitive biochip on a sandwich configuration strategy.

2. Materials

Biosafety Level II laboratory facilities are demanded when handling bacteria from the genus *Salmonella*. Good practices comprise restricted access when work is being conducted, the obligatory use of gloves and lab coats, dedicated laminar flow cabinets, and the use of proper containers for biological waste. All the materials used in bacterial culture/handling must be sterilized (e.g., autoclave).

S. typhimurium cells were kindly provided by the Water Analysis Laboratory of Instituto Superior Técnico from Technical University of Lisbon. Cells were provided in Ramback Agar (Merck), which is a chromogenic growth medium for the discrimination of *Salmonella* spp. cells from other *Enterobacteriaceae*. Specifically, *Salmonella* spp. will appear as red colonies, opposite for example to *Escherichia coli* that will originate blue colonies.

There are many alternatives for the necessary reagents, equipment, and suppliers. The following section describes the solutions implemented at our laboratory for the accomplishment of this protocol.

2.1. Buffers and Culture Media for Cell Culture

1. Sodium Phosphate, Dibasic (Merck, Cat: 567550).
2. Sodium Phosphate, Monobasic, Monohydrate, Molecular Biology Grade (Merck, Cat: 567549).
3. Tween® 20 detergent (Merck, Cat: 655205).
4. Sodium chloride (NaCl) (Panreac, Cat: 211659).
5. Sodium Acetate, Anhydrous, Molecular Biology Grade (Merck, Cat: 567418).

6. Tris, Hydrochloride, Molecular Biology Grade (Merck, Cat: 648317).
7. EDTA, Disodium Salt, Dihydrate, Molecular Biology Grade (Merck, Cat: 324503).
8. Sodium meta-periodate, $NaIO_4$ (Sigma-Aldrich, Cat: S1147).
9. BSA-Fraction V lyophilized powder (Sigma-Aldrich, Cat: 85040C).
10. Sodium cyanoborohydride solution 5.0 M in 1 M NaOH, $NaBH_3CN$ (Sigma-Aldrich, Cat: 296945).
11. Thiol-polyethyleneglycol, HO-PEG-SH b (RAPP Polymer, Cat: 103000-40).
12. DL-Dithiothreitol, DTT (Sigma-Aldrich, Cat: D0632).
13. Rappaport Vassiliadis Liquid Broth (BD, Cat: BA-257257.01).
14. LB-AGAR (Sigma-Aldrich, Cat: L2897).

2.2. Magnetic Particles

1. Amine (NH_2) coated magnetic nanoparticles, 250 nm (Micromod, Cat: 09-01-252).
2. Streptavidin coated magnetic nanoparticles, 250 nm (Micromod, Cat: 09-19-252).
3. Protein A coated magnetic nanoparticles, 250 nm (Micromod, Cat: 09-20-252).

2.3. Antibodies to Salmonella spp.

1. Rabbit polyclonal antibody (PAb) to *Salmonella* spp. (Abcam, Cat: ab35156).
2. Rabbit PAb to *Salmonella* spp. biotin conjugated (Abcam, Cat: ab69255).
3. Mouse monoclonal antibodies (MAbs) to *Salmonella* spp. (Abcam, Cat: 13633).

2.4. Chemical Cross-Linkers for Antibody Immobilization

1. Sulfo-LC-SPDP (Pierce, Cat: 21650).

2.5. Sample Loading and Salmonella Detection in the Magnetoresistive Biochip

1. Portable electronic platform for magnetic detection readout (INESC-ID—http://sips.inesc-id.pt/biochip/).
2. Syringes 1 mL (Codan, Cat: 621640).
3. Blunt tips 23 G (Small Parts, Cat: NE231PL25).
4. Teflon tubing 0.02 in. ID (Small Parts, Cat: B0013HMTI0).
5. Syringe Pump (New Era, Cat: NE-300).

2.6. Others

1. Refrigerated centrifuge (Hermle, rotor 220.72 for 15-mL centrifuge tubes and rotor 220.87 for 1.5-mL tubes).
2. Isopropanol (Sigma-Aldrich, Cat: 190764).

3. Glacial acetic acid (Sigma, Cat: 320099).
4. Magnetic concentrator Dynal MPC-S (Invitrogen, Cat: 120.20).
5. Sterile centrifuge tubes 15 mL (Corning, Cat: 430766).
6. Sterile Eppendorf tubes 1.5 mL.
7. Spectrophotometer (Hitachi U-2000).
8. Disposable plastic cuvettes (VWR-Int., Cat: 97000-586).
9. Disposable Petri dishes.
10. Disposable inoculating loops (VWR-Int., Cat: 1612-9358).
11. Orbital shaker (Aralab, Agitorb 2000).
12. Balance (Denver Instruments company, AA-250).
13. Microcon centrifugal filter devices 100 kDa cut-off (Millipore, Cat: UFC510096).
14. Coomassie Protein assay Kit (Pierce, Cat: 23200).
15. Clear 96 microwell plates (Greiner Bio-ONE, Cat: 650901).
16. Microplate reader (Molecular Devices, SpectraMax 340 PC).
17. UV-Ozone source (Jelight, Cat: 144AX-220).
18. End-over-end Eppendorf rotation with fixed speed at 12 rpm (Orbit, Cat: 3000445).

3. Protocols

3.1. Buffer Preparation

3.1.1. 0.1 M Phosphate Buffer (PB) pH 7.2

1. Prepare two stock solutions* as follows:
   Solution A: 0.2 M
   - Weight 35.61 g of Na_2HPO_4.
   - Bring the final volume to 1 L with dH_2O.

   Solution B: 0.2 M
   - Weight 27.6 g of $NaH_2PO_4 \cdot H_2O$.
   - Bring the final volume to 1 L with dH_2O.

   *These solutions, when autoclaved, are stable up to 1 year when stored in the dark at 4 °C.
2. Prepare 500 mL of 0.1 M phosphate buffer pH 7.2* as follows:

Component	Volume (mL)
Solution A	180
Solution B	70
dH_2O	250

*This solution, when autoclaved, is stableup to 1 year when stored in the dark at 4 °C

3.1.2. 0.1 M Phosphate Buffer, 0.02 % v/v Tween 20

1. Prepare 50 mL of 0.1 M phosphate buffer–0.02 % Tween 20 (v/v):

Component	Volume (mL)
Tween 20	0.01
0.1 M Phosphate buffer pH 7.2	Up to 50

3.1.3. 0.1 M Phosphate-Buffered Solution pH 7.2

1. Prepare 100 mL of 1.5 M NaCl solution as follows:

Component	Composition
NaCl	8.8 g
dH_2O	100 mL

2. Prepare 500 mL of 0.1 M phosphate buffered-saline solution at pH 7.2* as follows:

Component	Volume (mL)
Solution A	180
Solution B	70
1.5 M NaCl solution	50
dH_2O	200

*This solution, when autoclaved, is stable up to 1 year when stored in the dark at 4 °C

3.1.4. 0.1 M Sodium Acetate Buffer pH 5.5*

1. Prepare 100 mL of 0.1 M sodium acetate buffer pH 5.5 as follows:

Component	Composition
$NaC_2H_3O_2$	0.82 g
dH_2O**	100 mL

*This solution, when autoclaved, is stable up to 1 year when stored in the dark at 4 °C
**Before bringing the final volume up to 100 mL, adjust the pH to 5.5 with glacial acetic acid

3.1.5. 10 mM Tris–HCl, 1 mM EDTA (TE) pH 7.4*

1. Prepare 500 mL of 10 mM Tris–HCl, 1 mM EDTA pH 7.4 as follows:

Component	Composition
Tris–HCl	0.8 g
EDTA	0.15 g
dH_2O**	500 mL

*This solution, when autoclaved, is stable up to 1 year when stored in the dark at 4 °C
**Before bringing to the final volume, adjust the pH to 7.4

3.1.6. 20 mM Sodium Meta-periodate ($NaIO_4$) in 0.1 M Acetate Buffer, pH 5.5*

1. Prepare 1 mL of 20 mM sodium meta-periodate as follows:

Component	Composition
Sodium *meta*-periodate	4 mg
0.1 M acetate buffer, pH 5.5	1 mL

*This solution must be prepared immediately before use and protected from light

*3.1.7. 1 mg/mL Solution of Sulfo-LC-SPDP**

1. Prepare 1 mL of sulfo-LC-SPDP solution in phosphate buffer as follows:

Component	Composition
Sulfo-LC-SPDP	1 mg
0.1 M phosphate buffer, pH 7.2	1 mL

*This solution should be prepared immediately before use

3.1.8. 25 % (w/v) Bovine Serum Albumin and 0.05 % (v/v) Tween 20 in Phosphate Buffer

1. Prepare 10 mL of solution as follows:

Component	Composition
BSA	2.5 g
Tween 20	0.005 mL
0.1 M phosphate buffer, pH 7.2	Up to 10 mL

*3.1.9. 1 mg/mL SH-PEG**

1. Prepare 1 mL of solution as follows:

Component	Composition
SH-PEG	1 mg
TE buffer, pH 7.4	1 mL

*This solution should be prepared immediately before use

*3.1.10. 1 mM DTT**

1. Prepare 1 mL of solution as follows:

Component	Composition
DTT	0.15 mg
0.1 M sodium acetate buffer	1 mL

*This solution should be prepared immediately before use

3.2. Media Preparation

3.2.1. Rappaport Liquid Medium

1. Prepare 25 mL of Rappaport liquid medium* as follows:

Component	Composition
Rappaport	0.66 g
dH_2O*	25 mL

*Autoclave at 115 °C for 15 min. Protect from light and store at 4 °C

3.2.2. LB-Agar Medium

1. Prepare 500 mL of LB-Agar Medium* as follows:

Component	Composition
Rappaport	17.5 g
dH_2O*	500 mL

*Autoclave at 115 °C for 15 min. Store at 4 °C

3.3. Polyclonal Antibody Immobilization on the Magnetic Particles

3.3.1. Carbohydrates Oxidation Site-Directed Immobilization Chemistry

1. Filtrate 100–200 μL of PAb molecule at 1 mg/mL using the Microcon centrifuge filters (100 kDa cutoff) at 10,000 rpm for 15 min at room temperature to remove the previous buffer solution and exchange buffer to 100–200 μl of 0.1 M sodium acetate buffer, pH 5.5 (see Notes 1 and 2).
2. Add an equal volume of 20 mM sodium *meta*-periodate solution (final concentration 10 mM), mix well, and incubate the mixture for 30 min at 4 °C, protected from light.
3. Filtrate with the Microcon (100 kDa cutoff) using a refrigerated centrifuge with the temperature set to 4 °C, at 10,000 rpm for 15 min.
4. Remove the sodium *meta*-periodate solution with the Microcon filters (100 kDa cutoff) at 10,000 rpm for 15 min and exchange buffer to 100–200 μl of 0.1 M phosphate buffer, pH 7.2. The carbohydrate molecules are now oxidized and the PAb molecules are ready for the coupling reaction with amine groups (see Note 3).
5. Gently resuspend the amine-coated magnetic particles stock solution (5×10^{11} particles/mL stock concentration) in order to obtain a homogeneous solution (see Note 4).
6. Remove the necessary volume of particle stock solution to obtain a final concentration of 10^7 particles/μL in 0.1 M phosphate buffer pH 7.2 into a 1.5-mL Eppendorf tube (1:10 dilution). For example, for a final volume of 100 μL, remove 10 μL from the stock solution.
7. Place the 1.5-mL Eppendorf tube containing the nanoparticles in the magnetic concentrator and wait for 3–5 min, until the particles pellet is clearly visible on the tube sidewalls. With the aid of a pipette, gently remove the original buffer without aspirating the particles. Resuspend the particles in 100 μL of 0.1 M PB pH 7.2. Repeat this procedure at least twice in order to completely remove the original preservative buffer.
8. Prepare a 100 μL dilution of the oxidized PAb at 6.25 μg/mL in 0.1 M PB, pH 7.2.

9. Using the magnetic concentrator remove the buffer from the particles solution (see step 7) and replace it with the antibody solution prepared above.
10. Incubate the mixture for 3 h at 37 °C with moderate agitation (end-over-end rotator at 12 rpm) (see Note 5).
11. Using the magnetic concentrator, discard the supernatant and resuspend in 100 μL of 0.1 M PB, pH 7.2. Repeat this procedure twice in order to remove the non-bound antibody molecules.
12. Reduce the primary amines by adding directly from the stock solution 10 μL of cyanoborohydride. Allow to react for 10 min at room temperature (see Note 6).
13. Using the magnetic concentrator, wash the Ab-conjugated particles three times with 0.1 M PB, pH 7.2.
14. Finally, resuspend the particles in 1 mL 0.1 M PB (10^6 particles/μL). The Ab-conjugated particles are preserved up to 1 week at 4 °C in this buffer without loss of antibody activity (see Note 7).

3.3.2. Biotin–Streptavidin Affinity Binding

1. Filter 100–200 μL of biotin-PAb molecule at 1 mg/mL using the Microcon centrifuge filters (100 kDa cutoff) at 10,000 rpm for 15 min at room temperature to remove the previous buffer solution and exchange buffer to 0.1 M PBS, pH 7.2.
2. Gently resuspend the streptavidin particles stock solution (5×10^{11} particles/mL stock concentration) in order to obtain a homogeneous solution (see Note 4).
3. Remove the necessary volume of particles stock solution to obtain a final concentration of 10^7 particles/μL in 0.1 M PB, pH 7.2 to a 1.5-mL Eppendorf tube (1:10 dilution). For example, for a final volume of 100 μL, remove 10 μL from stock solution.
4. Using the magnetic concentrator, wash the particles three times with 0.1 M phosphate saline buffer pH 7.2, as indicated in Sect. 3.2.1, step 7, in order to remove the preservative buffer.
5. Prepare a 100 μL dilution of the biotin-PAb at 10 μg/mL in 0.1 M PBS, pH 7.2.
6. Using the magnetic concentrator, remove the buffer from the particles solution and replace it with the antibody solution prepared in step 5.
7. Incubate the mixture for 30 min, at room temperature, with moderate agitation.

8. Using the magnetic concentrator, discard supernatant and wash the Ab-conjugated particles three times with 0.1 M PBS, pH 7.2.
9. Resuspend the particles in 1 mL 0.1 M PB (10^6 particles/μL final concentration). The Ab-conjugated particles are preserved up to 1 week at 4 °C in this buffer without loss of antibody activity.

3.3.3. Protein A Affinity Binding

1. Proceed as described above for biotin/streptavidin immobilization but replace the biotin-PAb by the nonconjugated PAb molecule and use the Protein A-coated magnetic nanoparticles (5×10^{11} particles/mL stock concentration) (see Note 8).

3.3.4. Evaluation of Immobilization Efficiencies

Protein quantification methods can be used to determine the effectiveness of the immobilization chemistries described above. The following protocol is based on the Coomassie Protein assay kit and can be used to determine the number of immobilized antibodies per particle.

1. Standards: Prepare a serial dilution of the antibodies ranging from 0 to 100 μg/mL in a total volume of 100 μL (see Note 9).
2. Samples: Preserve the 100 μL supernatant and the washing solutions containing the unbounded antibody molecules (see above sections).
3. Transfer the standards and samples to a clear 96-microwell plate.
4. Add an equal volume of Coomassie reagent to each well, mix, and incubate in the dark for 10 min.
5. Transfer the plate to the microplate reader and collect the absorbance at 595 nm.
6. Plot the absorbance of the standards as a function of its concentration and use the standard curve to interpolate the concentration of each sample (see Note 10).
7. Calculate the amount of immobilized antibody as follows:

$$\text{Ab molecules} = \text{Concentration}\,(\text{g/L}) \times \left(100 \times 10^{-6}\right)^{*}\,(\text{L}) \times 6 \times 10^{23}\,\text{molecules/mol} / 150{,}000\,(\text{g/mol})$$

(*Reactional volume)

$$\text{Immobilized Ab molecules} = \text{Ab molecules}_{\text{Initial}} - \sum_i \text{Ab molecules}_{(\text{supernatant} + \text{washing buffers})}.$$

3.4. Monoclonal Antibody Immobilization on the Chip Surface

1. Wash the chip three times with isopropanol followed by a three-time wash in dH_2O. Dry under compressed air or N_2.
2. Treat the chip with the UV-ozone source setting for at least 15 min for the UV irradiation time (see Note 11).
3. Filter 50 μL of MAb molecule at 50 μg/mL using the Microcon centrifuge filters (100 kDa cutoff) at 10,000 rpm for 15 min to remove the preservative buffer and exchange buffer to 0.1 M PB, pH 7.2. Keep this solution at 4 °C until step 8.
4. Dispense 20–50 μL of sulfo-LC-SPDP solution over the chip surface and incubate at room temperature for 1 h in a humidified atmosphere in order to prevent evaporation (see Note 12).
5. Remove excess cross-linker by rinsing with 0.1 M PB.
6. Incubate the chip with 50 μL of DTT solution for 15 min. This step aims to reduce the remaining disulfide bridges on the cross-linker and further enhances Sulfo-LC-SPDP cross-linkage to the gold surface. Wash using a pipette with 0.1 M PB, pH 7.2.
7. Incubate the chip with 50 μL of MAb solution (from step 3) for 2 h at room temperature in a humidified atmosphere to prevent evaporation.
8. Remove the unbound MAb molecules by washing with 0.1 M PB, pH 7.2.
9. Block the chip surface with 50 μL of OH-PEG-SH solution in TE buffer for 1 h at room temperature in a humidified atmosphere (see Note 13).
10. Remove excess blocking solution by washing the chip two times with TE buffer. Rinse with PB. The chip is now ready to use.

3.5. *Salmonella typhimurium* Magnetic Capture

3.5.1. *S. typhimurium* Cell Culture

1. In an aseptic environment and with the aid of an inoculating loop, remove one isolated colony of *S. typhimurium* from the culture plate and inoculate 25 mL of Rappaport liquid medium.
2. Incubate for 16 h in an orbital shaker at 37 °C with 250 rpm agitation.
3. In aseptic conditions transfer the enriched medium to two sterile 15-mL centrifuge tubes. Centrifuge at 5,000 rpm for 5 min.
4. Discard supernatant into a proper biohazard container for further decontamination. Resuspend the cell pellet in 0.1 M PB, pH 7.2.
5. Using the spectrophotometer, measure the culture optical density at 600 nm (OD_{600}) using PB as blank. All contaminated waste must be discarded into a proper container.
6. The cell concentration is calculated as follows:

$$1OD_{600} \sim 5\times10^{8}\,\text{cells} / \text{mL}.$$

7. In aseptic conditions dilute the cell suspension to the desired concentrations (typically in the range of 0–10^6 cells/mL) in 0.1 M PB, pH 7.2.

3.5.2. Magnetic Capture

1. Prepare 100 μL (10^6 particles/μL) of the PAb-modified particles as described in Sect. 3.3.
2. Using the magnetic concentrator, remove the conservation buffer and replace it for 100 μl of a 5 % BSA–0.02 % Tween 20 solution. Block the particles for 30 min at room temperature with moderate agitation (see Note 14).
3. Discard blocking solution and replace it by the cell suspension (see Sect. 3.5.1, step 7). Incubate for 30–45 min at room temperature, with moderate agitation (see Note 15).
4. Using the magnetic concentrator, gently remove the supernatant. Repeat this procedure three times with 0.1 M PB. All waste should be discarded into a proper biohazard's container for further decontamination.
5. Perform a final suspension of the magnetically labeled cells in 10 μL of 0.1 M PB, pH 7.2.

3.5.3. Evaluation of Cell Capture Efficiencies

Evaluation of capture efficiencies can be performed to assess the efficacy of the capturing procedure. The following protocol is based on traditional colony plate assays coupled to mass balances and affords a mean to calculate the number of captured cells originated from cell suspensions not exceeding 10^4 cells/mL. Higher cell concentrations will result in difficulty in cell counting.

1. Autoclave LB-Agar medium for 15 min at 115 °C. Cool down approximately to 40 °C, assuring that the medium has not solidified.
2. In an aseptic environment, add approximately 20 mL of LB-Agar medium into each Petri dish. Wait until the culture medium completely solidifies at room temperature.
3. Preserve the supernatant and washing solutions from step 4 in Sect. 3.5.2 and 100 μL of the initial cell suspension. With the aid of an inoculating loop, plate the samples in the LB-Agar plates. Spread expansively in order to obtain isolated cells. Each isolated cell will originate an individual cell colony. Incubate the plates for 16 h at 37 °C.
4. Count the Colony-Forming Units (CFUs) obtained in each plate. The number of captured cells can be calculated as follows:

$$\text{Captured cells} = \text{No. CFU}_{\text{Initial cell suspension}} - \sum_i \text{No. CFU}_{(\text{supernatant} + \text{Washing buffers})}.$$

3.6. Magnetically Labeled Cell Detection

3.6.1. Sample Loading

1. Assemble the PDMS block (containing the U-shaped microfluidic channel) against the poly(methyl methacrylate) (PMMA) piece.
2. Press the piece against the chip surface (previously functionalized as depicted in Sect. 3.4) and tighten the screws in order to seal the microfluidic channel.
3. Using tweezers, insert the Teflon tubing in the inlet and outlet ports to form the complete device (see Fig. 6b).
4. Using the 1-mL syringe, carefully aspirate the solution containing the magnetically labeled cells (see Sect. 3.5.2, step 5), avoiding air bubbles.
5. Assemble the blunt end tip into the cell-containing syringe and connect it to the inlet tubing. Place the syringe in the syringe pump and set the flow rate at 5 μL/min.

3.6.2. Signal Acquisition

1. Signal acquisition starts with the connection of the device to a portable electronic platform, containing all the necessary circuitry for the sensors readout. Operational measurement conditions use an external magnetic AC+DC field set to 13.5 Oe rms+30 Oe. During the measurements, the sensors are biased with a 1 mA electrical current. The signal is captured using a gain of 40 and the sensors are sequentially read. Finally, the signal is digitally filtered with a 1-Hz-bandwidth filter and transferred through a Universal Serial Bus (USB) port to a laptop.
2. Start to acquire the baseline signal for 10 min ($V_{baseline}$). Initiate sample loading by plugging the syringe pump (see Sect. 3.6.1, step 5). The binding of the magnetic particles on the chip surface can be monitored in real time. Stop flow and allow the binding reaction to occur for 35 min.
3. Replace the sample-containing syringe with a new one containing 0.1 M PB pH 7.2–0.02 % Tween 20. Connect to the inlet tubing and start the flow at 5 μL/min in order to wash the non-bound particles. Stop the washing reaction when the reference sensor ($V_{reference}$) has returned to the baseline (see Note 16).
4. A positive binding signal ($V_{binding}$) is obtained when the signal of the bioactive sensor ($V_{positive}$) is higher than the signal in the reference sensor ($V_{reference}$).

4. Typical Results

4.1. Polyclonal Antibody Immobilization on the Magnetic Particles

For all the described chemical strategies, the successful immobilization of the antibodies on the surface of the magnetic particles was tested by varying the initial antibody concentration in solution

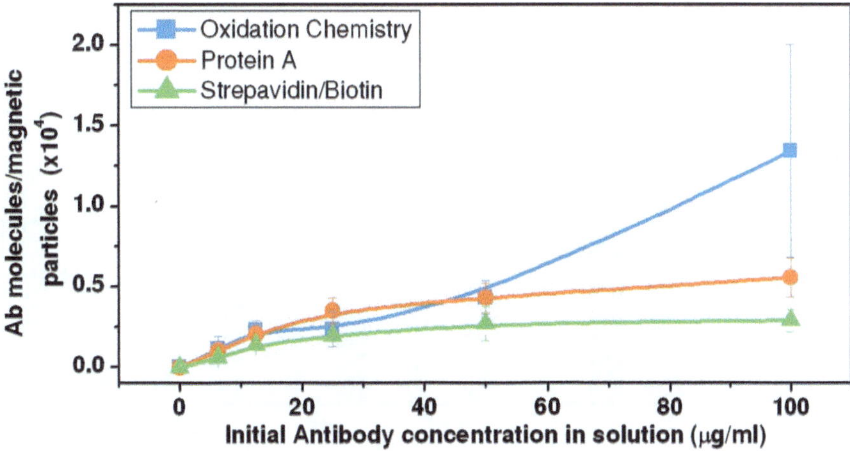

Fig. 7. Density of immobilized Abs as a function of immobilization strategy. Particle number was set to ~10^9 particles in a total volume of 100 μl.

and calculating the number of antibodies per particle, as described in Sect. 3.3.4. Results are depicted in Fig. 7.

Antibody immobilization could be confirmed for all the immobilization chemistries. Because the antibody molecules are immobilized on a solid surface, the antibody surface density is an important variable affecting capture efficiencies. A very sparse surface coverage will result in poor capture efficiencies. However, as surface density increases, concerns regarding steric effects hindering the antibody recognition event may emerge. For the affinity-based chemistries, biotin/streptavidin and Protein A, the antibody surface density follows a saturation profile, with maximum surface densities reaching 2,000 and 3,500 antibodies/particle, respectively. For the carbohydrate oxidation, a double-layer profile emerged. This trend suggests interactions between the immobilized antibodies. Consequently, a superior number of antibody molecules could be immobilized.

4.2. *Salmonella typhimurium* Magnetic Capture

The biological activity of the antibody-functionalized particles was tested by incubating the particles with a fixed number of *S. typhimurium* cells (10^3 cells) as described in Sect. 3.5.3. Results depicted in Fig. 8 show that it was possible to magnetically capture *S. typhimurium* cells from a solution, independently of the immobilization protocol in use. Apparently, biotin/streptavidin affinity binding affords higher capture efficiencies. However, it must be pointed out that the blocking protocol could not be performed (see Note 14) and therefore the nonspecific adsorption contributes significantly to the observed capture efficiencies. An inverse proportionality between antibody surface densities and capture efficiencies is observed. These results indicate that despite the fact that more antibodies can be loaded as initial concentration increases, biological functionality becomes compromised. This trend enabled us to define optimum surface densities (~1,000 antibody/particle) for each particular immobilization strategy.

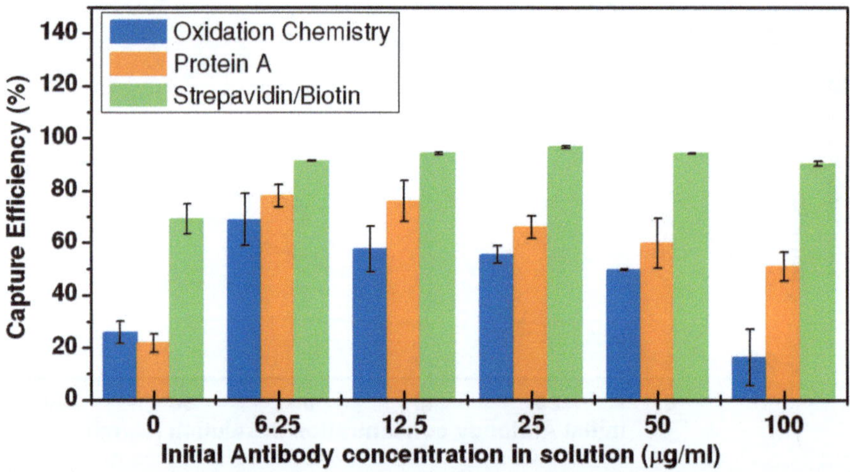

Fig. 8. Capture efficiencies of *S. typhimurium* cells (10^4 cells/mL) as a function of the different immobilization chemistries. Particle number was set to ~5×10^8 in a total volume of 100 μl. No blocking strategy (see Note 12).

Fig. 9. Optical microscope pictures (800× magnification) from sensors after chip incubation with Protein A particles. The sensor on the *left* is a positive control for MAb immobilization (gold pad), whereas the sensor on the *right* is a reference sensor, with no gold pad. The presence of protein A particles only on the positive control confirms the presence of immobilized MAb molecules.

Using the described approaches, it was possible to magnetically capture different cell concentrations (10^1–10^6 cells/mL) with efficiencies higher than 80 %.

4.3. Detection of Magnetically Labeled Cells with the Magnetoresistive Biochip

The magnetically labeled cells were detected using the magnetoresistive platform as described in Sect. 3.6. It was verified that aminated particles adsorbed strongly on the gold chip surface. On the other hand, Protein A particles do not adsorb on the surface but react strongly when the MAb is immobilized. In the latter case, this apparent disadvantage was indeed useful to confirm the presence of the MAb molecules on the surface of the chip. Results on Fig. 9 show that the presence of Protein A particles is restricted to the bioactive sensors, which possess gold pads modified with Mab molecules.

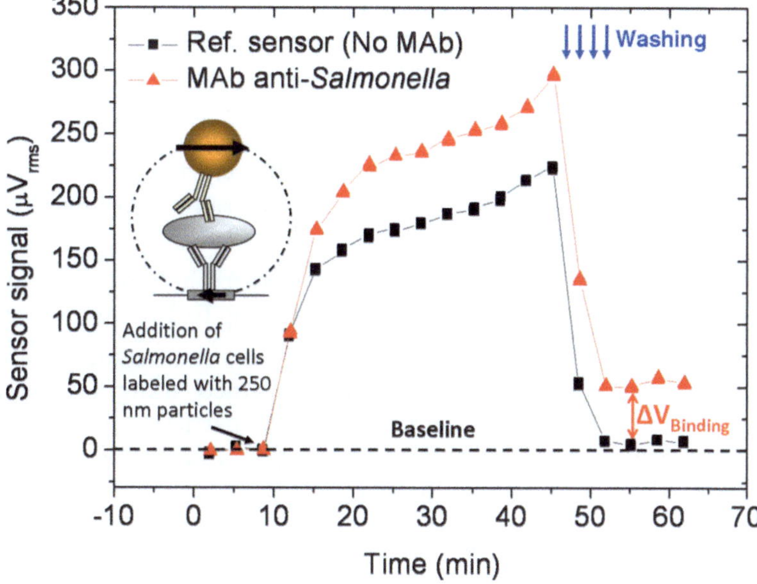

Fig. 10. Real-time sensing data from a spin valve sensor. Detection of antibody–antigen recognition event between immobilized monoclonal anti-*Salmonella* antibodies and *Salmonella* cells labeled with 250 nm streptavidin-coated magnetic particles. The signal of the bioactive sensor (MAb immobilized over the sensor surface) is 50 times higher than the signal from the reference sensor.

For the magnetoresistive measurements, cell capturing was performed using Ab–streptavidin-conjugated nanoparticles and a cell suspension containing 10^6 cells/mL. Real-time measurements were acquired as described in Sect. 3.6 and results are depicted in Fig. 10. At the end of the experiment, after the washing procedure, it was possible to observe that the signal from the bioactive chip (modified with MAb) was kept at 50 μVrms while the signal from the reference sensor (no MAb) returned near the baseline. These results validate the approach, confirming that the magnetically labeled cells were recognized by the MAb molecules and remained attached even after washing. Taking into consideration an initial concentration of 10^6 cells/mL (before capturing) and a reaction volume of 0.5 μL (channel capacity), approximately 1,000 cells were incubated over the chip surface. Since the chip has 28 bioactive sensors, the 50 μVrms signal corresponds to an average detection of ~30 cells.

5. Notes

1. Commercial antibodies are usually supplied in buffers containing preservatives and stabilizers such as sodium azide or BSA, which should be removed prior to any coupling reaction (Sect. 3.3.1, step 1). This step is mandatory in case of the presence of glycerol since this compound is an inhibitor of the oxidation reaction.

2. Avoid dryness of the filter membrane by over-centrifugation in order to prevent nonspecific adsorption and consequent antibody losses in the recovery step (Sect 3.3.1, step 1). The time and centrifugation speed depend on the rotor rotation angle. In the first usages, it is advisable to carefully monitor this step by stopping the centrifugation within short period of time to ensure that the sample volume is retained in the sample reservoir.
3. Aliquots of oxidized antibodies (1 mg/mL) were stable when stored at −20 °C up to 1 month. Once thawed, use immediately.
4. Do not centrifuge the magnetic particles. Particles collection must be performed using proper magnetic concentrators. Additionally, the particles cannot be frozen.
5. Alternatively, the reaction in Sect. 3.3.1, step 10, can be accomplished by incubating the mixture overnight at 4 °C.
6. Cyanoborohydride is toxic. Use protective personnel equipment and work inside a fume hood to perform protocol in Sect. 3.3.1, step 12.
7. Please note that MAbs may lack carbohydrate residues at the Fc region, which may compromise the applicability of the protocol in Sect. 3.3.1.
8. Proteins A and G mentioned in Sect. 3.3.3 are native and recombinant proteins of microbial origin that bind to mammalian immunoglobulins. Binding specificities and affinities of these proteins differ between source species and antibody subclass. Please consult Table 1 to select the antibody-binding protein that is best for your application.
9. The antibodies in Sect. 3.3.4, step 1, are diluted in different solutions according to the immobilization chemistry. For the carbohydrate oxidation chemistry the standard curve should be prepared in 0.1 M PB, whereas for biotin/streptavidin chemistry, 0.1 M PBS should be used.
10. Under our experimental conditions (Sect. 3.3.4, step 6), the sensitivity of Coomassie assay was in the order of 3 μg/mL. For applications requiring lower protein concentrations the Micro-BCA assay kit should be preferable.
11. The UV/ozone treatment (see Sect. 3.4, step 2) cleans the surface from organic compounds rendering it hydrophilic, which is crucial for a successful biochemical modification. The treatment efficiency depends on the distance between the sample and the UV lamp. A distance of 5 mm (achieved by adjusting the level of the tray inside the chamber) is recommended by the supplier as the optimal condition. Alternatively, oxygen plasma generating systems can be used. Corona discharge or chemical treatments (e.g., Piranha solution) can be used on

Table 1
Binding affinities of Protein A and G to the Fc region of antibodies from different mammalian source species

Species	Protein A	Protein G
Human	+++	+++
Mouse	+++	+++
Rat	+	++
Cow	+	+++
Goat	+	+++
Sheep	+	+++
Horse	+	+++
Rabbit	+++	+++
Guinea pig	+++	+
Pig	+++	+
Dog	+++	+
Cat	+++	+
Chicken	–	–

+++ strong binding, ++ medium binding, + weak binding, – no binding. Adapted from Pierce (http://www.piercenet.com/files/TR0034-Ab-binding-proteins.pdf)

surfaces not comprising electronic circuitry, since these treatments may irreversibly damage the structures.

12. The NHS ester in the Sulfo-LC-SPDP cross-linker is moisture sensitive and hydrolyzes easily in the presence of water molecules. Increasing the reaction time (see Sect. 3.4, step 4) will not increase the efficiency of the surface chemical modification. For this purpose, it is preferable to increase the linker concentration.

13. SH-PEG blocks unoccupied sites on the gold surface. Non-reacted NHS groups are blocked by the amine molecules of Tris solution (refer to Sect. 3.4, step 9).

14. The composition of the blocking solution was previously optimized with the aim of reducing nonspecific adsorption of *Salmonella* cells on the particles surface. Although effective for amine- and Protein A-coated particles, it was observed that streptavidin particles precipitate, forming clusters in an irreversible way. In view of this, no blocking procedure was included in the immobilization chemistry described in Sect. 3.5.2, step 2.

15. Exceeding the incubation time mentioned in Sect. 3.5.2, step 3, it will increase the cell nonspecific adsorption.
16. Different flow velocities may be used as stringent washing in order to increase discrimination between the positive and negative signals. We have assayed washing speeds as high as 50 μL/min (see Sect. 3.6.2, step 3).

6. Future Directions

In this work, a protocol for the magnetic capture and detection of labeled *S. typhimurium* was optimized in the scope of the development of an MR biochip-based platform. It is important to highlight that this protocol, despite being developed for particular application on MR chips, may be straightforwardly implemented on the functionalization of other types of gold surface-based immuno-analytical devices. On the contrary not all the gold-surface biochemistries available are suitable to use on the electronic integrated devices such as the MR chip, since they may damage the electrical structures of it.

Immunomagnetic capture coupled to magnetic detection, as it was demonstrated in this chapter, has the potential of becoming a valuable tool for bioanalytical analysis over a wide range of environmental samples. Indeed, the presence of a specific pathogen, *S. typhimurium*, in control samples could be detected within 1 h (capture + detection) total reaction time.

Although attractive for the perspective of point-of-use application, a major challenge will be the response of these systems to real samples with constraints regarding low target analyte concentrations that are usually dispersed on complex sample matrices. Therefore a few important issues have been identified, and already started to be addressed at our research group. For example scaling down the sensing units close to the target dimensions is expected to improve sensitivity and detection limits. We envisage the design and microfabrication of dedicated chambers for sample pretreatment (e.g., concentration). Microfluidic solutions for sample loading, preparation, and transportation inside the device are being designed, including microfluidic structures such as valves and pumps.

Finally, scalability of the detection features, seeking high-throughput applications and improvements on the electronic control of the system, will potentiate device's autonomy, portability, and automation.

Acknowledgements

The authors want to acknowledge the availability of a customized electronic platform developed by INESC-ID group, by personally thanking to José Germano, Leonel Sousa, and Moisés Piedade.

References

1. Butler JE (2000) Enzyme-linked immunosorbent assay (reprinted from Immunochemistry, pg 759–803, 1994). J Immunoassay 21: 165–209
2. Tsai W-C, Li I-C (2009) SPR-based immunosensor for determining staphylococcal enterotoxin A. Sensor Actuator B Chem 136:8–12
3. Kim G-H, Rand AG, Letcher SV (2003) Impedance characterization of a piezoelectric immunosensor part II: Salmonella typhimurium detection using magnetic enhancement. Biosens Bioelectron 18:91–99
4. Marquette CA, Blum LJ (2006) State of the art and recent advances in immunoanalytical systems. Biosens Bioelectron 21:1424–1433
5. De Palma R, Reekmans G, Laureyn W, Borghs G, Maes G (2007) The optimization of magnetosandwich assays for the sensitive and specific detection of proteins in serum. Anal Chem 79:7540–7548
6. Dittmer WU, de Kievit P, Prins MWJ, Vissers JLM, Mersch MEC, Martens MFWC (2008) Sensitive and rapid immunoassay for parathyroid hormone using magnetic particle labels and magnetic actuation. J Immunol Methods 338:40–46
7. Liu G, Wang J, Lin Y, Wang J (2008) Nanoparticle-based biosensors and bio-assays. In: Zhang X, Ju H, Wang J (eds) Electrochemical sensors, biosensors and their biomedical applications. Academic, San Diego, pp 441–457
8. Tansil NC, Gao Z (2006) Nanoparticles in biomolecular detection. Nano Today 1:28–37
9. Suter M, Butler JE (1986) The immunochem istry of sandwich ELISAS. 2. A novel system prevents the denaturation of capture antibodies. Immunol Lett 13:313–316
10. Domen PL, Nevens JR, Mallia AK, Hermanson GT, Klenk DC (1990) Site-directed immobilization of proteins. J Chromatogr 510:293–302
11. Endo N, Umemoto N, Kato Y, Takeda Y, Hara T (1987) A novel covalent modification of antibodies at their amino-groups with retention of antigen-binding activity. J Immunol Methods 104:253–258
12. Peluso P, Wilson DS, Do D, Tran H, Venkatasubbaiah M, Quincy D, Heidecker B, Poindexter K, Tolani N, Phelan M, Witte K, Jung LS, Wagner P, Nock S (2003) Optimizing antibody immobilization strategies for the construction of protein microarrays. Anal Biochem 312:113–124
13. Turkova J (1999) Oriented immobilization of biologically active proteins as a tool for revealing protein interactions and function. J Chromatogr B 722:11–31
14. Hoffman WL, Oshannessy DJ (1988) Site-specific immobilization of antibodies by their oligosaccharide moieties to new hydrazide derivatized solid supports. J Immunol Methods 112:113–120
15. Oshannessy DJ, Hoffman WL (1987) Site-directed immobilization of glycoproteins on hydrazide-containing solid supports. Biotechnol Appl Biochem 9:488–496
16. Brogan KL, Wolfe KN, Jones PA, Schoenfisch MH (2003) Direct oriented immobilization of F(ab′) antibody fragments on gold. Anal Chim Acta 496:73–80
17. Jung Y, Jeong JY, Chung BH (2008) Recent advances in immobilization methods of antibodies on solid supports. Analyst 133:697–701
18. Graham DL, Ferreira HA, Freitas PP, Cabral JMS (2003) High sensitivity detection of molecular recognition using magnetically labelled biomolecules and magnetoresistive sensors. Biosens Bioelectron 18:483–488
19. Ferreira HA, Feliciano N, Graham DL, Clarke LA, Amaral MD, Freitas PP (2005) Rapid DNA hybridization based on ac field focusing of magnetically labeled target DNA. Appl Phys Lett 87:013901–013903
20. Gijs M (2004) Magnetic bead handling on-chip: new opportunities for analytical applications. Microfluid Nanofluid 1:22–40
21. Janssen XJA, van Ijzendoorn LJ, Prins MWJ (2008) On-chip manipulation and detection of magnetic particles for functional biosensors. Biosens Bioelectron 23:833–838

22. van der Wijngaart W, Andersson H, Stemme G (2003) Handling of beads in microfluidic devices for biotech applications. In: Oosterbroek RE, van den Albert B (eds) Lab-on-a-chip. Elsevier, Amsterdam, pp 187–204
23. Wirix-Speetjens R, Fyen W, Xu K, De Boeck J, Borghs G (2005) On-chip magnetic particle transport: where physics, chemistry and biology meet. In: Magnetics Conference, 2005 INTERMAG Asia 2005. Digests of the IEEE International, pp 521–522
24. Tartaj P, Morales MP, González-Carreño T, Veintemillas-Verdaguer S, Serna CJ (2005) Advances in magnetic nanoparticles for biotechnology applications. J Magn Magn Mater 290–291:28–34
25. Grüttner C, Rudershausen S, Teller J (2001) Improved properties of magnetic particles by combination of different polymer materials as particle matrix. J Magn Magn Mater 225:1–7
26. Wu J-H, Ko S, Liu H, Kim S, Ju J, Kim Y (2007) Sub nm magnetite nanoparticles: synthesis, microstructure, and magnetic properties. Mater Lett 61:3124–3129
27. Martins VC, Cardoso FA, Germano J, Cardoso S, Sousa L, Piedade M, Freitas PP, Fonseca LP (2009) Femtomolar limit of detection with a magnetoresistive biochip. Biosens Bioelectron 24:2690–2695
28. Germano J, Martins V, Cardoso F, Almeida T, Sousa L, Freitas P, Piedade M (2009) A portable and autonomous magnetic detection platform for biosensing. Sensors 9: 4119–4137
29. Martins VC, Germano J, Cardoso FA, Loureiro J, Cardoso S, Sousa L, Piedade M, Fonseca LP, Freitas PP (2009) Challenges and trends in the development of a magnetoresistive biochip portable platform. J Magn Magn Mater 322:1655–1663

Index

Sonia M. Tiquia-Arashiro (ed.), *Molecular Biological Technologies for Ocean Sensing*, Springer Protocols Handbooks, DOI 10.1007/978-1-61779-915-0,

S

T

U

MIX
Papier aus verantwortungsvollen Quellen
Paper from responsible sources
FSC® C105338

If you have any concerns about our products,
you can contact us on
ProductSafety@springernature.com

In case Publisher is established outside the EU,
the EU authorized representative is:
Springer Nature Customer Service Center GmbH
Europaplatz 3, 69115 Heidelberg, Germany

Printed by Libri Plureos GmbH
in Hamburg, Germany